The Mammoth Bo
UFOs

Also available

The Mammoth Book of Ancient Wisdom
The Mammoth Book of Arthurian Legends
The Mammoth Book of Battles
The Mammoth Book of Best New Horror 2000
The Mammoth Book of Best New Science Fiction 13
The Mammoth Book of Bridge
The Mammoth Book of British Kings & Queens
The Mammoth Book of Cats
The Mammoth Book of Chess
The Mammoth Book of Comic Fantasy
The Mammoth Book of Seriously Comic Fantasy
The Mammoth Book of Dogs
The Mammoth Book of Erotica (New Edition)
The Mammoth Book of Endurance and Adventure
The Mammoth Book of Gay Erotica
The Mammoth Book of Haunted House Stories
The Mammoth Book of Heroic and Outrageous Women
The Mammoth Book of Historical Detectives
The Mammoth Book of Historical Erotica
The Mammoth Book of Historical Whodunnits
The Mammoth Book of How It Happened
The Mammoth Book of International Erotica
The Mammoth Book of Jack the Ripper
The Mammoth Book of Jokes
The Mammoth Book of Legal Thrillers
The Mammoth Book of Lesbian Erotica
The Mammoth Book of Lesbian Short Stories
The Mammoth Book of Life Before the Mast
The Mammoth Book of Locked-Room Mysteries and Impossible Crimes
The Mammoth Book of Men O'War
The Mammoth Book of Murder
The Mammoth Book of Murder and Science
The Mammoth Book of New Erotica
The Mammoth Book of New Sherlock Holmes Adventures
The Mammoth Book of Nostradamus and Other Prophets
The Mammoth Book of Oddballs and Eccentrics
The Mammoth Book of Private Lives
The Mammoth Book of Seriously Comic Fantasy
The Mammoth Book of Sex, Drugs & Rock 'n' Roll
The Mammoth Book of Short Erotic Novels
The Mammoth Book of Sports & Games
The Mammoth Book of Soldiers at War
The Mammoth Book of Sword and Honour
The Mammoth Book of Tasteless Lists
The Mammoth Book of the Third Reich at War
The Mammoth Book of True Crime (New Edition)
The Mammoth Book of True War Stories
The Mammoth Book of 20th Century Ghost Stories
The Mammoth Book of Unsolved Crimes
The Mammoth Book of War Diaries and Letters
The Mammoth Book of the Western
The Mammoth Book of the World's Greatest Chess Games

The Mammoth Book of UFOs

Lynn Picknett

ROBINSON
London

Constable Publishers
3 The Lanchesters
162 Fulham Palace Road
London W6 9ER
www.constablerobinson.com

First published in the UK by Robinson,
an imprint of Constable & Robinson Ltd 2001

Collection and editorial material
copyright © Lynn Picknett 2001

All rights reserved. This book is sold subject to the condition that it shall not, by way of trade or otherwise, be lent, re-sold, hired out or otherwise circulated in any form of binding or cover other than that in which it is published and without a similar condition including this condition being imposed on the subsequent purchaser.

A copy of the British Library Cataloguing in Publication Data is available from the British Library.

ISBN 1-84119-075-6

Printed and bound in the EU

TO CLIVE
the best friend anyone could ever have
love and honour

CONTENTS

ACKNOWLEDGMENTS

A great many people have helped me with this book, especially those who ensured, in innumerable ways, that I had the time and resources to finish it. And there have been some very special people who have kept my interest in the subject of UFOs alive over the years, who also deserve my admiration and gratitude. In particular I would like to thank the following:

Clive Prince, whose enthusiasm, knowledge, loyalty and unique friendship – not to mention a heightened sense of the absurd – was guaranteed to get me through the bleaker moments. I will always be indebted to him for his very practical support, despite the many other pressing demands on his time.

My late father, Cyril Picknett MBE, for his love, encouragement, and the strange story – which took some courage to share – of the "flying train".

My friend and colleague Dr Stephen Prior, not only for information about UFOs but also his patience, hospitality and great good humour over the lengthy course of my writing this book. Thanks too, to Francesca Norton, for looking after me so well and being such good company.

Nick Pope, for generously supplying information and sharing ideas on this complex and controversial subject.

Craig Oakley, whose unconventional views often helped enormously, and whose infectious humour never fails to lift my spirits.

Albert Budden, for taking time to discuss his hypothesis and supplying me with research material.

Brian Allan, of Scotland's Strange Phenomena Investigations, who generously supplied me with information about the Bonnybridge sightings, and for his infectious enthusiasm about the subject.

Bob Brydon, for his help, a memorable Hogmanay discussion

about "the stargate conspiracy" – and his patience, as this project took up an increasing amount of my time.

In some cases for their extraordinary insights and inspiration and in others for their support and friendship: Geoff Gilbertson; Manfred Cassirer; Marisa St Clair; the late Ralph Noyes; Ken Seddington; Robin Crookshank-Hilton; Georgina Bruni; Peter Brookesmith; Kevin McClure; Lionel Beer; John Spencer; Hilary Evans; Dr Jacques Vallée; Maggie Fisher; Professor Archie Roy; Malcolm Robinson; Amanda Nevill; Bob Rickard and Paul Sieveking; Mary Saxe-Falstein; David Bell; Sheila and Eric Taylor; Sarah Litvinoff; Moira Hardcastle and Simon Hinton.

And finally, the enthusiastic – and unusually understanding – Krystyna Green of Constable & Robinson Publishers, whose "non-threatening" notes kept me up to the mark, and who worked her magic on terrifying deadlines.

INTRODUCTION

On my first day as Deputy Editor of the weekly publication *The Unexplained* in late 1979, I was asked which subjects I would like to deal with. As an enthusiast about all matters paranormal and weird, I was only too happy to be spoilt for choice, but finally mentioned I had an abiding interest in Unidentified Flying Objects (UFOs) and would look forward to commissioning, editing and generally overseeing any articles on that subject. Assuming that I might have to fight for the honour, I was taken aback by the reaction of the team, who by the time of my arrival had already been working on the planning stages of the publication for some weeks. "UFOs?" said one doughty sub-editor scornfully. "You'd like to do UFOs? You won't last two weeks." I was astonished. Was this because of the workload? Perhaps the office was regularly overwhelmed with reports of strange flying craft. Or perhaps I had the look of someone whose psyche was too fragile to deal with case histories of such "high strangeness"? But no. The sub-editor said, "You'll go round and round in ever-decreasing circles. You'll hate it. The big problem is that there are no answers."

I didn't hate it. On the contrary, I found the subject enormously fascinating. But I do admit that my UFO job often suffered from phases of deep frustration, feelings that have often resurfaced over the years. Now, having delved considerably deeper into the subject, I disagree that there are "no answers". If anything, the reverse is true. Often it seems as if the real problem with studying UFOs is that there are far *too many* answers – often apparently mutually exclusive, but each appearing to carry virtually equal weight. But I agree that the subject, unless approached with an open mind and a taste – even a relish – for the absurd, can be maddeningly elusive, confusing and contradictory.

This book will attempt to present all sides of the great UFO debate, for while there are cases that seem to demonstrate that the strange objects seen in our skies are unknown craft (which may or may not be reassuring), others appear to be more apparitional, properly belonging to the paranormal realm of ghosts, fairies and vampires. Inevitably, given the elusive nature of the phenomenon, some cases straddle the borderline between the two categories – almost as if cheekily defying ufologists to pin them down in one category or another.

There are many thousands of sightings of anomalous craft in our skies: hard shiny discs, solid metal triangles, three-dimensional cigar shapes or any variation of these. They gleam in the sun, cast shadows, leave traces such as burn marks on the ground – and occasionally on people who have witnessed them. And sometimes, it is believed, they crash . . . They are witnessed by sane, sober and eminently sensible people, besides cranks – and also, infrequently, by those, like myself, whose self-appointed business it is to try to make sense of it all. UFOs are no respecters of persons.

These swooping, darting, very real objects were once considerably more people-shy than they are now: these days whole cities stop to witness their aerial choreography, as during the curiously intensive Mexico City sightings in the early 1990s, which continue sporadically to this day. And although it is a shame in some respects that the cities over which they choose to dance in great numbers so blatantly do not include London, New York or San Francisco – which would certainly make the world's media take the subject more seriously – the fact can no longer be ignored that *this phenomenon is real.*

It is true that many a promising sighting of an apparently unexplained aerial object comes to nothing, the UFOlogical equivalent of fairy gold. After all, 'UFO" is merely the acronym for "Unidentified Flying Object": disappointingly, most of those reported rapidly become IFOs – Identified Flying Objects, be they misidentified aircraft, satellites (still a last guess to a surprising number of people), hot air balloons or advertising airships, some sort of unusual atmospheric or meteorological phenomenon such as a lenticular cloud, sunlight bouncing off flocks of birds or a bright planet. Recent research has shown that even some of the most famous sightings may have a mundane explanation. Yet there is still a small number – about ten per cent – of all sightings that remain stubbornly unexplained, the "true" UFOs.

Many people automatically assume that they are extraterrestrial craft come to reconnoitre, and they may be right, at least in some cases. Even if there are other, no less disturbing explanations, the very presence of strange craft in the Earth's air space might reasonably be expected to ruffle the feathers of any responsible government. Moreover, strange craft that shape-shift and transmute from hard, shiny discs into amorphous blobs of dancing light should prompt intensive investigation by the think-tanks of the world's top brains. There is massive, and ever-accumulating, evidence that our skies are routinely, even insolently, being buzzed by UFOs that show no greater fear of us than an elephant shows for a dung beetle. Even so, most people ignore the subject, relegating it at best to a fringe interest all too often associated with cranks. In academic circles UFOs are *bad taste*, and a sure way of having university research grants withdrawn (at least in the UK) is to admit an interest in them. Astronomers may debate the existence of extraterrestrial life, but any suggestion that aliens may already be here is to invite ridicule and derision. Who wants to be known as anoraks who believe in "little green men?"

Perhaps, as increasing numbers of conspiracy theorists believe, the world's governments are not only as ignorant of them but also as arrogant about them as they appear to be, and the only people who worry about the shiny craft skipping across the twenty-first century are we superstitious peasants. Then again, perhaps, as many have claimed – including those who should know, such as the Ministry of Defence's Nick Pope – the only conspiracy, if it can be called that, is not one of secret knowledge, but one of *ignorance*. The powers that be maintain their lofty disregard for UFOs because they refuse to admit that not only have they no idea what the craft are, where they come from or why they're here, but also have no means of dealing with them.

Certainly, governmental obsession with secrecy makes UFO research extremely difficult and confusing, especially in Britain (top UK UFOlogist Jenny Randles tells how it is *illegal* to reveal what brand of teabags are used in the Ministry of Defence's canteen, for example, and any breach of that extraordinary law is punishable by imprisonment), but what of the Americans, with their Freedom of Information Act, and the Australians with theirs? Have their files finally resolved the mystery of the UFOs? Unfortunately not. In fact, if anything the vast mountain of paper generated by the release of previously classified information seems only to cloud the issue further, although individual items – some of which will be dealt with later in the book – may seem exciting confirmation of one particular theory, or another . . . And of course most truly sensitive or revelatory documents are still routinely withheld, or released with so many words blacked out as to be worse than useless – except to fuel the never-ending controversies yet further.

However, those tons of paper prove *something*. They show that despite long campaigns of denial that UFOs are anything other than weather balloons, the planet Venus or the tooth fairy (and all the other fatuous non-explanations), governments have taken them seriously enough to write reports, convene committees (behind very firmly closed doors) and cover their tracks. If the American government has behaved in this way since at least the end of the Second World War then you can be sure that Her Majesty's Government has followed suit, and that there are somewhat weightier matters being obscured by the Official Secrets Act than the use of PG Tips teabags in the MoD canteen.

The usual line is that governments are keeping lips firmly buttoned about UFOs because they have no wish to cause mass panic. Perhaps this paternalism has a point: who needs another 1930s *War of the Worlds* fiasco? The loss of the last shred of faith in our governments when we realized they know virtually nothing about the UFOs and worse, and are helpless against them, would be a traumatic blow, possibly with terrible results. No Hollywood blockbuster would have prepared us for the moment when the US President tells the world that the Reta Reticulans or aliens from Sirius are about to take over. But is that really so likely? Some believe that the "aliens", while immensely powerful, do not come from the stars at all, but from much closer to home.

It does seem that the unknown craft have a strong kinship with the paranormal, despite their three-dimensional appearance. They shape-shift from hard-edged metal to luminous, amorphous lights, travel in ways that defy the known laws of physics (possibly simply to show that it can be done) and leave witnesses in a state of turmoil, beset with a kaleidoscope of emotion ranging from fear to awe and an almost religious sense of wonder. And in this, I speak from experience.

One insomniac night in 1983 I gazed idly out across the South London sky and saw what I thought to be the lights of two planes, one behind the other, flying slowly into view from behind the grey mass of the South London Polytechnic (now the University of South London). It only occurred to me much later that it would have been very unusual for two planes to follow the same flight path so closely, especially at three in the morning when the sky was almost completely clear of air traffic. But at the time I simply stared in fascination, soon feeling compelled to concentrate hard, because the two lights suddenly diverged, skipping about like maddened insects, in great loops within the rectangle of sky framed by my window. They kept up this wild inchoate dance for perhaps two minutes before suddenly colliding and coalescing into one great, bright light, which then zoomed off out of sight at a phenomenal speed. Then there was nothing, just the dark sky and dull rooftops of Borough Road, SE1.

There is no doubt that I saw a UFO – or two – that night, if only technically, in the sense that I witnessed something that was both unidentified and flying, although whether it was truly an object in the usual sense of the word will never be known. It seemed not to be, being more of a blob of energy.

Whenever I tell the story I am inevitably asked whether I reported it. With apologies to UFOlogists who are keen to do everything by the book, I can only answer that it never occurred to me to report my strange nocturnal sighting of the dancing lights. Nor, indeed, did it occur to me even to tell anyone about it for several days, and even then I deliberately played it down, not so much in case I was ridiculed (after all, I *was* working on *The Unexplained*: weirdness was our daily bread) but because the experience seemed somehow too *precious* to discuss. The dancing lights were my connection with a numinous world, a place beyond the normal rules of life, and the allegedly inviolable laws of physics where lights do not dance in the sky – certainly not above the South Bank Polytechnic, anyway.

But had I really seen a spaceship from some distant star system? Or was it a misidentification of some, to me, unknown natural phenomenon, an artefact of my own idling mind, an intrusion from a magical realm, or just something elusively and unknowably *other*?

The second UFO I saw, some years later, had a considerably less powerful effect on me, and is perhaps a cautionary tale of the danger of falling in love with the excitement and awe of the UFO experience. At about six o'clock on a late spring evening I was in the back of a black cab as it drove from Hampstead to St John's Wood in north London where I live. Somewhere close to the area of Swiss Cottage we stopped at a red light. And waited and waited and waited . . . even though the lights changed several times. Yet curiously the cab driver did not lean out of his cab and shout some colourful expletives at a daydreaming driver in the time-honoured fashion. I thought it odd that there was not a huge impatient queue backed up behind us at that busy spot, and it seemed strangely quiet for what was still rush hour. Then I looked up.

A massive craft almost blotted out the sky. It was dark grey with a row of illuminated portholes, at first appearing to be cigar-shaped, but as I continued to look, awestruck, I realized it was actually a round craft tilted on its side, creating the illusion of being cylindrical. Hoisting my jaw up from the floor, I glanced hurriedly around, fully expecting to see throngs of passersby staring up at the sky, or at least the cabbie looking upwards. Nothing. A couple sauntered past deep in conversation. The cab driver stared stolidly ahead. Plucking up courage I pointed upwards, asking him what he thought "it" was. He just said: "Can't see anything." When I looked up again, there was nothing there. We drove off.

This time, although once again I had the sense of witnessing something awesome and special, I couldn't wait to tell the few good friends who would take me seriously. Telling even the most understanding of pals, however, about such an experience can only ever be partially satisfying. They weren't there and you were. Even though they believed me and were supportive and excited for me, I still felt obscurely that I was wrong in mentioning it, making a pact with myself not to tell anyone else – which, as things turned out, may have been just as well.

A day or so afterwards, a colleague casually mentioned that the Goodyear blimp was flying over London that week on some advertising stunt . . . Instantly my stomach turned over. I was filled with terrible humiliation and self-doubt. I tried my hardest to recall exactly what I had witnessed, seeing again in my mind's eye the tip-tilted roundness, the yellow porthole lights, feeling again the strange silence. I interrogated myself to try and find some answers. If it was the Goodyear dirigible why hadn't the cabbie seen it? Answer: perhaps he couldn't have seen the same patch of sky I could see from his window. Why weren't other people looking up at it? Answer: why should they? Maybe they were much too world weary and blasé to bother to look up at the Goodyear ship. This was London, after all.

Did I really see a gigantic round spaceship with porthole lights? Perhaps the Goodyear ship looks like that when seen from underneath. I was sorry I'd shared my experience even with close friends, and hot with horror at the thought I could have done so to a considerably wider audience – after all at that time I had a weekly slot on a local radio programme (on the Clive Bull Show on LBC). The very thought brought me out in a cold sweat.

Suddenly, I was willing, even strangely eager, to explain away my "sighting" as the Goodyear blimp, even though in the deepest recesses of my mind I knew nothing would change what I saw – or, at least, *what I thought I saw.* And as I write this, a decade or so later, I can still recall the massive "UFO" with its lights above my cab in Swiss Cottage.

That experience perfectly illustrates the extraordinary difficulty of dealing with reporting UFOs at all objectively. There is something about seeing unknown things in the sky that can create powerful feelings of wonder, almost of being "chosen", while sometimes the very suggestion of an alternative explanation can provoke such intense self-doubt that the original sighting is denied.

But at least my own experiences, which were kept at an anecdotal level and never reported to UFOlogists, the RAF, police or the military, have made me wonder how many other people have never so much as mentioned their own sightings? I suspect that the UFO phenomenon must be *massive.*

Over the past decade I have been involved in working with Clive

Prince on several books, all of which, in their way, are about the power of belief. *Turin Shroud: In Whose Image?*, (1994), argues that Christianity's greatest relic, the Shroud of Turin – long believed to be the winding sheet of Jesus, miraculously imprinted with his image – was in fact a cynical fake by that tormented genius, Leonardo da Vinci. Even though the cloth was conclusively shown by carbon dating to be at least 1,000 years too young to be the miraculous cloth, there are those to whom any suggestion that it is a fake is tantamount to an attack on their entire religion. This is, surely, a sad reflection on the depth of their faith, just as challenges to the Roswell story provoke intense anger in certain UFOlogists. Interesting though the Shroud and the Roswell Incident are in their very different ways, why are they so important, when both religion and UFOlogy contain so much else that is exciting, challenging and rewarding?

Following on from our Shroud book, Clive and I decided to follow Christianity back to its roots, totally ignoring all preconceptions and prejudice – beginning, as it were, from scratch. This resulted in *The Templar Revelation, Secret Guardians of the True Identity of Christ* (1997), which revealed that the religion actually began in ancient Egypt.

The last Picknett and Prince book was an appropriate starting point for *The Mammoth Book of UFOs. The Stargate Conspiracy: Revealing the truth behind extraterrestrial contact, military intelligence and the mysteries of ancient Egypt* (1999), dug deep into the background of New Age channelling cults and famous twentieth century prophets, revealing the amazing ease with which quite outrageous concepts can be accepted unquestioningly by thousands of people. As a species, it seems we would rather believe something than nothing, and believe so intensely that those who do not agree with us are the objects of derision or worse.

Nowhere is that more apparent than in UFOlogy, which for many people is nothing less than a substitute religion, with articles of faith, prophets – and heretics. For example, where those who believe in the Extraterrestrial Hypothesis (ETH) are concerned, to question the nuts and bolts reality of the craft is tantamount to blasphemy. UFOlogy is an uncomfortable place to be at the best of times, but to question the sacred cornerstones of the religion is to be cast into outer darkness or publicly vilified. Equally, to those who fervently believe that all UFOs are paranormal in origin, caused by mental aberration, or simple misidentifications, to suggest otherwise is to rock their world, too.

As someone who has never been anything other than a commentator on the outside of the UFOlogical community, my own observations have led me to conclude that the phenomenon is so fluid and changeable that *no one is always right*. Conversely, this means that everyone is sometimes right, which may be little comfort to the zealots of any persuasion!

Acting as Devil's Advocate, I have attempted to present all the major theories, challenging them vigorously if they appear to deserve it. Neither am I afraid to attack the great sacred cows of UFOlogy – such as the Roswell incident – if the evidence is lacking. We do no one any favours by avoiding the truth, unpalatable though it may seem. Being over-selective with the data in order to make one particular hypothesis look good is threatening to bring the whole field into disrepute. What is needed now is complete open-mindedness and a glasnost between the various factions.

Hopefully, this book will demonstrate to those already in the UFO world that there is enough room for everyone, and to those who are new to the subject, perhaps it will excite, intrigue and inspire them to do their own research. Between us we may find that, of all the possible answers, there is one that makes sense of all the nonsense, taking us forward into a less dogmatic era. And somewhere along the way, our individual lives may be transformed by the phenomenon we have chosen to investigate.

Lynn Picknett
London, 2000

Chapter 1

UNIDENTIFIED FLYING OBJECTS: THE PREQUEL

"We know that . . . this world was being watched closely by intelligences greater than man's and yet as mortal as his own. We know now that as human beings busied themselves about their various concerns they were scrutinized and studied, perhaps almost as narrowly as a man with a microscope might scrutinize the transient creatures that swarm and multiply in a drop of water. With infinite complacence people went to and fro over the Earth about their little affairs, serene in the assurance of their dominion over this small spinning fragment of solar driftwood, which by chance or design man has inherited out of the dark mystery of time and space. Yet across an immense ethereal gulf, minds that are to our minds as ours as to the beasts in the jungle, intellects vast, cool, and unsympathetic, regarded this earth with envious eyes and slowly and surely drew their plans against us."

The opening words of Orson Welles'
War of the Worlds broadcast, 30 October 1938.

Today the UFO phenomenon is generally thought of as beginning in 1947 when Kenneth Arnold saw nine objects flying in formation that he described, memorably, as "flying saucers", but history has recorded strange objects being seen in the skies and encounters with otherworldly entities for many centuries before that classic sighting. Perhaps what we call UFOs have been observed by people from all times – even when human beings lived in caves – the only difference being the theories put forward to account for them.

Almost all Bible scholars and church-goers refuse to accept that UFOs have any connection with their beliefs, but a great many UFOlogists have come to accept that countless descriptions of heavenly appearances may actually have been misinterpreted alien activity. For example, in Chapter 2 of the Second Book of Kings in the Old Testament, we read how Elijah guides Elisha into the desert where a chariot of fire comes down from heaven to carry him off to safety. However, in the original Hebrew text, the "chariot" was actually a fish-shaped vessel, with flames spewing from its tail – perhaps an uneducated description of a spaceship?

Certainly, Soviet physicist Professor M. Agrest had no doubts about what really happened to the infamous cursed cities of the plains – Sodom and Gomorrah, which God is said to have destroyed because of their wickedness. Agrest believes that they were annihilated by an atom bomb,[1] and that Lot's wife was not turned to salt, as in the Biblical account, but to *ash*. He also made the astonishing claim – made famous by the Swiss writer Erich von Daniken[2] – that ancient Baalbek was the "Cape Kennedy of its day", the launchpad for alien craft, citing the vast numbers of "tektites" (pieces of glass, created by fierce heat) found at the site, which are usually found after atomic blasts.

During the Exodus from Egypt,[3] the Children of Israel, led by Moses, are believed to have been preceded by a pillar of fire in the sky (Exodus Chapter 13). And when Moses went up to the top of Mount Sinai to converse with God, he disappeared into a cloud that kept him from sight for forty days and nights. As some commentators[4] have pointed out, it seems there was more to this "cloud" than meets the eye. Even though the phrase "forty days and forty nights" is no longer taken literally by theologians – who see it as a figure of speech intended simply to imply a long time – clearly it was an impressive length of time to be enshrouded in cloud on top of a bare mountain top. Was Moses taken away from Sinai during this lengthy period? Did he actually meet God, not merely on the top of Mount Sinai, but way beyond it, in his alien spaceship? Even though belief in the ancient astronaut theory has long since peaked, there are still those who believe this to be the case, and that the Bible and other ancient texts are records of alien intervention in human lives many millennia ago.

Erich von Daniken and others believe that certain ancient texts, limited by the writers' knowledge and experience, actually describe the arrival of astronauts from elsewhere, who were then worshipped as gods by the ignorant primitive humans, just as the GIs with their luxury goods were worshipped by the "cargo cults" of the South Seas in the mid-twentieth century. Adherents of the "ancient astronaut" school of thought believe, for example, that when the ancient Egyptians wrote of the god Osiris bringing the secrets of civilization to their ancestors, they were really describing the bestowal of knowledge by a superior being from elsewhere in the universe.

Even the young Dr Carl Sagan,[5] intrigued by Professor Agrest's speculations, suggested that the old myths should be re-analyzed in the light of modern knowledge about space travel, citing the Old Testament story of the prophet Enoch, who dreamed of "two men, very tall, such as I have never seen on earth. And their faces shone like the sun, and their eyes were like burning lamps . . . They stood at the head of my bed and called me by name. I awoke from my sleep and saw clearly these men standing in front of me." What we would call "bedroom visitors", these "men" carried Enoch off on a trip around the "seven heavens". Although Enoch wrote about his incredible experience widely,[6] his works are not well known today.

The New Testament, too, some believe, contains references to contact with extraterrestrials. John A. Keel suggests that the lurid prose of the Book of Revelations – which either defies interpretation or only too often provides fuel for the more extreme Fundamentalist vision – may actually contain startling new information about the intervention of aliens in human affairs. In Chapter 4 we are told that "a door was opened in heaven" and in Chapter 10 that the writer, the visionary John of Patmos, "saw another mighty angel come down from heaven clothed with a cloud; and a rainbow was upon his head, and his face was as it were the sun, and his feet as pillars of fire". Speaking for many other UFOlogists of the 1970s, John A. Keel wrote:[7] "It sounds as if he were describing a brilliantly glowing sphere surrounded by vapors ("cloud") and colored lights ("a rainbow") and two beams of light or flame were jetting down beneath it ("pillars of fire")." Although there are other, perhaps more valid, interpretations of Revelations,[8] it would be a mistake not to consider all the possibilities.

Open-minded researchers inevitably come across strangely compelling stories from history that appear to be tales of alien contact, although obviously the terms of reference, and therefore the mode of describing them, are very different from our own.

While researching for an article on the Rosicrucians for the Dutch magazine *Bres*, Ludo Noens of Antwerp, Belgium, came upon a fascinating story in a book entitled *Le Comte de Gabilis* (*Count Gabilas*), by the Abbé Montfaucon de Villars in 1670. This described one of the revelations of the mysterious Count Gabilas, an initiate into much esoteric lore (including the Kabbalah), who wrote of a strange event from Charlemagne's time, c. AD 800:

> . . . it so happened that one day in Lyon, people saw three men and a woman descending from these Aerial Ships; the whole town assembled around the place, crying that these people were Magicians, sent by Grimoald, Duc de Bennevent, Charlemagne's enemy, to destroy the French harvest. The four innocents defended themselves by saying that they were from the country as well, and that they had been kidnapped shortly before by miraculous Men who had shown them unheard-of wonders and who had asked them to tell the world about it . . .[9]

As we will see in the Explanations chapter, many of these early stories were intimately bound up with belief in fairies or little people, and a host of phenomena were described that are now associated with UFO sightings, such as mysterious moving lights. One such tale tells how, when the sixteen-year-old German poet Goethe[10] was on his way with two fellow travellers from Frankfurt to the University of Leipzig, they sometimes had to walk beside the coach because rain made its progress difficult. It was during one of those periods in their journey that the poetic genius noticed something very strange a little distance away – some kind of mysteriously shining object. He described what happened next:

> All at once, in a ravine on the right-hand side of the way, I saw a sort of amphitheatre, wonderfully illuminated. In a funnel-shaped space there were innumerable little lights gleaming, ranged step-fashion over one another, and they shone so brilliantly that the eye was dazzled. But what still more confused the sight was that they did not keep still, but jumped about here and there, as well downwards from above and *vice versa*, and in every direction. The greater part of them, however, remained station-

> ary, and beamed on. It was only with the greatest reluctance that I suffered myself to be called away from the spectacle, which I could have wished to examine more closely. The postilion, when questioned, said he knew nothing about such a phenomenon, but that there was in the neighbourhood an old stone-quarry, the excavation of which was filled with water. Now whether this was a pandemonium of will o'-the-wisps, or a company of luminous creatures, I will not decide.[11]

Without more information it is impossible so long after the event to decide whether these strange lights were indeed, as Goethe himself suggested, caused by natural gases – will o'-the-wisps – or by a phenomenon such as earthlights (see page 451), or by some kind of supernatural event, possibly associated with non human entities. It may be significant that Goethe himself was given to psychic experiences, at least once seeing his own double, for example. Yet this story has a timeless ring to it, evoking the classic scenario of modern UFOlogy where weird lights suddenly appear to travellers on a lonely road, heralding a brush with "high strangeness".

However, more recent witnesses to anomalous things seen in the sky, while being just as confused and scared, are less hampered by a lack of vocabulary or convenient point of reference. In the late 19th century, when technology had progressed enough to produce hot-air balloons and prototype airships, those were what people saw in the skies as UFOs.

Yet before the great 'scareship' flaps of the late nineteenth and early twentieth centuries, a formation of *circular* objects was actually photographed. On 12 August 1883 Mexican astronomer José Bonilla was looking through his telescope at the Sun at his observatory at Zacatecas when suddenly a formation of 143 unknown objects slowly flew across his field of vision, which he then captured for posterity with his new technological toy – a camera. The developed photograph revealed solid, spindle- and cigar-shaped objects flying at an estimated altitude of 200,000 miles (320,000km). Intrigued and perhaps a little excited, Professor Bonilla despatched a copy of the photograph and his notes about the event to the French scientific publication *L'Astronomie*. Although it was published, this astonishing event made no impression on the scientific community, who hastened – only too characteristically – to expunge it from their minds. Bonilla had obviously upset them with his "damned data", countless examples of which filled the files (and books) of the great American collector of anomalies, Charles Fort.[12]

(No doubt there will be those eager to declare that what Professor Bonillo had actually witnessed was a flock of geese flying across his field of vision, and that he had misjudged the distance involved. While there is something to be said for this kind of interpretation in other cases,[13] it would be unfair and arrogant to suggest that a well-respected astronomer was as bad at calculating speeds as lesser mortals might be, not to mention being incapable of interpreting what he saw through a telescope.)

A few years before Bonilla's uncomfortable observation, Texan farmer John Martin was reported in the Dennison *Daily News* as seeing a large round UFO pass overhead – very fast.[14] He described it as a "saucer", thus anticipating Kenneth Arnold's famous phrase by nearly eighty years.

Then the airships began to arrive in force. For example, in November 1896 crowds of people around San Francisco witnessed a huge, black cigar-shaped object pass over their heads, illuminating them as it did so with its powerful searchlights. Although this mystery airship disappeared from sight for the first three months of 1897, suddenly it was the centre of attention again, this time with many thousands of witnesses as it flew over the Midwest, causing such wild speculation as to its origins and intentions that it verged on hysteria.

Able to fly against the wind, it had turbine wheels and a glass-covered portion through which strange entities could be seen: it was like nothing known to have been designed, certainly anything that was capable of flight, at that point in history. Then in March the airship suffered a spectacular mishap when its trailing anchor caught in the clothes of farmer Robert Hibbard near his home outside Sioux City, Iowa, pulling him into the air for many yards before he fell back down to earth. Unlike modern UFOs, which on the whole avoid flying over cities, the "scareship" seemed positively to wallow in overflying large conurbations where huge crowds gathered to watch its progress. (Of course in those days there was no danger of its being chased in the air or shot down.) But like some reported UFOs of the modern era, the mysterious airship is said to have "dropped probes", while exhibiting the same tendency to change course without warning and manoeuvring with astonishing speed and dexterity. But what of the crew of the airship? Were they merely human, or were they truly alien in appearance and behaviour?

Unfortunately, we will never know for certain. Some witnesses described them as being "hideous people", a group of two men, one woman and three "children", all constantly jabbering in an unknown tongue among themselves. Of course in those more circumscribed and uneducated times, most unrecognizable foreign languages would have seemed barbarous to the good people of the Midwest, but it is clear that many witnesses truly believed the occupants of the scareship not to be of the human race (although there were reports of crews that looked like "Japs", prefiguring the modern Men In Black phenomenon by decades). Several people reported that they had communicated with the visitors, sometimes having proper conversations with them, as the story of Captain James Hooton, an Iron Mountain railroad conductor, reveals.

After hunting for a few hours in the countryside around the area of Homan, Arkansas, he had begun to make his way back to the train station when:

> . . . my attention was attracted by a familiar sound, a sound for all the world like the working of an air pump on a locomotive.
>
> I went at once in the direction of the sound, and there in an open space of some five or six acres, I saw the object making the noise. To say that I was astonished would but feebly express my feelings. I decided at once that this was the famous airship seen by so many people about the country.
>
> There was a medium-sized looking man aboard and I noticed that he was wearing smoked glasses. He was tinkering around what seemed to be the back of the ship, and as I approached I was too dumbfounded to speak. He looked at me in surprise, and said: 'Good day, sir, good day.' I asked: 'Is this the airship?' And he replied: 'Yes, sir,' whereupon three or four other men came out of what was apparently the keel of the ship.
>
> A close examination showed that the keel was divided into two parts, terminating in front like the sharp edge of a knife-like edge, while the side of the ship bulged gradually towards the

> middle, and then receded. There were three large wheels upon each side made of some bending metal and arranged so that they became concave as they moved forward.
>
> 'I beg your pardon, sir,' I said, 'the noise sounds a great deal like a Westinghouse air brake.'
>
> 'Perhaps it does, my friend: we are using condensed air and aeroplanes, but you will know more later on.'
>
> 'All ready, sir,' someone called out, when the party all disappeared below. I observed that just in front of each wheel a two-inch tube began to spurt air on the wheels and they commenced revolving. The ship sprang forward, turning their sharp end skyward, then the rudders at the end of the ship began to veer to one side and the wheels revolved so fast that one could scarcely see the blades. In less time than it takes to tell you, the ship had gone out of sight.

Making a detailed sketch of the airship, Captain Hooton remarked that he was greatly shocked to discover that the ship had no evidence of a bell, because he thought "every well regulated air locomotive should have one." But bell or no bell, the scareship was clearly so far ahead of known human technology as to be almost incomprehensible – *almost*, but not quite. People at the time knew about airships, even though they were rarely seen. But significantly, they did not see flying saucers or space-shuttle-style UFOs, either because their minds would not have allowed them to see something so beyond their experience (the vagaries of human perception are discussed in the Explanations chapter), or because whoever or whatever was in control of the phenomenon were only one step ahead of known terrestrial technology themselves. When human technology produced airships, the aliens (whoever or whatever they were and are) also produced airships, although more advanced and awe-inspiring than the human sort; when the Space Age began to dawn, the aliens were slightly ahead of that technology, too.

It is interesting that when the nineteenth century witnesses refer to the UFO being a "ship" they meant an airship, just as today we mean a spaceship. But in the eleventh or twelfth century a ship seen in the sky was just that – a real ship, one that might otherwise sail on the sea quite happily. An Irish tale of about AD 1211 tells how:

> There happened in the borough of Cloera, one Sunday, while the people were at Mass, a marvel. In this town is a church dedicated to S. Kinarius. It befell that an anchor was dropped from the sky, with a rope attached to it, and one of the flukes caught in the arch above the church door. The people rushed out of the church and saw in the sky a ship with men on board, floating before the anchor cable, and they saw a man leap overboard and jump down to the anchor, as if to release it. He looked as if he were swimming in water. The folk rushed up and tried to seize him, but the Bishop forbade the people to hold the man, for it might kill him, he said. The man was freed, and hurried up to the ship, where the crew cut the rope and the ship sailed out of sight. But the anchor is in the church, and has been there ever since, as a testimony.

Another version of this story[15] puts the scene at Gravesend in Kent, England, when the trailing anchor of the poetically-named "cloudship" became entangled in the gravestones, requiring one of its crew to slide down a rope and cut it free before returning to the safety of the ship, which then flew off, never to be seen again by the people of that area.

The Middle Ages may not have been hospitable enough for the celestial mariners – there was always the danger they would have been captured, condemned as diabolical in origin and treated to the usual fate reserved for all such unfortunates – but the nineteenth century seemed to appeal more to their apparently sociable natures. In April 1897 the *Houston Post*[16] reported two astonishing episodes of human/"scareship" crew interaction within four days of each other. On 22 April it reported:

> Rockland: Mr John M. Barclay, living near this place, reports that last night at about 11 o'clock, after having retired, he heard his dog barking furiously, together with a whining noise. He went to the door to ascertain the trouble and saw something, he says, that made his eyes bulge out and but for the fact that he had been reading of an airship that was supposed to have been in or over Texas, he would have taken to the woods.
>
> It was a peculiar shaped body, with an oblong shape, with wings and side attachments of various sizes and shapes. These

were brilliant lights, which appeared much brighter than electric lights. When he first saw it, it seemed perfectly stationary about five yards from the ground. It circled a few times and gradually descended to the ground in a pasture adjacent to his house. He took his Winchester [rifle] and went down to investigate. As soon as the ship, or whatever it might be, alighted, the lights went out. The night was bright enough for a man to be distinguished several yards [away], and when within about thirty yards of the ship he was met by an ordinary mortal, who requested him to lay his gun aside as no harm was intended. Whereupon the following conversation ensued: Mr. Barclay enquired:

'Who are you and what do you want?' 'Never mind about my name, call it Smith. I want some lubricating oil and a couple of cold chisels if you can get them, and some bluestone. I suppose the saw mill hard by has the two former articles and the telegraph operator has the bluestone. Here is a ten-dollar bill: take it and get us these articles and keep the change for your trouble.'

Mr Barclay said: 'What have you got down there? Let me go and see it.' He who wanted to be called Smith said: 'No, we cannot permit you to approach any nearer, but do as we request you and your kindness will be appreciated, and we will call you some future day and reciprocate your kindness by taking you on a trip.'

Mr Barclay went and procured the oil and cold chisels, but could not get the bluestone. They had no change and Mr Barclay tendered him the ten-dollar bill, but same was refused. The man shook hands with him and thanked him cordially and asked that he not follow him to the vessel. As he left Mr Barclay called him and asked him where he was from and where he was going. He replied, 'From anywhere, but we will be in Greece day after tomorrow.' He got on board, when there was again the whirling noise, and the thing was gone, as Mr Barclay expresses it, like a shot out of a gun. Mr Barclay is perfectly reliable.

Just 30 minutes after this extraordinary encounter took place, Mr Frank Nichols of a neighbouring town had his own fifteen minutes of fame[17] as the local newspaper revealed:

Josserand: Considerable excitement prevails at this writing in this usually quiet village of Josserand, caused by the visit of the noted airship, which has been at some many points of late. Mr Frank Nichols, a prominent farmer living about two miles east of here, and a man of unquestioned veracity, was awakened night before last near the hour of twelve by a whirring noise similar to that made by machinery. Upon looking out he was startled upon beholding brilliant lights streaming from a ponderous vessel of strange proportions, which rested upon the ground in his cornfield.

Having read the despatches, published in the [*Houston*] *Post* of the noted aerial navigators, the truth at once flashed over him that he was one of the fortunate ones and with all the bravery of Priam at the siege of Troy [sic] Mr Nichols started out to investigate. Before reaching the strange midnight visitor he was accosted by two men with buckets who asked permission to draw water from his well. Thinking he might be entertaining heavenly visitors instead of earthly mortals, permission was readily granted. Mr Nichols was kindly invited to accompany them to the ship. He conversed freely with the crew, composed of six or eight individuals about the ship. The machinery was so complicated that in his short interview he could gain no knowledge of the workings. However, one of the crew told him the problem of aerial navigation had been solved. The ship or car is built of newly-discovered material that has the property of self-sustenance in the air, and the motive power is highly condensed electricity. He was informed that five of these ships were built at a small town in Iowa. Soon the invention will be given to the public. An immense stock company is now being formed and within the next year the machines will be in general use. Mr. Nichols lives at Josserand, Trinity County, Texas, and will convince any incredulous one by showing the place where the ship rested.

Unfortunately, as the history of UFOlogy has repeatedly shown, it takes more than pointing to the "place where the ship rested" to convince most people of the truth of a frankly incredible story. Yet there is little reason to doubt Mr Nichols, or most of the others who came forward with their airship stories (although some are more dubious – see below). The tales do have a kind of consistency, although not one of rationality but rather of *absurdity*: what was the "newly-discovered material that has the property of self-sustenance in the air", for heaven's sake? And although some form of "condensed electricity" could, in theory, drive an airship, it is highly unlikely that such a power was harnessed in those days, as the greatest scientist and innovator of his day, Thomas Alva Edison[18] pointed out, saying of the scareship:

> You can take it from me that it is pure fake. I have no doubt that airships will be successfully constructed in the near future but . . . it is absolutely impossible to imagine that a man could construct a successful airship and keep the matter a secret. When I was young, we used to construct big colored paper balloons, inflate them with gas, and they would float about for days. I guess someone has been up to that fine game out west.
>
> Whenever an airship is made, it will not be in the form of a balloon. It will be a mechanical contrivance, which will be raised by means of a powerful motor, which must be made of a very light weight. At present no one has discovered such a motor, but we never know what will happen. We may wake up tomorrow morning and hear of some invention which sets us all eagerly to work within a few hours, as was the case with the Roentgen rays. Then success may come. I am not, however, figuring on inventing an airship. I prefer to devote my time to objects which have some commercial value. At the best, airships would only be toys.[19]

And in the late 1960s, the British expert Charles H. Gibbs-Smith, stated: "Speaking as an aeronautical historian who specializes in the periods before 1910, I can say with certainty that the only airborne vehicles, carrying passengers, which could possibly have been seen anywhere in North America in 1897 were free spherical balloons, and it is highly unlikely for these to be mistaken for anything else. No form of dirigible [a propelled gasbag] or heavier-than-air flying machine was flying – or indeed *could* fly – at this time in America."[20]

Nor is there any evidence of five such ships having been built at "a small town in Iowa" – although some might take this as an indication that there was a nineteenth-century equivalent of Area 51, where strange craft were built in great secrecy, perhaps even through reverse engineering from captured alien airships. But if so, why has no hint ever come out about the existence of such a place over the last hundred years or so – even through the delving of the most rabid conspiracy theorists? And, once again, why is the aliens' technology only one step ahead of our own?

It seems that the crew and/or controllers of the mystery airships were either from a world or dimension that was very similar to ours, but a sort of Looking Glass Land, where science and technology, while recognizable up to a point, have an air of high strangeness about them, a surreality that inevitably seems to smack of fantasy. And their fine words are totally insubstantial – even deliberate lies – told, if not simply to invite ridicule on the heads of the poor deluded witnesses, then perhaps for the pure joy of lying. Could it be that these "aliens" are some kind of race of tricksters, whose descendants now haunt our own skies and darken our dreams?

In his masterly *Passport to Magonia,* Jacques Vallée recounts several cases of peculiar encounters with UFOnauts, few of which, superficially at least, seem to make any sense (indeed, some might think all alleged contact with aliens suffers from the same marked air of complete mystification underpinned with a hefty wadge of nonsense). Vallée[21] cites the case of Monsieur Carreau from Chaleix, Dordogne, who, on 4 October 1954, saw a "cauldron-shaped" object, the size of a "small truck" land in his field, out of which came two "normal" – European-looking – men wearing brown overalls. They shook hands with the startled Frenchman and asked abruptly: "Paris? North?" Tongue-tied with shock, Garreau could not answer. After stroking his dog, the men climbed back into their craft and flew off. The whole exercise was completely pointless. Why ask a question – to which presumably they really did need an answer – and not wait for the man to recover his wits? And why would "men" who flew around in such superior craft need to ask a farm worker the way to Paris? One assumes they had navigational aids on board, such as good old-fashioned maps. The whole event seems somehow *staged,* as if the point of the exercise was simply to cause consternation, or to flag up the very existence of the weird. Perhaps with the coming of modern sophistication the Otherworld needs to reassert its reality in the minds of increasingly sceptical human beings. But in that case, why appear to a farm worker in the back of beyond – something "aliens" are very fond of doing, even today? Why not stage-manage an appearance to ultra sophisticated city dwellers in their thousands?

A similar case involved forty-year-old Lazlo Ujvani just sixteen days after M. Carreau's brush with the world of the unexplained. At 3 a.m. he was on his way to work near Raon-L'Etape in the Vosges area of France when he came across a man in a grey jacket with strange insignias on it, wearing a motorcyclist's helmet and carrying a gun. The man said something to him in an unidentified language. Knowing a little Russian, the Czech-born worker tried it out on the stranger, who replied immediately, "in a high-pitched voice": "Where am I? In Italy? Or Spain?" He also asked how far he was from the border with Germany and what the time was. On hearing it was 2.30, the stranger looked at his own watch, which said 4 o'clock, then brusquely instructed Ujvani to get on with his business. After walking on for a while, the worker encountered a bizarre craft in the middle of the road, shaped like "two saucers glued together, about five feet [1.5 m] in diameter and three feet [900 cm] high". The stranger told him not to get any closer, and then the UFO flew up vertically "with the noise of a sewing machine" before disappearing into the sky.

Again, little of this makes any sense. A stranger in a weird assortment of clothes – which, despite the unknown insignia, were recognizably European-style – and who hails from a very advanced craft, accosts a man in the early hours of the morning and asks whether he is in Spain or Italy, when he is actually in France. His watch is wrong. But the strangeness is not all on the visitor's side. At first Ujvari does not recognise the language the alien speaks, but after trying Russian, they successfully have a brief conversation. Was the stranger speaking Russian to start with? And if so, why did Ujvari not recognize the language? If it was some other language, it was mighty convenient that the stranger also spoke Russian. What a coincidence. Like a great many other alien encounter stories it seems like a practical joke, albeit rather pointless. But who are the jokers? And what is their game plan – if any?

Sometimes, however, it is the humans who appear to be the hoaxers, as may be seen in the classic nineteenth-century scareship story that became a cult classic when rediscovered in the 1960s. The mystery began on 19 April 1897, when the *Dallas Morning News* carried the following story:

> Aurora, Wise County, April 17 – About 6 o'clock this morning the early risers of Aurora were astonished at the sudden appearance of the airship which has been sailing throughout the country. It was travelling due north, and much nearer the earth than before. Evidently, some of the machinery was out of order, for it was making a speed of only ten or twelve miles an hour, and gradually settling towards the earth. It sailed over the public square and when it reached the north part of town collided with the tower of Judge Proctor's windmill and went to pieces with a terrific explosion, scattering debris over several acres of ground, wrecking the windmill and water tank and destroying the judge's flower garden. The pilot of the ship is supposed to have been the only one aboard, and while his remains are badly disfigured, enough of the original has been picked up to show that he was not an inhabitant of this world.
>
> Mr T.J. Weems, the US Signal Service officer at this place and an authority on astronomy, gives it as his opinion that he was a native of the planet Mars. Papers found on his person – evidently the records of his travels – are written in some unknown hieroglyphics, and cannot be deciphered. The ship was too badly wrecked to form any conclusions as to its construction or motive power. It was built of an unknown metal, resembling somewhat a mixture of aluminum and silver, and it must have weighed several tons. The town today is full of people who are viewing the wreckage and gathering specimens of strange metal from the debris. The pilot's funeral will take place at noon tomorrow. Signed F.E. Hayden.

This story was resurrected in the January/February 1967 issue of *Flying Saucer Review* in an article co-authored by Donald B. Hanlon and Dr Jacques Vallée, entitled 'Airships Over Texas', together with a postcript to the effect that Dr J. Allen Hynek had despatched a friend to Aurora, Texas, to check out the details. Very quickly it transpired that the original newspaper story was riddled with errors: Judge Proctor had no windmill; T.J. Weems was not an officer in the Signal Service, but was the local blacksmith, and there was no evidence of unidentified graves in the cemetery. More damningly, according to Hynek's informant, one Oscar Lowry, the journalist responsible for the story who had been eleven at the time of the alleged incident, was known to have been greatly concerned at the ill-luck that plagued the town – an epidemic, a fire and destroyed crops, besides being left off the railway route – and sought to make Aurora a tourist attraction. Having heard of the interest generated by other mystery airship stories, Hayden decided to put his beloved town on the map with the most sensational of them all.

Then when, in the late 1960s, local historian Etta Pegues raked over the scareship story, interviewing some of the senior citizens who had been alive at the time, none of them could remember the famous airship. Ms Pegues wrote: "Cliff D. Cates would have included it in his *Pioneer History of Wise County* which was published in 1907. It would have sold him a billion copies. Also, if it had been true, Harold R. Bost would have included it in his *Saga of Aurora.* It would have been the highlight of his theme. But neither men [sic] mentioned it because it had been forgotten as any other piece of fiction would have been forgotten." She added that, "It was all a hoax cooked up by Hayden and a bunch of men sitting around in the general store", and that the journalist was well-known for his hoaxes, although some thought Judge Proctor himself was behind it – one old man saying that his father had remarked that the Judge had "outdone himself this time".

So was the phantom airship of Aurora merely a hoax, or at best an attempt to exploit the scareship phenomenon as a public relations exercise for the town? Renewed interest in the story in the 1970s seemed, for a while at least, to cast doubt on the negative verdict of the 1960s. In March 1973 Bill Case, aviation writer for the *Dallas Times Herald* interviewed some of the old timers of Aurora for a series of articles about the crash, including Brawley Oates, resident on the Proctor farm since the end of the Second World War, who said readily that "I've heard this story all my life", adding that he had found a cache of metal under the windmill. An eighty-six-year-old claimed his father had seen the airship moments before it crashed, while Mrs Mary Evans, aged ninety-six, said her parents refused to let her see the crash site at the time, but did tell her about it afterwards. More spectacularly, ninety-eight-year-old G.C. Curley claimed that two of his friends had seen the crash site and the "torn up body".

Case's articles unleashed the floodgates. Within two months, one Fred N. Kelly, who styled himself "a scientific treasure hunter", claimed to have found something of which he said: "I've never seen any metal like that in twenty-five years of experience" – a scoop that Case was not slow to splash under banner headlines such as "UFO alloy unknown back in '97". Having allegedly located the UFOnaut's grave with a metal detector – and declared it bore "a unique handmade headstone" – certain UFOlogists sought a court order to enable them to have the body exhumed.

While the big UFO organizations argued vehemently about the story, matters were taken out of their hands by an unknown grave robber who broke open the alleged tomb of the alien in the Aurora cemetery. While it is still possible that there was a lone alien airman lying in the tomb, photographs of the headstone showed nothing more unusual than bad carvings and a crack, and when the International UFO Bureau tracked down Case's witnesses the story fell apart. It transpired that they either said nothing of the sort (as reported by Case) or were never even interviewed in the first place.

It took a while for the tale of the Aurora crash to lose its appeal. As Frank X. Tolbert, a Dallas journalist wrote in June 1973: ". . . I understand the yarn of the 1897 visitor from another planet rivals Watergate for space in European publications."

While the Aurora incident may have been a hoax, there were many other strange sightings of roughly that time that must have had other explanations, if only we knew what they were. The year 1908, for example, brought a wave of airship sightings to the same area around Tacoma, Washington, that was to play host to the Maury Island hoax of 1947 (see page 49). A brilliant reddish craft – described as "two or three times as bright as Jupiter" flew over Kent, Washington on 1 February and again twenty-four hours later, witnessed by whole crowds, some of whom said it was "cigar-shaped".[22] The Tacoma newspaper, the *Daily Ledger* of 4 February wrote: "During the same week, on clear nights, colored lights were displayed at high altitudes, and on one occasion a rocket was discharged high in the air, it is asserted."

Because the Russian-Japanese war of three years earlier was still fresh in the minds of the populace – and the so-called "Yellow Peril" a source of continuing paranoia – some columnists suggested that the dirigible was an unknown Japanese surveillance craft. This reinforcement of underlying bigotry and war hysteria was to surface in Europe within a few years when countries on both sides of the English Channel were visited by their own "scareships".

On the night of 13 or 14 October 1912 an aircraft was heard flying close to Sheerness and the Naval Flying School, Eastchurch, in Kent, England. At that time night flying was virtually unknown – not surprisingly, for daytime aviation was dangerous enough, but to fly those flimsy craft in the dark was usually asking for trouble. It was all very odd. But, as Nigel Watson admits in his article on the subject in *Strange Magazine*[23] the matter would have been quickly forgotten had the Member of Parliament for Brentford, Mr Joyston-Hicks, not asked Winston Churchill (then First Lord of the Admiralty) for a statement about the mystery. Churchill replied that flares had been lit that night at the Flying School but nothing had actually landed there. Privately, he let it be known that he suspected that the phantom craft was German in origin, although he said nothing of this in public. Suspiciously, the Admiralty Air Department requested a report on the subject only two weeks later, so obviously in their eyes at least it was neither serious nor urgent.

The report stated that a Lieutenant Fitzmaurice said he heard the aircraft as he walked through Sheerness, but it was only the next day, when the matter escalated, that he thought any more of it. A crowd grew in Sheerness High Street to watch as a strange light, with a dark object accompanying it, flew eastwards over their heads. The *Sheerness Guardian*[24] reported that the "mysterious light" travelled at great speed towards the harbour, where it "turned sharply and went off back in pretty much the same direction as it came. The sound of an aerial engine was heard from the direction in which the light (was there more than one light?) appeared."

Although this engine was fast, implying it was one of the early airplanes, most people at the time assumed it was an airship, or dirigible, of some kind. At that time Germany had a few advanced Zeppelin airships, but only one of which was considered capable of having overflown Sheerness – the *L.I. (LZ14)*, which at the time had just begun a marathon flight across Germany – perhaps it could have been blown off course? Or perhaps it could have "accidentally on purpose" strayed over the east coast of Britain? At that time there was no love lost between the two countries: the First World War was just over a year away, and tension, not to mention suspicion amounting almost to mass paranoia, was high.

The Editor of *The Aeroplane*, Mr C.G. Grey, noted that the airship may have been the *Hansa,* privately hired by Prince Henry of Plesse and a party of ten German officers with the intention of visiting "a certain nobleman" (presumably British), but the weather had made a landing impossible.

Then another unknown dirigible was spotted on the afternoon of 3 December over Portsmouth, flying the White Ensign and apparently carrying four passengers. Rumours spread and excitement grew, but disappointingly this "phantom" airship turned out to be the British Army airship *Beta II* on its way on a return journey from Farnborough to Portsmouth.

Then on 4 January 1913 three witnesses – one John Hobbs and two policemen – heard the sound of a powerful engine overhead and witnessed a huge dark object, together with a light, as before, venturing inland. One of the constables reported admiringly: "It could only have been a very powerful engined aircraft to have flown in such a wind, and daring airmanship was also involved in the flight."

The previous day, the *Clément-Bayard IV/Dupey-de-Lôme* of France was known to have been in service just over the Channel, so it was wondered if it had been blown off course. There were, however, several other suggestions, including the Zeppelin *Hansa* and – much more bizarrely – a new motor boat showing off, although that fails to take into account the anomalous accompanying light and the fact that this was an *aerial* phenomenon.

Although there were other sightings, the "scareship" flap only really took off on that magical date, 17 January[25] when unexplained aircraft were seen in several areas of Britain, but mainly South Wales, which continued to act like a magnet for the phenomenon for a while. In early February it was the turn of the north-east and Yorkshire, and, after a brief period when nothing happened, the mystery dirigibles were back in force. Many hundreds of people saw something in the sky that they interpreted as German Zeppelins. Their feelings at that particular time can only be imagined, particularly if they knew just how unprepared the British Government and military were for any show of Teutonic aerial might. In *The Standard* on 25 February 1913 a Royal Flying Corps officer from Aldershot had warned about the pathetic state of British preparedness, writing:

> We are helpless. We have neither aeroplane nor dirigible capable of coping with these vessels in the air . . . The fact is that we have been hopelessly left behind in military aeronautics, and there seems little prospect of any advance being made as long as responsible Ministers give public utterances of their being content to wait to pick the brains of foreigners.

Hoaxes, pranks and the inevitable "explanations" of misidentified stars and planets, searchlights, balloons and so on, seemed to mark the end of the scare, although there were a few reported from time to time for a few more months. The populace remained firmly convinced that the Germans were to blame, which did nothing to quell the suspicions of that nation when they suffered their own airship scare in March 1913 over the Eastern provinces. The London *Daily Telegraph*[26] reported that two women witnessed the progress of a strange craft, which:

> . . . had two cars, and while they were watching it a black cloud of smoke suddenly rose from one of these. Then flames appeared, and quickly enveloped the hull of the airship, which began to fall rapidly towards the earth. Just before it reached the tree-tops one of the cars became detached, and the vessel, thus lightened, soared rapidly upwards.

Firemen, a large detachment of troops, ambulances and police raced to the area, but failed to locate the wreckage. Although there was a rumour that someone had smelt gas from the punctured infrastructure, this was dismissed because airships used odourless hydrogen, which in any case would have dispersed rapidly. The only rational explanation was that the two women had seen sparks from the exhaust of a test plane under the command of one Lieutenant Zwickan.

It is interesting that the scareships were interpreted by the British (and, incidentally, the French and Belgians) as being German – probably carrying spies – while the Germans thought the anomalous craft hailed from England, from perfidious Albion. At least, as Nigel Watson points out, in both countries the scares had the effect of galvanizing the build-up of air defences, not to mention death-dealing aerial weaponry. But although the pre-war tension, rapidly escalating into something approaching full-blown paranoia, may have coloured the perceptions of the scareship witnesses, the origins of the craft remain unknown to this day. As Watson muses: "Are ambiguous lights in the sky interpreted according to prevailing social pressures and concerns allied to individual fears of invasion and attack?" Then he asks the great Fortean question: "Or is there something or somebody out there playing games with us?"

Like many other UFOs, these craft seemed to taunt and distress those on the ground with scraps of evidence of their existence – noise, lights, shadows, sparks, and so on – seemingly proving their hard, solid reality. But then they disappear without trace, leaving a bewildered and frightened population believing their skies were being overflown by representatives of a nation that was shortly to become their enemy.

Perhaps the scares were cleverly organized by British and/or German Governments to be tantalizing glimpses of something terrible looming in order to prepare the people for the imminent conflict, although in the case of Britain's woeful unpreparedness, that seems highly unlikely. Perhaps the rising war-hysteria induced false perceptions: ensuring that people took perfectly ordinary experiences to be extraordinary because emotion had temporarily hard-wired their unconscious, dreaming minds to their consciousness, so what they saw had a dream-like quality. After all, history has repeatedly demonstrated that hysteria is a powerful creator of the most convincing – and terrible – special effects.

But hysteria, even if it could explain a few of the sightings of mystery airships, by no means explains them all. Unfortunately, as with many aspects of this phenomenon, even the most likely-sounding explanations turn out to conceal mysteries of their own. The favourite explanation for the airship flaps in the United States in the 1890s was that they were the creations of a mystery inventor, whose hangars lay hidden in remote rural areas, and who took them for outings where they were seen. At first glance, there did seem to be hard, checkable facts to support this theory, for one such "secret inventor" actually came forward and announced himself. Yet although that might seem to have clinched the matter once and for all, things became considerably weirder . . .

In September 1909 Wallace E. Tillinghast, the Vice President of the Sure Seal Manufacturing Company of Worcester, Massachusetts, approached the newspapers with an astonishing tale to tell. An eminent and reputable local man, he claimed he had invented and flown a dirigible, "capable of carrying three passengers with a weight limit of 200 pounds [90 kg] each, a distance of at least 300 miles [480 km] without a stop to replenish the supply of gasoline, and if necessary, at a rate of 120 miles an hour [192 km/h]."

Tillinghast claimed that on 8 September he had flown around the Statue of Liberty and then on to Boston before returning to New York without landing once. The newspaper reports went on:

> Another part of this trip is still more wonderful. Mr Tillinghast says that when near Fire Island [off Long Island], one of the cylinders of the flier ran irregularly, so the motor was stopped, with the machine 4,000 feet [1,200 m] in the air, and sailed forty-six minutes, while two mechanics repaired it in midair, the engine being started again when the airplane was near[27] enough to land to be seen by a member of the lifesaving crew patrolling the beach.[28]

Yet as such a machine and such a flight were completely impossible at that time, there is only one conclusion: Tillinghast had fabricated the entire story, although he stood to lose everything with such a bare-faced lie. None of it makes sense, but as John Keel points out in his *Operation Trojan Horse*, Tillinghast's apparently crazy announcement came at a very interesting moment – "when a massive UFO flap was *about* to inundate the New England states".[29] Thousands of witnesses reported mysterious airships during December 1909, but they were *preceded* by Mr Tillinghast's extraordinary boast. The newspapers reported his claimed that he had made "over 100 successful trips, of which 18 have been in his perfected machine. His latest airplane is so perfect and adjusted so correctly that upon being taken from the shop it immediately made uninterrupted trips covering 56 miles [89 km]."[30]

In December 1909 the flap really began. Just after midnight on 19/20 December, the inhabitants of Little Rock, Arkansas, were astonished by a "cylindrical shaft of light, which, arising from the southeast horizon, stretched athwart the firmament far to the east."[31] Local astronomers confessed themselves completely bewildered by the phenomenon. That same morning the residents of Boston, Massachusetts saw "a bright light passing over" which may have been "an airship of some kind".[32] And it must be remembered that this was the alleged airship supremo, Mr Tillinghast's, home town. Was he flying a new, improved version of his innovative airship? Indeed, when the flap erupted all over the eastern United States, with powerful searchlights scouring the countryside below, not unnaturally, reporters wasted no time in doorstepping Mr Tillinghast's home, seeking for confirmation from the great inventor himself. Unfortunately, he was not there, although Mrs Tillinghast said rather repressively, "My husband knows his business. He'll talk when the proper time comes."

The next night, anomalous lights were observed in many places, flying against the wind – which ruled out balloons. The phenomenon was succinctly described by the Providence, Rhode Island, *Journal* of 24 December 1909:

> As on Wednesday night, the light was first reported passing over Marlboro about 6:45 o'clock. The light, which was at a height so great as to make impossible a view of its support, disappeared to the southwest in the direction of Westboro and Worcester.
>
> It was traced from North Grafton, not far from Worcester, through Grafton, North Grafton, Hopedale, and Milford, and then after being lost sight of reappeared in Natick about 7:30 o'clock, going in the direction of Boston. Observers are positive that it was a searchlight.
>
> At 7:45 it was seen from Boston Common, by the testimony of several persons, among them men who were at a prominent clubhouse on Beacon Hill.

> At Northboro and Ashland, early in the evening, the population turned out en masse to watch the light pass overhead.
>
> Observers at several points report that while the light was generally steady, occasionally it flashed, and once or twice it disappeared entirely.

Was this the local hero (or anti-hero) Mr Tillinghast up to his old tricks? Apparently not, for when local reporters finally tracked him down, he told them:

> I was out of Worcester last night. Where I was is my own business. It may be that I flew over the city, but that is my own business, too.
>
> When I said recently that I had flown from Boston to New York and returned, I said nothing but what was true. I have an airship which will carry three or four persons and will make the speed I claimed for it – that is, about one hundred twenty miles an hour.
>
> When I get ready, I shall speak fully and not until then.[33]

As the furore grew, Tillinghast's lips became even more firmly buttoned. As the Providence, Rhode Island, *Journal* commented:

> Tillinghast is absolutely incommunicado. The notoriety that has followed him since the mysterious lights were seen has seriously interfered with his business and with his homelife. He has not been permitted an hour's peace. At his office there are constantly two or three persons who want to know something. At the door of his place of business and at his home he is closely watched by mysterious men. When he is home, his telephone rings constantly. As his wife has only recently recovered from an illness, the constant clangor is not conducive to his good nature.

Who were the "mysterious men" who "closely watched" Mr Tillinghast? And surely an experienced man of business like himself would have expected such a response from the public after making such astonishing claims. One wonders if he was suffering from some kind of mental problem, perhaps a *folie de grandeur*, or a nervous breakdown. Perhaps he was just quietly mad. Yet there is considerably more to this peculiar little story than meets the eye.

John Keel reports[34] that an unidentified "staff correspondent for the United Press" was reportedly arrested for trespassing when he tried to get to the bottom of the Worcester mystery . . ." The journalist, acting on a tip-off, arrived at the estate of one John B. Gough, just outside the city of Worcester, Massachusetts, where he found an 100-foot long shed hidden in thick forest land. His syndicated UPI report has this to add to the growing mystery of the airships:

> Fourteen men in the employ of the Morgan Telephone Company of this city were at work there on some secret occupation. Paul B. Morgan, head of the telephone company, is a close friend of Wallace E. Tillinghast, who is supposed to be the inventor of the mysterious flying machine . . . Morgan has been interested in aviation for several years, and two years ago he spent $15,000 trying to perfect a machine invented by a Swedish aviator. The Swedish invention, however, proved unsatisfactory and was abandoned . . . John D. Gough, on whose estate the shed was found, is an old-time temperance lecturer and is friendly with Tillinghhast and Morgan . . . .
>
> The secrecy maintained at the Gough estate and the careful manner in which the shed discovered today is being guarded lends new weight to the belief that a marvelous ship has been constructed.

The article added a warning note: "The correspondent was taken before the justice summarily today, and the swift manner in which he was prosecuted for trespassing is believed to have been employed as a warning for others who might attempt to invade the secrecy of the airship plant."

But *what* "airship plant"? To this day, nothing more has emerged about its very existence, let alone its production of cutting-edge dirigibles. So what on earth was the whole Tillinghast saga about? Assuming that he – and his associates Paul B. Morgan and John D. Gough – were not mad, what was going on? If it was some kind of confidence trick on their part, it was singularly unsuccessful, for not only is there no evidence that they ever made a cent out of the airship flap, conversely there is abundant evidence that Tillinghast, at least, was damaged by it, both professionally and personally.

John A. Keel theorizes that Tillinghast – "one of the most prominent and reputable members of his community with a track record as an inventor, was approached by a man or a group of men who offered to take him for a ride in a marvelous new "secret" aircraft. Mr Tillinghast was a man of science, and he was far too curious to reject such an opportunity. He went to an isolated field and climbed aboard the machine he found there. His hosts kept their promise and flew him around the countryside, perhaps even to Boston and back."

The mysterious pilots, Keel suggests, put a proposition to Tillinghast: "They struck a bargain (which they had no intention of keeping), and perhaps they offered him a large interest in the profits from their flying machine, provided he did exactly as they ordered during the next few months."

Keel's theoretical reconstruction of the deal goes on: "They explained that they needed a responsible, respectable man to front for them while they ironed the bugs out of their invention. They appealed to his ego, saying that they were interested only in giving their airship to the world, and they didn't care if he took full credit for it. After the machine was fully tested, they promised, they would turn it over to him, and he could make all the arrangements for manufacturing more of them. He could also claim full credit for inventing it. They, the real inventors, would happily remain behind the scenes."

What a golden opportunity for the enthusiastic inventor Mr Tillinghast! There is no doubt that he would have seen this as an offer he could not refuse. Keel suggests that he "accepted the proposition, visions of glory dancing in his brain. The machine had been *proven* to him. He was convinced of the reality of the trip he had taken. When reports of mystery airplanes started to filter into the press in early December, his mysterious friends called upon him and told him that it was time to disclose the existence of the invention. Tillinghast dutifully appeared before the reporters, revealed that he had already made a number of flights, and that the invention would be fully unveiled at an appropriate time in the near future."

Messrs Morgan and Gough may have been party to a similar

approach, and may even have made the same sort of agreement with the mystery airship inventors. But it all came to nothing. As Keel says sombrely, "Like so many of the modern UFO contactees, they were *used*." (My italics.) It certainly appears that Tillinghast was taken for a ride in more ways than one, but by whom? Who were the genuine secret inventors who hid behind him – and eventually threatened to ruin his reputation? Where did the airships come from?

As we have seen, the scareship flap just before the outbreak of the First World War prompted profound fears of superior enemy technology (which as it happened, was non-existent) and here we have prominent citizens being taken for fools, while impossibly advanced airships continued to fly about the skies, faster and more agile than anything on Earth . . .

Who or what was gleefully playing with us, as a cat toys with a mouse? If they were the same entities who later transmogrify into UFOnauts in their flying saucers, first preaching brotherly love and then, a generation later, allegedly abducting and abusing vulnerable humans, the over-riding question is, surely, *why* . . . . (Perhaps one should bear in mind that cats do not merely play with hapless birds or mice and then let them go. After having their fun, they go in for the kill.)

But while the mystery airships had apparently had their day with the coming of the First World War in 1914, that was by no means the end of the phenomenon of strange things seen in the sky, and alien entities associated with them encountering humans – usually to the latter's ultimate bemusement, distress or even ruin. Anomalous blobs of light darted around – and occasionally even in – fighter bombers of the Second World War, where, just like the scareships of the 1900s, the Axis Powers blamed the Allies, and the Allies blamed the Axis powers. But it was as if the phenomenon was looking for an image, an instantly-recognisable brand-name as it were, in order to strike a chord deep within the human psyche. And that image, perhaps born of a poetic turn of phrase, was not long in coming. Soon, very soon, more unidentified flying objects would haunt our skies, disturb – perhaps even shape – our dreams and infect us with a longing to make contact with something Other and further. Barely was the apocalypse of the Second World War over when, parallel with the coming Cold War, came the shadow of the saucers, but whether they brought hope and balm after the great scarring of the nations' souls, or another kind of deadly threat, remained to be seen.

Notes

1. In the *Literaturnaya Gazeta* of Moscow, 1969.
2. In a series of books, such as *Chariots of the Gods.*
3. Which may or may not have actually taken place. So far archaeologists have uncovered no evidence to support the idea that the Israelites were slaves in Egypt.
4. See for example, p. 70 of John A. Keel's classic *Operation Trojan Horse* (London 1971).
5. The late American astronomer, who although relatively open-minded – he suggested that there might be something to the "pyramids on Mars" theory – was basically a mainstream scientific rationalist. However, one of his legacies was the script of the movie *Contact*, starring Jodie Foster, about an astronomer who has a profound mystical experience as a result of apparent contact with extraterrestrials.
6. Apparently out of no fewer than 366 books, only one of which, *The Secret Book of Enoch*, is available today. Only a few lines about his experiences appear in the Bible (Genesis, Chapter 5).
7. P. 73 *Operation Trojan Horse.*
8. It seems like a drug-induced vision – which may indeed have been the case. Eating hallucinogenic mushrooms or contaminated grain (ergot poisoning) could have induced such apocalyptic visions, or any of the many available techniques for inducing a shamanic trance, such as prolonged fasting or whirling to a drum beat. Shamanistic practices were commonplace in the ancient world: there is evidence (see Picknett and Prince, *The Stargate Conspiracy*) that the ancient Egyptian college of Heliopolis was based on shamanic practices. See also Jeremy Narby's *The Cosmic Serpent.*
9. From a letter to *Fortean Times* No. 127. Translated from the original French:

   "*. . . il arriva qu'un jour entr'autres, on vit a Lyon decendre de ces Navires Aeriens, trois hommes et une femme; toute la Ville s'assemble a l'entour, crie qu'ils sont Magiciens et que Grimoald Duc de Bennevent, ennemi de Charlemagne, les envoye pour perdre les moissons des Francois. Les quatre innocents ont beau dire pour leur justification qu'ils sont du pays meme, qu'ils ont ste enleves dupuis peu par des Hommes miraculeux qui leurs ont fait voir des merveilles inouies, et les ont pries d'en faire le recit . . .*"
10. Johann Wolfgang von Goethe, 1749–1832.
11. From the sixth book of Goethe's autobiography, as pointed out to Dr Jacques Vallée by American cult film-maker Kenneth Anger.
12. See the Chapter on Explanations.

13. Even, disappointingly, for the Kenneth Arnold classic. See page 42.
14. On 24 January 1878.
15. See Gervase of Tilbury's *Otis Imperialia.*
16. As recounted, for example, in Jacques Vallée's *Passport to Magonia,* pp. 144-145.
17. As reported in the *Houston Post* of 26 April 1897.
18. 1847-1931.
19. Quoted on p. 34 of Keel.
20. *Ibid,* p. 103.
21. See Vallée, pp. 146-147.
22. See p. 109 of Keel.
23. *Strange Magazine,* No. 13, Spring 94. This was the second of a two-part article on the scareship mystery.
24. 23 November 1912.
25. The date of 17 January is associated in esoteric circles, in particular the controversial occult/Johannite group the Priory of Sion, with completion of the alchemical "Great Work", when the alchemist transcends mortality and is spiritually, and it is said, physically, transfigured. Some esotericists and conspiracy theorists claim that cabals within the governments of world powers ensure that key events take place on or around 17 January, such as the outbreak of the Gulf War. For a discussion of the alchemical theory about this date and the Priory of Sion, see *The Templar Revelation* by Lynn Picknett and Clive Prince.
26. 14 March 1913.
27. Keel, p. 112.
28. *Ibid* p. 113.
29. Portland, Oregon, *Journal,* 23 December 1909. Quoted in *Ibid,* p. 114.
30. The Arkansas *Gazette,* 20 December 1909.
31. The New York *Tribune,* 21 December 1909.
32. Keel, p. 110.
33. *Ibid.*

CHAPTER 2

THE INVADERS

Cases of UFO sightings from different places and times in the modern era.

Even in a book this size there is not enough space for all the classic cases of UFO sightings, let alone a truly representative selection of lesser-known, but perhaps equally fascinating, stories from around the world. Therefore, apart from the essential stories – such as Rendlesham Forest and, of course, Roswell, which has a chapter of its own – I have chosen to present a very varied selection. Some are well known, at least to UFO buffs, but others will be completely new to everyone because they are the stories of people who have exclusively confided them to me (with strict instructions to keep their identities secret). Some of these deeply personal tales came from family – my own father's story is included – and friends, while others originally came to my notice through radio phone-in shows, after lectures or through my former column in *Woman* magazine. In one or two cases I have blended two stories in order to disguise the identity of the witnesses, but otherwise they are exactly as originally told.

Although such cases are purely anecdotal – and therefore probably deeply irritating to UFOlogists who consider themselves more "purist" or scientific – I still feel they have much to offer because of their freshness and very lack of conformity to any particular hypothesis, influence or trend. These people had no interest in UFOs before – and often even after – their experience. They recount what they thought they saw with a refreshing honesty and directness, and, as far as I could ascertain, have no reason to fabricate their claims. Like many of those involved in the more famous cases, these are ordinary people caught up in extraordinary events, and the stories they tell are all the more fascinating because they have never been told before.

I have also included witnesses' statements culled from magazines and the like, because although they are transient, and were often never investigated or followed up, such clippings are, to my mind, an intrinsic part of the overall UFO story.

Where the classic cases are concerned, I have presented a fair sprinkling, from the controversial Polaroid pictures of UFOs taken by "Mr Ed" at Gulf Breeze, Florida, to the involvement of the US military at Rendlesham Forest and the flap that ended with demonic creatures terrifying the people of Varginha, central Brazil in more recent times. Inevitably, though, some readers may search in vain for their favourite case, and some that are presented as honest testimony may, by the time this book is published, have been proved to be hoaxes – or, indeed, *vice versa*. The world of UFOs is very fluid, and feelings can and usually do run high among both the "True Believers" and sceptics.

This book's ambition – which some may think an impossible dream – is to appeal to both the committed UFO buff and the general reader, those for whom this is their first-time dip into the UFOlogical pool. The notes and references will enable researchers to trace information to the original sources, and the list of books for further reading (the tip of a truly mammoth iceberg) will enable the reader to decide for him- or herself whether the truth really is "out there" – or in here . . .

One thing is for sure: there is enough mystery, conspiracy and magic in the subject to capture the imagination of anyone with a questing mind and a desire to understand humanity's place in the universe.

THE PRE-1947 SIGHTINGS

This brief section presents some of the strange phenomena that were reported before the landmark sighting of 1947, the prelude to today's greater UFOlogical drama.

13 AUGUST 1917, FATIMA, PORTUGAL

Although this has become a classic story of twentieth-century Catholicism, a case of visionaries and the "miracle of the dancing sun" at Fatima, Portugal in 1917, the testimony of some of the original witnesses suggests very strongly that this was *also* – or perhaps even exclusively – a multiple-witness UFO sighting, which was misinterpreted as a vision due to the intensely religious frame of reference of those involved.

About 70,000 people gathered in the meadows of a place known as Cova da Iria near Fatima, Portugal on 13 October 1917 to witness a prophesied miracle. There were newsmen in abundance, some with the movie cameras of the day: under a lowering sky and squally rain, all eyes were turned to the tiny figures of three children standing in the distance amid the churned-up mud of the field, under a small tree. The crowd waited. The children waited. Then just after midday, one of them – ten-year-old Lucia – let out an audible gasp, and raised her face to a vision of rapture, invisible to the multitude. This heralded the arrival – as promised – of "the Lady", who announced that she was "the Lady of the Rosary". A clear indication that she was the Virgin Mary, this was the first time that she had identified herself. She told the children that the Great War would end very soon (although in fact it lasted until November 1918).

Then something terrifying happened. The crowd screamed as one in horror as a *massive silver disc-shaped object* swooped out of the clouds, scattering fine strands of the mysterious material known as "angel hair" on the scene beneath, although the substance melted almost instantly. This has become the more acceptable "petals" in descriptions of the event in religious literature. The (to be literal) Unidentified Flying Object danced around the sky, and as it did so changed colour rapidly, keeping this up for about ten minutes as the crowd watched, fascinated despite their palpable fear.

Some of those who were there committed their memories of that amazing day to paper; one of whom, Professor Almeida Garrett of Coimbra University, wrote:

> It was raining hard . . . suddenly the sun shone through the dense cloud which covered it: everybody looked in its direction

> . . . It looked like a disc, of very definite contour; it was not dazzling. I don't think it could be compared to a dull silver disc, as someone said later at Fatima. No. It rather possessed a clear, changing brightness, which one could compare to a pearl . . . It looked like a polished wheel . . . This is not poetry; my eyes have seen it . . . This clear-shaped disc suddenly began turning. It rotated with increasing speed . . . Suddenly, the crowd began crying with anguish. The "sun", revolving all the time, began falling toward the earth, reddish and bloody, threatening to crush everybody under its fiery weight . . .[1]
>
> But the sun did not crash onto the earth. Instead, a wave of intense heat enveloped the crowd, instantly drying their sodden clothes – which was almost miracle enough, one might think. Some reported miraculous healings, which they ascribed to the beneficent presence of the "Lady".

Soon the story had become transmogrified into a modern myth. The UFO had become the sun, the angel hair rose petals. The children who appeared to act as focus, or magnet, for the phenomenon went the way of all visionaries: while Lucia was put away in a nunnery for life, her two little companions were to die not long after the miracle.

Yet even those who revere the children tend to forget the prelude to the extraordinary event known as "the dance of the sun", airbrushing some inconvenient details out of the true picture in their haste to create a modern myth of a religious miracle.

It happened like this: during the summer of 1915 four little shepherdesses of Cabeco, Portugal, saw a strange, white-shrouded entity hovering in the air. "It looked like somebody wrapped in a sheet. There were no hands or eyes on it," they told their parents, who tried to dismiss it as nonsense. One of the girls, Lucia Abobora, went on to have other, more momentous encounters the following year, together with two friends of roughly the same age. They observed a light moving just above the tree-line, which approached, transmuting into a "transparent young man", who seemed to be in his mid-teens. Shining radiantly like iridescent crystal, he bobbed down onto the ground and said: "Do not be afraid. I am the Angel of Peace. Pray with me." Obediently the children knelt and prayed with him until he simply disappeared into thin air. The angel reappeared some weeks later and followed the same procedure, establishing a strong bond with the children by praying with them.

A month after the United States entered the war,[2] Lucia and her two friends, seven-year-old Jacinto Marto and Franciso Marto, nine, were with their sheep close to the area of Cova da Iria when they all noted a peculiar "flash of lightning", although the sky was perfectly clear. Fully expecting a torrential downpour, they dashed off for cover under an oak tree, but someone had got there before them . . . Hovering just above a nearby tree was a radiant sphere of light, in which was a being clad in a white robe, with a countenance so brilliant that "it dazzled and hurt the eyes", as the children said later. This angel seemed to be feminine, saying in a modulated, musical voice: "Don't be afraid. I won't hurt you." Utterly over-awed, the children nervously asked her where she came from, to which she replied: "I am from heaven. I come to ask you to come here for six months in succession, on the thirteenth day at this same hour. Then I will tell you who I am, and what I want. And afterward I will return here a seventh time."

Full of their astonishing news, the children ran home and told their families, who declined to believe them. Even so, the story had spread, and the local people – no doubt recalling the story of Bernadette of Lourdes – gathered at the appointed place at the time of the next contact with the Lady. The children, eyes fixed on an invisible being, abruptly knelt and began talking. A woman called Maria Carreira claimed later that although she saw nothing, she heard a sound like the buzzing of a bee, while another witness was to report hearing a noise like a horsefly inside an empty pot – presumably an angry buzzing sound. (As we will see, buzzing is frequently reported by UFO contactees and witnesses, possibly because there is a distinctly electrical or electromagnetic element to the phenomenon.)

In his 1971 classic *UFOs: Operation Trojan Horse,* John Keel uses the Fatima story as an example of his "ultraterrestrial" theory, the idea that the power behind the UFOs was, and is, a realm of shape-changing beings who are normally invisible to us but who occasionally choose to interpenetrate our reality. Morally ambiguous – sometimes they appear to be "good", but more often seem considerably less altruistic, they seek to guide our civilization, especially when our actions threaten them in some way. Other writers, notably Jacques Vallée, Hilary Evans and Patrick Harpur have taken a similar line.

Keel believes that little Lucia had been singled out by the beings for contact because she had shown a natural talent for what would in other circumstances be called trance mediumship – acting like a gateway for the "miraculous" phenomena to manifest. The middle of the First World War, it was a time of great global upheaval and the entities needed to "correct our course",[3] choosing to use the time-honoured method of 'signs in the sky' to do so.

The entities gave three predictions to the children, the third of which has allegedly been revealed to concern the attempted assassination of Pope John Paul II, although – especially as he survived – it seems curiously low-key for the massive build-up bestowed on it over the years. (And it is said it caused one pope to faint dead away, which also seems very odd under the circumstances.) Predictions are often given to "contactees", but usually only the first few come true: the last and most apocalyptic seem designed to bring ridicule and ruin upon the "prophet". In this case, two of the children died very soon after the visions, while Lucia spent her life hidden away in a convent, which may or may not be considered a bad fate.

While the "ultraterrestrial" theory remains controversial – and is little short of anathema to many who hold the Extra Terrestrial Hypothesis dear – it does provide many of the answers to some of the more complex and paranormally-tinged cases, which are arguably the majority.

It is also interesting that Whitley Strieber, author of *Communion* and perhaps the world's best-known abductee, believes the Fatima incident to have involved a genuine UFO.

AUGUST OR SEPTEMBER 1938, NEW APPOLLONIA, THESSALONIKI COUNTY, NORTHERN GREECE

This story, although unashamedly a "foaf" (friend-of-a-friend) account, and may therefore have lost or gained something in the retelling over the years, seems to have its roots in a genuine experience, and is all the more interesting because it predates the post-1947 UFO age.

Recounted in *Strange* magazine by Greek contributor Thanassis Vembos,[4] the anonymous narrator – who Vembos calls "A.I." – tells how his father, "P.I.", encountered two bizarre non-human entities and an unknown craft near the village of New Appollonia, in northern Greece in August or September, 1938.

This is his account:

> . . . My father had gone to Ladja, an area near Volvi Lake, in order to cut down trees and collect dry leaves. He left before sunrise with his donkey. They had reached Ladja, and both were crossing an area with bushes. Suddenly, he saw two people in a clearlng. He was astonished and approached them carefully. There was enough light, as the Sun was to rise in a couple of minutes, so my father could see them clearly. These two "men" were tall, taller than an ordinary man, with big heads and short hair. Their eyes were red and were staring at him in a very peculiar way.
>
> . . . Both had dark skin, not brown like a negro's skin, but red, bronze, suntanned. Their faces seemed sunburned and bloated. They were wearing uniforms, something like English military outfits. The two "people" were standing in front of a large "thing." The object was egg-shaped and had a height of about 3 meters [9 feet]. Its width was enough for two standing occupants and the craft was standing upon 3 or 4 "legs". Apparently it was half metallic and half glass (its upper half was transparent). There was an opening on its metallic portion, like a manhole, and a little ladder with 3 or 4 steps. P.I. started shouting, calling the two "people".

However, as he was carrying an axe, they seemed to be afraid of him, so,

> They stepped back, climbed the ladder, and entered the craft. Now they were clearly visible within the crystal half of the craft. One of them was staring at the witness and the other one was doing something unknown. They pulled the ladder inside and

> suddenly there was a noise, like something that was activated. Then on the top of the craft, something like a balloon appeared, or rather inflated, and the craft took off vertically. During its ascent no flames or smoke were visible, and no noise was heard. P.I. thought that one of the craft's occupants waved once at him. Then the witness took some steps and when he looked into the sky again, the object had become as small as a bird. P.I., upon recovering from the shock of the incident, examined the area of the landing. He found a bottle of about half a liter. But he was afraid to touch it, so he pushed it with his axe. The bottle fell and its cork blew off. A thick liquid poured out. The liquid combusted spontaneously, and set fire to the nearby grass and bushes. Some drops of the unknown liquid fell on the axe's stick and immediately set it on fire. P.I. put out the fire by burying the axe in the sand. Then he carefully took the half-burned axe and went back home. Later he threw away the axe stick, fearing more unpleasant influences.

The witness's son recalls that, "When he told his story in the village, everybody laughed at him. I don't know if anybody went to the site of the incident, to check out his claims. But I was his favourite son, and I believed everything he said. And he trusted me."

Thanassis Vembos added that A.I. was twenty years old at the time of the incident, while his father was then in his mid-fifties. He was a well respected local shepherd who inspired something like awe with his ability to hypnotize snakes and make them wrap themselves round his waist like a belt. Unfortunately, P.I. has since died, so there is no way of corroborating the story.

This account is particularly interesting because both the craft and its occupants are unlike today's more familiar disc-shaped UFOs and the tiny Grey aliens or even the tall blond Venusian types that enjoyed a vogue in the 1950s. The egg-shaped craft, with its half metallic and half glass superstructure, seemed to be just large enough to take two upright occupants, who were tall and sunburnt, with bloated faces, wearing uniforms similar to "English military outfits". The craft took off vertically – something beyond the capabilities of 1930s technology – making no sound. The contents of the bottle they left behind, although unknown, were extremely flammable, causing the grass and axe handle to catch fire, which if nothing else indicates that this episode was no mere hallucination or dream.

These aliens do not appear to have chosen to appear to P.I. On the contrary, he seemed to have found them accidentally, startling them. Nor had the craft crashed or, apparently, broken down. Perhaps they had merely stopped briefly to stretch their legs or to reconnoitre. Also, they seemed friendly enough – P.I. thought he saw one of them waving as they took off – and hardly the abducting, terrifying sort of non-human entity. But who were they and where did they come from?

AUGUST, 1939. GUISBOROUGH, CLEVELAND, NORTH-EAST ENGLAND

"Jack Quinn",[5] now a retired businessman from Newcastle-upon-Tyne, was ten years old in August 1939. On the day of his strange experience, he and his mother "Annie" took a walk because it was too hot and airless in their cramped terraced house. They left the little town of Guisborough, Cleveland, behind them, climbing up the local beauty spot, Rosebery Topping, which commands a stunning view of the local countryside.

"We walked on over the top," Jack recalls, "where there is a sort of field that dips down over the other side of the hill, full of ling and heather that covers some nasty unexpected holes. Being local, we knew of the dangers underfoot and were careful to pick our way, holding hands and concentrating hard. Our aim was to get down to a track that ran round the hill and down to the bottom again. But suddenly Mum just froze. She motioned to me to keep quiet, and whispered: 'Listen'."

Jack went on:

> There was a strange humming sound, apparently coming from the ground under our feet, which seemed to be vibrating. We'd never heard of anything like that. At that age you usually take your cue from grown-ups, so after one look at Mum's face I was scared, too, although she was trying to be brave for my sake. She grabbed my hand and pulled me across the heather, but in our

haste I slipped and hurt my ankle down one of the natural ditches. As Mum knelt down to have a look, the noise got louder and there was a strong smell of something burnt, like burnt paper, then a big white globe of light suddenly appeared from nowhere, there in the middle of the field. I couldn't run and Mum wouldn't leave me, but what wouldn't I have given to just leg it out of there!

The sphere of light didn't move at first, just sat on the heather about four or five yards away. I suppose it was about six feet [1.8 m] across – quite large enough to terrify us half to death. Then two little men appeared on the brow of the hill. Even though they were about thirty yards [27 m] away, we could see they were tiny, maybe about three feet [900 cm] high. They were wearing some kind of shiny material, a sort of light greenish colour, and had close-fitting helmets made of a similar kind of material. They were jabbering away to each other in high-pitched voices. Then they looked across at us and pointed, getting excited. The globe of light moved closer to us and Mum – never a keen church-goer – held me close and started praying. I honestly had no idea what was happening but didn't like it much. I must have been crying because afterwards there were tear stains on my cheeks.

The light rolled right round us very slowly, as if taking a good look, while we clung to each other and cried. I guess it circled round leaving a gap of only about four feet or so between us and it. Then it stopped, although the humming sound got louder under us in the earth. Mum was saying "Oh God, Oh God, what's happening" over and over again between sobs, but she didn't shut her eyes. She wanted to see what was going on.

Then suddenly everything stopped. The humming noise and vibrating disappeared and the globe of light and the little men seemed to dematerialize. Puf! Just like that, into thin air, as if it had never happened. We got up, brushed ourselves down, and then Mum helped me to hobble across the heather.

However, that was not quite the end of their problems on Rosebery Topping that day. As they made their slow and painful progress across the field to the track, they distinctly heard some weird high-pitched laughter, which seemed to come from the air. Then there was a sudden, unnaturally complete stillness. As Jack says, "Most of the time, even in rural areas, there is some kind of background noise – a dog barking or an engine running somewhere. Even just birds. But up there that day all that just stopped. The light turned strange – sort of nasty thick yellow, and there was that ugly burning smell again. We both felt very tired, dog tired, and were tempted just to fall asleep up there on the hill among the heather. God knows what would have become of us if we had."

> But Mum was a fighter. Something made her realize we'd be lost if we lay down and slept. So she struggled to keep me awake and alert – and it was a real struggle, my eyes were closing and my limbs felt like lead – and gradually I came awake again. The thick yellow light had gone and we set off again downhill to the track. But then we couldn't find the way. I can't tell you how odd that was. We'd been up there dozens of times and anyway once you get to that point you can actually see the track, so there's no possibility of getting lost. But somehow we did. Neither of us wore watches, but we must have stumbled around in that field for something like half-an-hour – at least that's what it felt like. We were both panicking, although Mum tried to keep our spirits up by chatting about our forthcoming holiday (which we never had, as things turned out, because of the war).

Suddenly, as if by magic, the Quinns found themselves on the track – but not at its highest point, as might be expected, but much further down, within yards of the bottom, where the hill met the road proper. They seemed to have been transported hundreds of yards down the side of a steepish hill. Once reality started to bite once more, they recognizd other anomalies.

"We'd gone up the hill at about two in the afternoon," explains Jack. "But although on our reckoning it should only be about four at the outside when we got back to the road, it was six thirty. Dad was scouring the neighbourhood for us – it was a rare thing for his tea not to be waiting for him! We tried to explain what had happened, but although he was sympathetic he thought we'd just fallen asleep and dreamt it. I've never heard of two people having the same dream – and anyway, that must be the longest dream on record!"

Telling their tale was a sobering experience. They realized that nobody would believe them, although they did ask others if they had seen anything odd up on the hill. One neighbour, a policeman, said he and two mates on their day off had encountered what looked like a small shiny airship that made a disproportionately loud humming sound but vanished into thin air before they could investigate, and a twelve-year-old girl said she'd seen an "angel" among the heather (and been slapped by her father for telling lies).

Jack told me the story when I visited my aunt, now deceased, in the area some years ago, while investigating the paranormal. He insisted on absolute confidentiality, because as a successful businessman and a leading light of the local Conservative Club he thought having what he called – possibly with more accuracy than he knew – his "fairy story" publicized would do his reputation no good at all. Unfortunately his mother was by that time in a nursing home and would not be able to recall the event. Like many others with an unexplained story to tell, he said "I'm glad I've got that off my chest. I've never told anyone about that, not even my wife. It's a huge weight off my mind."

(It would be interesting to know if any other older inhabitants of that area have a similar story to tell.)

Although this experience happened on the very eve of the outbreak of the Second World War, neither Jack nor his mother ever associated the incident with the German menace. Nor, as far as could be ascertained, did the policeman friend and his mates, even though they saw a small *airship* up on Rosebery Topping. This episode had no tinge of war-hysteria about it. It seems that there is something much older and more atavistic involved, despite the machine-like globe that seemed as if it was monitoring the Quinns. Perhaps a clue lies in the high-pitched laughing that they heard, followed immediately by total loss of a sense of direction, even though they knew the place extremely well. In parts of Britain this phenomenon is known as being "*pixie-led*" . . .

While the fairy realm is discussed in the Explanations section of this book, suffice it to say here that a great many UFO/alien cases involve classic elements from fairy stories, such as missing time, absurdities, and the sense of being played with. As we will see, fairies are believed to be shape-shifters who change their appearance to match the times, the better to astound, bemuse and – perhaps ultimately – to lure mankind to its doom. They are mischief makers writ large, who delight in riddles, paradoxes and the wrecking if not of an individual's sanity, then in the loss of his reputation. Few walk away from a brush with the fairies totally unscathed, although neither Jack nor his mother suffered any ill effects (apart from a mildly sprained ankle and general bewilderment). Perhaps the fairy folk – if indeed, that is what they were – took pity on them, or maybe they just got bored.

24 JUNE 1947: MOUNT RAINIER, WASHINGTON. THE KENNETH ARNOLD SIGHTING – THE DAY FLYING SAUCERS WERE BORN

The historic day of 24 June 1947 began with private pilot Kenneth Arnold, thirty-two-year-old father and head of the Great Western Fire Control Supply company, piloting his own light aircraft to Pendleton, Oregon from a job[6] at Chehalis Air Service, Washington State. Finding himself near the location where a missing aircraft was reputed to have crashed – and with the prospect of claiming the £5,000 reward for locating it – he built time into his schedule to look for it. As he approached Mount Rainier at an altitude of 9,200 feet (2,760 m), he saw a bright flash then, a "screwy formation . . . a chain of nine peculiar looking aircraft" flying from north to south at approximately 9,300 feet (2,790 m) elevation and going, seemingly, in a definite direction at an angle of about 170 degrees. The objects approached Mount Rainier at a very high speed, every few seconds a few of them dipping or changing direction – as they did so, glinting brightly in the Sun. At first he thought they might be geese flying in formation, but realized they were flying much too fast to be mere birds.

Arnold tried to determine their precise speed between Mount Rainier and Mount Adams, only managing to work out that it was "pretty fast". Breaking his journey at Yakima, he told his old friend Al Baxter, manager of Central Aircraft, about his sighting of the fast-flying formation, who then gathered a few pilots around to discuss the matter. As Arnold recalled in his 1952 book *The Coming of the Saucers*:

> I proceeded to gather my scattered wits together, got back in my airplane, and took off for Pendleton, Oregon. I remembered that I had forgotten to mention the fact that some of these craft looked different from the rest, was darker and of a slightly different shape, and that I hadn't told the Yakima boys that I had clocked the speed of this formation within fairly accurate limits. While flying to Pendleton I took my map from its snap holder on the extreme edge of my instrument panel, grabbed a ruler, and began figuring mathematically, miles per hour. Figuring and flying my airplane at the same time was a little confusing, and I thought my figures were wrong and that I had better wait until I landed at Pendleton to do some serious calculating.
>
> When I landed at the large airfield at Pendleton there was quite a group of people to greet me. When I got out of my plane no one said anything. They just stood around and looked at me. I don't recall just how the subject came up in those first minutes after I landed, but before very long it seemed everybody around the airfield was listening to the story of my experience. I mentioned the speed I had calculated but assured everybody that I was positive that my mathematics were lousy.[7]

However, his calculations seemed in order: to everyone's astonishment, the objects appeared to have moved at a speed of at least 1,200 miles per hour (1,931 km/h), double that of any contemporary aircraft, a fact that greatly disturbed those Cold War Americans.

The very early newspaper reports are telling. The *East Oregonian* of 26 June (two days after the sighting) had Arnold describing the objects as being "flat like a pie pan and somewhat bat-shaped", and the *Oregon Journal* of 27 June reported him as saying, "They looked like they were rocking. I looked for the tails but suddenly realised they didn't have any. They were half-moon shaped, oval in front and convex in the rear" – very different from the traditional UFOs, such as those described by George Adamski (see page 281). But in a radio interview for KWRC Arnold makes an interesting slip (if, indeed, that is what it was), by describing the objects as "ships" – as if he already had spaceships in mind . . .

Revealing a memorable talent for sound-bites, Arnold also described the objects as resembling "the tail of a Chinese kite" as it might fly in the wind, adding that "they flew like a saucer would if you skipped it across the water" – and so one of the most evocative and pervasive phrases of the modern era was born. Flying saucers had arrived.

In drawings he made at the time, it is clear that one of the objects had a darker wing-tip, which, some claim, may prove extremely telling (see below) – not quite the "flying saucer" shape we have come to associate with Arnold's sighting. In fact, the modern idea of what he saw is closer to the completely round UFO George Adamski claims he witnessed and in which he travelled the universe.

At first, news of the sighting was treated as a hoax, but Arnold's impeccable record and obvious integrity soon prompted people to take him more seriously. Soon, too, newspapers were full of stories about "flying disks" and "flying saucers" – which others were reporting, especially from the same area in Oregon. On Independence Day, 4 July, four harbour patrolmen, several police officers and others had seen disc-shaped objects "shaped like chrome hubcaps" moving fast and revolving as they flew.

On the same day, the Army Air Force issued a statement to the effect that the saucers had nothing to do with them – they had no secret weapon that would account for them, while the Air Materiel Command put forward their own suggestions as to what the objects could have been: the Sun reflecting off cloud; meteor splinters reflecting the Sun or large, flat hailstones gliding through the air. However, all those suggestions were roundly ridiculed by respected scientists in the Press.

Almost immediately after this, Captain E. J. Smith and his co-pilot of United Airlines Flight 105 saw five disc-shaped objects flying in a loose formation, which flew off, only to have four more take their place. A stewardess also witnessed the discs. Captain Smith later became a friend of Kenneth Arnold.

Saucer fever gripped America: everyone was talking about them. If not secret weapons, what were they? Where did they come from? The 6 July edition of the *New York Times* suggested various possibilities, adding, "They may be visitants from another planet launched from spaceships anchored above the stratosphere."

Within a year of his sighting, Arnold had also come round to the Extra-Terrestrial Hypothesis (ETH), writing an article for Raymond Palmer's *Fate* magazine entitled "Are Space Visitors Here?", suggesting that the discs were spaceships – the first magazine article to suggest such a possibility. (Arnold's article for the first edition of *Fate*, entitled "I *Did* See the Flying Disks!" included the detail that he had had "an eerie feeling" on seeing the saucers – something he had omitted from his original report.) He collaborated with Palmer on the 1952 book, *The Coming of the Saucers*, fuelling the ETH further among an increasingly saucer-hungry public.

But what had Arnold seen? Immediately after the event he hypothesized that they might be some form of secret weapon or Soviet aircraft invading US airspace, an idea that was reinforced by a former USAF officer who later said to Arnold:

"What you have observed, I am convinced, is some type of jet or rocket-propelled ship that is in the process of being tested by our government or even it could possibly be by some foreign government".

However, the authorities denied that they had any weapon that matched Arnold's description – and although such denials may make the conspiracy theorists' eyes light up, it does seem that on that occasion, at least, they were telling the truth.

In the intervening half century or so, there have been many attempts to explain Kenneth Arnold's watershed saucer sighting, but perhaps British UFOlogist James Easton has come closer than most to discovering the true nature of what Arnold saw all those years ago. In a *tour de force* of research,[8] he has set out the stages of his detective work, revealing the possibility that the "screwy formation . . . of nine flying disks" were, in fact, a flock of *white pelicans* flashing in the sunlight.

At first, this seems ridiculous. How can relatively small creatures look like large flying discs? And how on earth can they travel at such enormous speeds – twice the speed of sound? Easton's research doggedly deals with these problems.

His interest had originally been kindled by a comment from UFOlogist Martin Kottmeyer in July 1997, who suggested that

Arnold could have been deluded by a flock of swans.[9] Intrigued, Easton made inquiries about the likelihood of such an explanation among American ornithologists,[10] and discovered that Arnold's description was familiar to them – "suddenly," says Easton,[11] "we had a prime candidate for his sighting . . . a formation of American White Pelicans . . . Indigenous to Washington, [it] is the largest bird in North America and among the largest in the world. It weighs up to 33 lb (15 kg) and its massive wingspan can extend to 10 ft (3m) or more . . ."

"Predominantly white, with black wing-tips, these birds are highly reflective, often described as 'sparkling' or 'flashing' in the far distance, even when effectively lost from sight. They are also 'tail-less' with a 'bat-like' profile."

Easton remarked that Arnold was clearly "not familiar with the unusual undulating flight – like a roller coaster or 'skipping across water' – and 'kite-tail' appearance of pelicans in formation." In the course of his research, he discovered the following extract from science-fiction writer Arthur C. Clarke's non-fiction essay *Things in the Sky* (1958) about an experience he had in Brisbane, Australia some years before:

> The sun was low in the horizon – and moving slowly above it from north to south was a line of brilliant silver discs.
>
> They looked like metallic mirrors, and they were oscillating or flip-flopping with a regular see-saw motion . . . I could not guess their size or distance; they were so bright and tiny against the darkening sky that it was almost impossible to decide their shape, but gave the impression of being ellipses.

The future bestselling author admitted: ". . . in the few minutes before they came closer I felt myself wondering if the Martian invasion had started; this was the only time I have ever seen a fleet of textbook flying saucers.

"In this case, the explanation turned out to be something I already knew – and didn't believe. Many UFO sightings . . . were due, I'd read, to birds reflecting sunlight under unusual conditions of illumination. This theory seemed so absurd that I dismissed it contemptuously, but it is perfectly correct."

Clarke then revealed: "The lights I saw flipping across Brisbane were nothing more than seagulls, the undersurfaces of their wings acting as mirrors. Though I have lived beside the sea for a quarter of my life and am doing so now, this is the only time I have ever witnessed this phenomenon, and I would never have credited it without the evidence of my own eyes. The effect of oscillating metallic discs was absolutely realistic; it would have fooled anyone."

Ornithologist Michael Price of Vancouver told James Easton: "Given the location, 25 miles [40 km] off Mount Rainier's glacial sides, ice would be a great substitute reflector and would easily blast enough sunlight back up onto the birds' underwings to make them reflect very brightly. Just look at the excruciating whiteness of the underwings of an adult white-headed gull [. . .] flying over snow on a sunny winter day."

Price added: "I'd submit that the hypothesis of a small southbound flock of [. . .] American White Pelicans observed by someone unfamiliar with underwing reflectivity would provide the same phenomena and be at least as good an alternative possibility than seeing artefacts from another planet."

Price was no doubt speaking for many when he added: "Darn it!"

Several prominent ornithologists joined the debate, remarking that even with their experience they had often been perplexed when seeing flocks of pelicans from a distance, especially in strong sunlight and against a highly-reflective background, such as snow-capped mountains. Bearing this in mind, it is interesting to note that Arnold wrote:[12] "Another characteristic of these craft that made a tremendous impression on me was how they fluttered and sailed, tipping their wings alternately and emitting those very bright blue-white flashes from their surface." But was he really describing the action of spaceships, or the characteristic dipping and fluttering of a flock of American White Pelicans?

Easton also points out the little-known fact that Arnold had another sighting just five days afterwards in the same area – at La Grande airfield, Oregon – when he saw "a cluster of about twenty to twenty-five brass coloured objects that looked like ducks". Clearly being unaware that he had probably hit on the correct explanation, he added: "I was a little bit shocked and excited when I realized they had the same flight characteristics of the large objects I had observed on June 24." But to him they were still a mystery, because, as he said, "I knew they were not ducks because ducks don't fly that fast."[13]

In fact, both ducks and White Pelicans are capable of astonishingly high speeds, as Mike Havener, a glider pilot, confirmed,[14] recalling one incident in which he "was flying at 52 mph [84 km/h] between thermals and these birds [in this case, White Pelicans] were staying with me." However, this is considerably slower than twice the speed of sound – which Arnold estimated to be the speed at which the discs were travelling. How could even the fastest birds compete with that sort of speed?

James Easton suggests that Arnold had made grave errors in his calculations – indeed, as Arnold himself had suspected at the time – and had mistaken his landmarks.[15] In other words it could be that the mystery objects had flown a considerably shorter distance in the time he clocked, in fact flying at speeds much more in keeping with a flock of birds.

The irony is, as American historian Ed Stewart discovered, that a newspaper report from 12 July 1947 had already reached pretty much the same conclusion, writing:

> A veteran Northwest Airlines Pilot who has flown over the Pacific northwest's 'flying saucer' country for 15 years today took all the glamor out of the mystery of the flying discs. All that people gave been seeing, he said, are pelicans. Or maybe geese or swans.
>
> Capt. Gordon Moore disclosed that he and his co-pilot, Vern Kesler were saucer-hunting last Wednesday on a regular flight between here [Spokane, Washington] and Portland, Ore. Kesler was sure he had seen some flying saucers on July 2, and the pilots were armed with movie cameras and binoculars for another encounter.
>
> "Suddenly we spotted nine big round discs weaving northward two thousand feet [6,000 m] below us," Moore related.
>
> "We investigated and found they were real all right – real pelicans."[16]

So is that really the end of the great classic sighting that ushered in the modern era of flying saucers? From the evidence, it certainly looks as if it might be – one of the greatest anticlimaxes in the history of the unexplained, perhaps even a body blow to the whole subject of UFOlogy. Yet only Arnold was *there*: he may well have misidentified flocking pelicans, but no one can ever be sure what he saw. It is all retrospective conjecture – of greater or lesser probability.

In any case, his story left an unprecedented legacy, perhaps through the strange process of "artefact induction", where an error is taken to be a genuine anomaly, which then somehow kick-starts a wave of inexplicable phenomena (see page 470). Or perhaps, coincidentally, space travellers really were in our skies at the same time that a flight of pelicans inadvertently made history, being seen – not by Arnold – but by others in the following weeks and years.

A cultural phenomenon had picked precisely the right moment to happen: it was as if Arnold's story had caught the populace at the perfect time to catch the collective imagination. Two days later a sensational story erupted: a flying disc had crashed close by a USAF base in New Mexico and a rancher had retrieved some strange debris. Moreover, it was official: the public relations officer said so. However, the next day he retracted his statement about a crashed flying disc, explaining that it had only been a weather balloon. The story was forgotten, but was to make a sensational comeback many years later. The Roswell story was born . . .

21 JUNE 1947, MAURY ISLAND, TACOMA, WASHINGTON

Allegedly predating Arnold's "flying saucer" sighting by three days, Harold A. Dahl, a harbour patrol man at Tacoma, his fifteen-year-old son, two crewmen and his dog were in his boat off Maury Island at 2 p.m. on 21 June 1947. Suddenly they saw five "doughnut-shaped" objects with rows of portholes circling round a sixth, which seemed to have problems, hovering about 2,000 feet (600 m) above the water at Puget Sound. The witnesses estimated that the UFOs were about 100 feet (30 m) wide with a 25-foot (7.5 m) hole in the centre, like a doughnut.

Landing on the beach at Maury Island, Dahl took several

photographs of the UFOs, watching as one of them touched the one in the middle as if attempting to repair it. There was a thudding explosion, and the centre object emitted dark, almost molten, rock-like material and sheets of extremely light metallic stuff, apparently consisting of silver and aluminium. Being directly under this shower, Dahl's dog was killed, his son hurt his arm, and the boat was damaged.

After the UFOs headed out to sea, Dahl tried to get help through his radio, but it was dead, so he and the others collected samples of the rock and metallic sheeting and took it back to base at Tacoma. Dahl's boss, Fred Lee Crisman, was initially sceptical about the story of how the boat had been damaged (perhaps not surprisingly). However, almost immediately he became convinced that UFOs had been responsible.

Hearing about the Maury Island incident, Kenneth Arnold took on the task of investigating it, although he had initial doubts about Dahl and Crisman's credibility. However, Arnold soon received back-up for the investigation – Captain William Davidson and Lieutenant Frank M. Brown of Military Intelligence flew out to join him, which may have had some bearing on a weird little episode that befell Arnold as he tried to find accommodation in the town. Everywhere was booked solid, but when he finally tried the best hotel, he was staggered to discover a room already booked in his name . . . It has been suggested that this had been arranged so that it could be "bugged".[17]

Dahl confided in Arnold about a visit from a MIB (Man in Black), who described the UFO sighting in uncanny detail, saying mysteriously – the MIB's stock-in-trade – "What I have said is proof to you that I know a great deal more about this experience of yours than you will want to believe." The MIB also said, in classic "B" movie fashion, that "if he loved his family and didn't want anything to happen to his general welfare, he would not discuss his experience with anyone."

When Arnold and his new friend, the airline pilot Captain Smith (see above), examined the specimen of the debris they found it to be similar (if not identical) to volcanic rock, offering to show it to the intelligence agents Davidson and Brown. Strangely, however, they seemed not the slightest bit interested in it, implying heavily that the whole thing was a hoax. Finally they loaded it on the B-25 that had been loaned to them for the occasion, which was piloted by Master Sergeant Elmer L. Taff, and had a "hitch-hiker" serviceman on board (apparently a common occurrence in those days). Shortly after taking off, the plane exploded – the pilot and hitchhiker parachuting to safety, while the two intelligence officers perished. The next day the *Tacoma Times*'s headline was: "Sabotage hinted in crash of army bomber . . .", and the article claimed that the B-25 had been "sabotaged . . . to prevent shipment of flying disk fragments" and that "the ill-fated craft had been carrying 'classified material,".

In fact, the authorities discovered a burnt exhaust had set a wing on fire, and the "classified material" was merely a stack of reports that had no connection with flying saucers.

It has been speculated[18] that the MIB was an agent of the Atomic Energy Commission, checking out the witnesses to an illegal dumping of radioactive slag over Puget Sound, Maury Island. However there may well have been no such person. A close scrutiny of the characters of Dahl and Crisman suggest that the whole thing was probably a hoax. Their stories are suspiciously inconsistent – even the date of the incident varies depending on the date of the report – and the manner in which Kenneth Arnold was roped into the case is particularly suggestive. Raymond Palmer, editor of *Amazing Stories* magazine, had written in the October 1947 edition: "On June 25 (and subsequent confirmation included earlier dates) mysterious supersonic vessels, either space ships or ships from the caves,[19] were sighted in this country! A summation of facts proves that these ships were not, nor can be, attributed to any civilization now on the face of the Earth."[20]

It was Palmer who, having been contacted by Dahl and Crisman, asked Arnold to investigate the story. In fact, Crisman was already known to Palmer, having written to him the year before claiming he had been involved in a battle with the "Deros" – strange entities who lived in caves and tunnels underground, according to *Amazing Stories* stalwart Richard S. Shaver. It was not an auspicious start for a serious investigation.

When the Army Air Force checked out Dahl and Crisman they discovered that far from being official harbour patrolmen, they were in fact salvage men, using scarcely seaworthy vessels to pick up interesting – or profitable – bits of flotsam and jetsam that might come their way. The official report was damning:

> Both [the men] admitted that the rock fragments had nothing to do with flying saucers. The whole thing was a hoax. They had sent the rock fragments to [Palmer] as a joke. One of the patrolmen [sic] wrote to [Palmer] stating that the rock could have been part of a flying saucer. He had said the rock came from a flying saucer because that's what [Palmer] wanted him to say.[21]

In the aftermath of the publicity, Dahl took care to distance himself from the story, as can be seen from a report in the San Francisco *News* of 4 August 1947: "Mr Dahl went to the *United Press* Bureau at Tacoma and denied he had any part of a flying disc. He exhibited metallic stones, which he said he picked up on the beach at Maury Island shortly before the flying saucer craze swept the country."

But like it or not, Dahl and Crisman's Maury Island hoax was to add to the fomentation of that same "flying saucer craze". White pelicans and pebbles had succeeded in unleashing a torrent of UFO sightings, rupturing the veil between the mundane world and the realm where anything is possible.

However, there is a possible twist to this story. As John Keel reports: "Slag fell out of the sky over Darmstadt, Germany, on June 7 1846, according to Charles Fort.[22] Slag! Preposterous, of course, Why, slag could no more fall out of the sky in Germany than it could over Puget Sound a century later".[23]

So was there at least a thread of genuine mystery, a tiny glint of real paranormality, running through this famous "hoax" after all?

UFOS: THE MODERN ERA

Sightings after the watershed of June 1947.

7 JANUARY 1948, FRANKLIN, KENTUCKY

This is the classic story of the mysterious death of Captain Thomas F. Mantell, a pilot with the Kentucky Air National Guard (ANG) – who is still widely considered to be UFOlogy's first martyr.

On 7 January 1948 Captain Mantell, 1st Lieutenant R.K. Hendricks, 1st Lieutenant A. W. Clements and 2nd Lieutenant B.A. Hammond set off on a training flight from Marietta Air Force Base (AFB) in Georgia, to Standiford AFB in Kentucky. They did not request that their F-51s be serviced with oxygen – something that was to prove extremely important later – because as they would be flying at a low altitude, they had no need of it.

At about 1.15 p.m. the State Police told the Fort Knox Military Police that "an unusual aircraft or object . . . circular in appearance approximately 250-300 feet [75–90m] in diameter" had been observed in the air over Mansville, Kentucky. The UFO began to move south slowly. One witness, at the Godman AFB control tower, Stanley Oliver, later described it – very suggestively, as it turned out – as ". . . the resemblance of an ice cream cone topped with red".[24] At that point the UFO seemed to be motionless, just hanging in the sky, bearing a strong resemblance to a parachute surrounded by a red glow. The operations officer, Captain Carter, also observed the UFO, describing it as "[appearing] round and white (whiter than the clouds that passed in front of it) and could be seen through cirrus clouds".

As the four training planes grew closer to Godman AFB, the control tower asked Mantell to investigate. Although Hendricks flew on to Standiford as planned, Clements and Hammond joined Mantell in his quest, climbing rapidly – but it was Mantell who was ahead of the others, already much higher than them. As he passed the 15,000 feet (4,500 m) mark at 3.15 p.m., he reported: "The object is directly ahead of and above me now, moving at half my speed. . . . It appears to be a metallic object, and is of tremendous size", adding: "I'm still climbing, the object is above and ahead of me moving at about my speed or faster. I'm trying to close in for a better look."

Reaching 22,000 feet (6,600 m) Hammond and Clements gave up the chase due to lack of oxygen. Their last view of Mantell was of his F-51 climbing even further, levelling off at 30,000 feet (9,000 m). At that point the plane began to dive, out of control, in a sickening spiral motion. It is estimated that when the pilot reached 25,000 feet (7,500 m) he lost consciousness. When the plane had plummeted to about 15,000 feet (4,500 m) it began to disintegrate.

Wreckage landed on a farm – 1,000 feet (300 m) away from the bulk of the debris lay the fuselage: inside what was left of the plane was the body of Captain Mantell, his watch having stopped at 3.18 – just three minutes after his last transmission.

Meanwhile, 1st Lieutenant Clements spotted the UFO again as he neared Godman AFB, describing it over the radio as "[appearing] like the reflection of sunlight on an airplane canopy". Landing at Standiford AFB he took oxygen on board before flying off to investigate further, but although he flew as high as he dared, he saw nothing more. Reports of a mysterious object began to pour in from a line of bases further south – at Madisonville, Kentucky, one witness who observed it through a telescope declared it to be nothing more than an air balloon. As the afternoon progressed, one Dr Seyfert, an astronomer at Vanderbilt University, Nashville, Tennessee, watched the UFO float past through binoculars, describing it firmly as "a pear-shaped balloon with cables and a basket attached."[25] Then, just as the Sun was setting, several airfield control towers reported observing a "flaming object" for about 20 minutes before it sank below the horizon.

Then the rumours began, rising to a pitch of hysteria, largely thanks to headlines such as the *Louisville Courier's*: "F-51 and Capt. Mantell Destroyed Chasing Flying Saucer". Among the unsubstantiated claims were the notions the wreckage was radioactive, and that Mantell's body was riddled with bullets or bore some sinister marks – or even that there was no body, which was why, (it was claimed) his coffin was closed at the funeral. However, the official investigation claimed to find no radioactivity in what was left of the F-51, and of course there are very good reasons for closing the coffin of a plane-crash victim.

Project Sign investigators (see the Conspiracies section) interviewed everyone involved, concluding (at least for public consumption) that the UFO had been the planet Venus, which – bizarre though it may seem – tallied with the control tower's description. The slow rate of the UFO's movement across the sky would have matched that of Venus, which is 15 degrees an hour – but at that time the planet was not particularly bright, and would have been difficult to spot in sunlight. However, as researcher Curtis Peebles concludes: "the sightings made across the midwest during the evening of January 7 were definitely Venus. The flaming appearance was due to its light traveling through the thick and turbulent atmosphere near the horizon."[26]

However, despite their sceptical public announcement, privately the investigators of Project Sign had become persuaded that UFOs were genuine alien craft and that Mantell's death was in some way suspiciously connected with tangling with a flying saucer. It was the Mantell case that prompted Project Sign to bring in a committee of experts, whose task it was to sort out the UFOlogical wheat from the chaff. One of the consultants was Dr J. Allen Hynek, one of the biggest names in the history of UFOlogy, who was to begin his involvement as a sceptic, and end as the famous believer who was consultant to Steven Spielberg for his 1977 movie, *Close Encounters of the Third Kind*. But for the moment, as the next case demonstrates, Hynek was more inclined to the sceptical approach.

24 JULY 1948, MONTGOMERY, ALABAMA

In the early hours of 24 July 1948 – six months after Mantell's death – Captain Clarence S. Chiles and co-pilot First Officer John B. Whitted were flying Eastern Airlines Flight 576 to Montgomery, Alabama. Then Chiles saw a strange reddish light. Thinking it was another aircraft, he said: "Look, here comes a new Army jet job",[27] but his certainty changed to alarm as the unknown object suddenly drew closer, apparently about to crash into them. Fortunately, however, whatever it was shot past them within just (210 m) 700 feet, giving them just enough time to take in some features: a cigar-shape with a double row of windows and flames emitting from its end. A passenger saw a flash of light. Within ten seconds it had vanished from their sight.

Arriving at Atlanta, the pilot and co-pilot duly reported their only-too-close encounter to the authorities, describing the UFO as being about 30 feet (90 m) across and 100 feet (300 m) long – something like a B-29 fuselage. They had heard nothing, neither was there any turbulence associated with the sighting.

Dr J. Allen Hynek declared that the UFO was a particularly bright meteor – indeed it proved that there were a great many of them around that night. He countered the rather obvious objection that meteors do not have rows of windows by saying: "It will have to be left to the psychologists to tell us whether the immediate trail of a bright meteor could produce the subjective impression of a ship with lighted windows."[28] Chiles and Whitted had already rejected this hypothesis, saying the UFO appeared to be under "intelligent control", while even the Air Force project officer said dismissively "It is obvious that this object was not a meteor . . . [the incident] remains unidentified as to origin, construction and power source."[29]

THE 1952 WASHINGTON UFO FLAP

In June 1952 there were dozens of reports of flying saucers in the skies over the east coast of the United States, and the "flap" continued through July, when, on the 10th, a National Airlines crew observed an anomalous light that was "too bright to be a lighted balloon and too slow to be a big meteor" close to Quantico, Virginia, not far from Washington, DC. Three days later another crew in the same vicinity witnessed a strange light that rose to their height, keeping pace with the plane for some minutes before shooting vertically upwards and vanishing. Then the next day, a Pan American Airlines crew saw a formation of eight flying saucers not far from Washington . . . Were the saucers closing in on the US capital? Certainly a scientist connected with Project Blue Book (see the Conspiracies section) seemed to think so. He said[30] – with either astonishing or suspicious prescience : "Within the next few days, they're going to blow up and you're going to have the granddaddy of all UFO sightings. The sighting will occur in Washington or New York, probably Washington."[31]

Around midnight on 19 July 1952 what has become known as the "invasion of Washington" began. Radar at the Air Route Traffic Control (ARTC) at Washington's National Airport picked up eight unknown – and therefore unauthorised – "targets" heading at around 120 miles per hour (192 km/h) straight for the White House, although nothing could be seen with the naked eye.

Staff at nearby Andrews AFB witnessed an orange-red ball of flame with a revolving tail, which was zooming around "at an unbelievable speed", followed by something very similar that arced through the sky and then disappeared. Then, at roughly the same time, a Capital Airlines pilot spotted another UFO, followed by six more, some speeding along and a few merely hovering.

By 2 a.m. flaming objects were seen all over the Washington area. Sometimes they were picked up on radar, sometimes they were simply seen with the naked eye. On at least one occasion the UFO appeared to be following a commercial airliner for a few minutes. But what were these things? Was Washington really in any danger from hostile flying saucers?

Gradually, over a period of hours, doubt set in. An Air Force captain at Andrews AFB noted that the mysterious light was stationary, concluding that its movement was merely an optical illusion, and another member of staff followed suit, ascribing the apparent movement of a UFO instead of an ordinary motionless star to the "power of suggestion".

Perhaps the most extraordinary aspect of the radar-visual flap that night was that nobody seemed to care about it very much, despite the apparent threat to the seat of American government. Worried, the controller of ARTC called the Air Force Command, saying later: "They were doing nothing about it so I asked if it was possible for something like this to happen, even though we gave them all this information, without anything being done about it. The man who was supposed to be in charge and to whom I had been talking, said he guessed so. Then another voice came on who identified himself as the Combat Officer and said that all the information was being forwarded to higher authority and would not discuss it any further. I insisted I wanted to know if it was being forwarded tonight and he said yes, but would not give me any hint as to what was being done about all these things flying around Washington."[32]

Incredibly, the Combat Officer told him that they "were not really concerned about it anyway . . . someone else was supposed to handle it." But despite this curious reluctance to do anything about the UFOs, they continued to be picked up on radar. At nearly 6 a.m. seven of them appeared on radar screens. It is quite extraordinary that it was only two days later that Air Force Intelligence heard of the flap – and even then it was only through a Washington newspaper! Were the military being criminally lazy and incompetent, or did they already have intelligence about the objects, knowing they were essentially harmless? Some of the UFOs had been travelling at about 7,000 mph (11,200 km/h), and had intruded into the prohibited airspace over the White House itself.

A week passed, during which the USAF were swamped with UFO reports – more than forty a day – of which over 30 per cent remained in the "unknown" category, a very high percentage. Then on 26 July 1952 at about 10.30 p.m., the ARTC controllers began to pick up another flotilla of UFO targets on radar, slowly moving over an area of about 2 miles of night sky. Many aircraft reported seeing white, orange or red lights flash – and this time a USAF B-25 was directed to intercept the intruders, but without success. All other air traffic was hastily cleared out of the way as, finally, at midnight, serious action was taken. Two F-94s were despatched, but could find nothing, even though they flew through "a batch of radar returns". Only one pilot, Lieutenant William Paterson, reported seeing anything: four lights then just one. He told the press:

> I tried to make contact with the bogies [UFOs] below 1,000 feet [300 m], but [the ARTC controllers] vectored us around. I saw several bright lights. I was at my maximum speed, but even then I had no closing speed. I ceased chasing them because I saw no chance of overtaking them. I was vectored into new objects. Later I chased a single bright light which I estimated [to be] about 10 [16 km] miles away. I lost visual contact with it [at] about 2 miles [3.2 km].[33]

Once again, it was the Press that alerted Major Ruppelt of Project Blue Book, the USAF's UFO investigation team. But although a Navy electronics expert, who they had with them, declared that he had seen "seven good solid targets" on the radar scopes for himself, they seemed to vanish almost immediately. A few very weak echoes were left, which were put down to temperature inversion.

Not unexpectedly, the newspapers loved it. (One headline said it all: "FIERY OBJECTS OUTRUN JETS OVER CAPITAL – INVESTIGATION VEILED IN SECRECY FOLLOWING VAIN CHASE."[34]) Frantically, the authorities tried to make some sense of it all, while more and more UFO reports flooded in. The most high-profile Press conference since the Second World War was called, with Major General John Samford, director of Air Force Intelligence at its head, saying that the Air Force had received hundreds of reports from "credible observers of relatively incredible things", but took care to add that there was no evidence that the UFOs posed any threat to the security of the country. The official explanation was that the radar echoes had been caused by temperature inversion, although Samford appeared to be keeping quiet about something, as if he was suppressing certain key facts. (It has been suggested[35] that Samford's manner was not so much evidence of a cover-up as of ignorance. He had not had time to be fully briefed on the matter. Later, the magazine *Washington Life*'s article "Washington Blips" suggested that the USAF knew "more about the blips than it admitted" – which certainly appears to tally with the scientist's prediction about the "granddaddy of all UFO sightings".

Experts brought in to solve the mystery for the authorities found that air inversion could account for most, if not all, of the sightings and radar targets. It was pointed out that the area had been in the grip of "rather peculiar" weather at the times concerned. The days had been hot and very humid, but after the Sun went down, the moisture and heat "radiated away, causing both temperature inversions and a drop in humidity with altitude." For example, at 10 p.m. on 19 July, there was a 3.1 degree Fahrenheit surface inversion. At 12,575 to 14,000 feet [3,772 to 4,200m] there was a layer formed by overlying moist air. The humidity went from 84 per cent at the ground to 20 per cent at the layer's base, then climbed to 70 per cent at the top of the layer . . . "Such conditions would cause false radar targets."[36]

But even if anomalous weather conditions produced phantom radar targets, did they also cause the strange coloured lights, some of which appeared to keep pace with commercial aircraft? And how could mere meteorology have prompted that curious prediction about UFOs buzzing the White House?

JUNE 1952, SPITZBERGEN, NORWAY

According to an article in the German newspaper *Zeitung*, in June 1952, Norwegian military planes had suffered radio interference and then seen a massive blue disc-shaped UFO on the snow-covered ground below. This was the start of a rumoured crash retrieval story.

Investigated by the Norwegian Air Force, the UFO was allegedly 125 feet (37 m) across, with what seemed to be a plexiglass dome on top. There were no dead aliens inside – only a mass of remote control equipment. Apparently, forty-six jets were arranged to make the disc rotate. According to the newspaper, the UFO was taken to pieces and shipped to Narvik for analysis where it was found to carry explosives and have a flight range in excess of 18,000 miles (28,800 km). Some reports claim that there was writing on parts of the craft – in the Russian alphabet, which may be a clue as to its origins. However, subsequent research had suggested that the whole story may have been a hoax by the German newspaper, although some still believe it to be genuine.

AUGUST 1954, TANANARIVE, MADAGASCAR

Waiting for the air mail to arrive from Paris, Edmond Campagnac, head of Technical Services of Air France, suddenly saw what seemed to be a green meteorite rushing earthwards. Watched by others on the Avenue de la Libération, the object disappeared from sight behind the mountains to the south of Tananarive.

It was early evening, and dozens of home-bound workers stopped to watch the sky, as a second green object appeared over the hills near the old Queen's Palace, but this one was not behaving like a meteorite at all – moving horizontally and much slower than the first. The green sphere had soon descended to almost roof-height and was heading towards the spectators on the Avenue de la Libération.

As it approached they could see that the light was in fact two objects, the first being a "lentil-shaped device"[37] enveloped in "electric-green luminous gas". About 100 feet (30m) behind this came a 130 foot (40m) long cigar-shaped metallic object, which some of the witnesses later described as reminding them of the fuselage of "contemporary Constellation aircraft shorn of fins, elevators, wings and engines"[38] (although how they could possibly have visualized this is hard to imagine). The UFO reflected the weak light of the sunset, while orangey-red flames could be seen coming from behind it. It was travelling at an estimated speed of 185 mph (300 km/h).

Everything was silent: the crowds standing open-mouthed, while the UFOs themselves moved completely noiselessly. Then as the objects flew over the city, the electricity failed, leaving the buildings in darkness. As soon as they had gone, all the lights came back on again.

The UFOs headed out towards the airport, but swung abruptly round to the west, where they swooped low over the zoo, panicking the animals, which broke through perimeter fences, causing chaos until they could be rounded up some hours later.

But what were the objects? How could they have violated the Madagascan air space with such ease? An official enquiry was ordered by General Fleurquin, head of the Air Force, which was carried out by Father Coze, director of the local observatory – from where he had witnessed the incident. But although Father Coze and his colleagues managed to talk to at least 5,000 of the estimated 20,000 witnesses to the event, their report seemed to vanish, only surfacing years later in an account by M. Rene Fouere of the *Groupément d'Etude de Phénomènes Aeriens* (GEPN) in the British journal *Flying Saucer Review*. Incredibly, the French only heard of it in 1974 – *twenty years after the event*- when Jean-Claude Bourret published details in *The Crack in the Universe* (1977).

28 NOVEMBER 1954, CARACAS, VENEZUELA AND CARORA, LARA, VENEZUELA, AND 18 DECEMBER 1954 SAN CARLOS DEL ZULIA, ZULIA, VENEZUELA

In the early hours of 28 November 1954 Gustavo Gonzalez and José Ponce were driving their van along a road near Caracas, Venezuela, when they encountered a luminous ball, which they estimated to be about a yard (1 m) across, hovering roughly 6 feet (2 m) above the ground. Stopping the van, Gonzalez got out to investigate but as he drew close to the mysterious globe, he was suddenly pushed over by a small dwarf-like creature, which was covered with bristly hair, and had glowing eyes. The terrified man pulled his knife out and lashed at the entity, but the knife merely bounced off its body.

A second being appeared on the scene, dazzling Gonzalez with an extremely strong light. At this, José Ponce rushed from the van to give assistance and saw two more entities coming towards them from the side of the road. They were carrying rocks and seemed to mean business, but at the last minute all of the creatures jumped into the hovering globe and disappeared.

Considerably shaken by their experience, Gonzalez and Ponce reported the incident to the police. The doctor who examined them had, as luck would have it, actually witnessed the scuffle from some distance away, so he knew that *something* had attacked the two young men. He found that they were in shock, while Gonzalez had a deep scratch down one side.

That event seemed to trigger yet more weirdness in the area. Less than two weeks later, on 10 December 1954, two hunters, Lorenzo Flores and Jesus Gomez were out in the countryside near Carora, Venezuela, when they came across a strange three foot (1 m) wide craft, "like two wash basins on top of one another", that hovered above the ground, emitting flames. The witnesses said that four creatures – small, dark and very hairy – leapt out of the UFO and jumped on them, trying to drag Gomez away, but when his companion lashed out with his (unloaded) shotgun, it broke in half on contact with the rock-like body of the entity. After a fierce and desperate struggle, the men managed to flee to a police station, where they were found to be in great distress and covered in scratches and bruises, their clothes ripped to shreds.

Nearly a week after this, on 16 December 1954, Jesus Paz was a passenger in a car travelling through the town of San Carlos del Zulia, Venezuela when he had the urgent need to urinate. Stopping the car, he went into some nearby bushes, where shortly afterwards he was heard to scream in fear and pain. Rushing to his aid, his friends discovered him lying unconscious, bleeding from long scratches. As they bent over him, a "small humanoid" suddenly appeared, scuttling into a disc-shaped UFO that was hovering close by. As soon as the creature was safely inside, the craft took off with a whistling sound.

Paz, although traumatized and suffering from several bad contusions, later recovered.

What does one make of these three Venezuelan cases in which young men suffer deep scratches after encountering strange, hairy creatures with rock-like bodies that disappear in shiny UFOs? Certainly, the entities seem to be nothing like the ubiquitous Grey aliens of the post-Strieber age, and there seems to have been no attempt to abduct the witnesses. It appears that they fell foul of the creatures because they accidentally disturbed their activities – whatever they might have been. Yet, although the scuffles were deeply shocking and unpleasant, the injuries inflicted were not life-threatening – unlike some reminders of contact with UFOs and their occupants, as we will see.

FEBRUARY, 1955, BROADLANDS, HAMPSHIRE, ENGLAND

In February 1955 the Queen's uncle Lord Louis Mountbatten discovered that a UFO was reported to have landed on his estate at Broadlands, Hampshire, leaving marks in the snow. This distinguished investigator was nobody's fool – Lord Louis was Earl Mountbatten of Burma, Admiral of the Fleet, and had been Supreme Allied Commander of South East Asia Command in the Second World War, besides being the Chief of the British Defence Staff and the last Viceroy of India – but was very open-minded about the unknown.

The UFO had been witnessed by one of the bricklayers on the estate, Frederick Briggs, of whom Earl Mountbatten said: "He did not give me the impression of being the sort of man who would be subject to hallucinations, or would in any way invent such a story. I am sure from the sincere way he gave his account that he, himself, is completely convinced of the truth of his own statement."[39]

Earl Mountbatten saw the marks in the snow where the UFO was alleged to have landed for himself, but of course they disappeared when the snow began to melt. He maintained his interest in all matters UFOlogical and paranormal to the end of his life,[40] occasionally inviting writers on such subjects, together with psychics and healers to Buckingham Palace for soirées with other members of the Royal family and their staff.[41]

25 DECEMBER 1955, MANCHESTER, ENGLAND

On Christmas morning eight-year-old Lisa Knottley[42] woke early, very excited at the prospects of presents and seasonal jollity. She jumped out of bed and got dressed hurriedly, then her father knocked on her door. They exchanged hugs and said "Happy Christmas!" to each other before standing, holding hands, by the window, looking out over the sea of drab, *Coronation Street*[43]-style terraced houses that stretched as far as the eye could see. Then, in the middle of the grey winter sky, they saw a small bright light.

"Look, pet, there's a plane," Lisa's father said. In those days and in that area a plane would have been a novelty. "Fancy having to fly on Christmas Day!" But as they stood and watched, the "plane" suddenly changed into a huge ball of "orange-yellow" light, which expanded rapidly into a massive diamond-shaped "star" hanging in the sky.

Lisa (now a senior media figure), recalls: "It was absolutely huge. It was no one particular colour, but an ever-changing rainbow, favouring the orange-yellow end of the spectrum. It just hung there in the sky, lighting up the dull slate rooftops. I clung on to my father's hand very hard, I can tell you! Then a thought occurred to me. I whispered to him, 'Daddy! It must be the Star of Bethlehem because it's Christmas', and he just laughed and said it must be."

When the "star" faded from view, after about five minutes, father and daughter went downstairs to breakfast, saying nothing. "We didn't tell Mum about it for some reason, nor anyone else. At least I know I didn't. I tried to talk about it at school about three or four years later but the kids just said it must have been Telstar, the satellite, but of course there were no such things at the time."

Many years afterwards, as Lisa became interested in the unexplained, she tried to talk to her father about their "sighting", but he looked at her blankly, saying he had no idea what she was talking about. She later wrote to American author John A. Keel about her experience and he replied that he, too, had had a similar "vision" at about the same age with his father – who also forgot all about it almost instantly.

Lisa was a religious child, which may explain her immediate identification of the UFO with the 'Star of Bethlehem', but in all other respects, the experience seems like a classic UFO sighting. Of course there are some who believe that the original Star of Bethlehem was itself a UFO, but that is unlikely, for several reasons.[44]

As a child, Lisa was more open to psychic experiences, which she seems to have been able to communicate, albeit temporarily, to her father. But as a grown-up, he forgot it almost immediately, just as the adults in *Peter Pan* forget how to fly, and about all of their childhood magic.

13 AUGUST 1956, BENTWATERS/LAKENHEATH, SUFFOLK, ENGLAND

One of the most sensational cases of a radar-visual UFO contact took place on 13 August 1956 at the joint RAF/USAF establishment at Bentwaters/Lakenheath in Suffolk, England.

Three ground-based radars vectored on the unknown objects and were confirmed both by witnesses on the ground and on aerial radar in a Venom night fighter. The UFOs – which were travelling at an estimated 4,000 mph (6,400 km/h) – were captured on gun camera film, but unfortunately the resulting photographs were too indistinct to be of any evidential use.

Even the usually tight-lipped Condon Report (see page 412), noted in its conclusion: "The apparently rational, intelligent behaviour of the UFO suggests a mechanical device of unknown origin." To this day the Bentwaters incident is still the subject of hot debate among UFOlogists, who seize upon the "coincidence" of the more recent – and more controversial – case of the sightings by USAF personnel of UFOs at Rendlesham Forest, which is just three miles away from Bentwaters/Lakenheath. Why do UFOs home in on this small rural area? Are they checking out the military hardware and the technology that drives it? Do they know something about those sites that remains a closely guarded secret? Or do they have a more intimate relationship with the military, one that is known only to those with the highest security clearance, those classified as having "the need to know"?

7 JULY 1957, ROME, ITALY

Just after lunchtime on 7 July 1957, Signore Luciano Galli, a businessman from the suburbs of Rome, was on his way back to work when a Black Fiat drew up. Although a small man with delicate features was driving, it was the tall, swarthy passenger, a man with piercing black eyes, who said "Do you remember me?"

This provoked a strange response. Suddenly, Galli recalled having seen the stranger in the streets of Rome and had felt oddly drawn to him, but he had vanished into the throng. Galli said, "I remember you."

Sitting in his black car, the man asked, "Would you like to come with us?" When the witness asked where to, the man smiled, replying, "Have confidence. Nothing will happen to you." At that, Galli climbed into the car, which drove off to the Croara Ridge on the outskirts of the Italian capital where they found a saucer-shaped UFO. A "cylinder" dropped down from underneath, which then opened up like a door. After the tall man took the witness inside, the cylinder ascended into the bowels of the UFO, and two intense lights flashed in his face.

"Don't be afraid," said the stranger with a laugh. "We have just taken your picture."

Then the UFO shot away from the ground, which Galli could see disappearing fast beneath them through a sort of lens-like window in the floor. Almost immediately they were beyond the Earth's atmosphere, rapidly approaching a massive cigar-shaped UFO that Galli estimated to be at least 2,000 feet (600 m) long, and had an intensely bright light at one end, with portholes through which yet more UFOs could be seen leaving and entering.

The alien said, "This is one of our spaceships." They entered the "mothership" where they disembarked. Galli stared around him, saying later, "There were no less than four or five hundred people there . . . standing and walking around."

The aliens took him on a tour of the ship, showing him the control room, the lounges, a comprehensive library and the captain's personal quarters. Then, not quite four hours later, he was back on Earth, where he had started from. Despite the sensational nature of his experience, Galli said stoutly, "I don't care what anybody says. The story is true. You can believe it if you wish."

Assuming that the story *was* largely true in that Signore Galli was not deliberately lying, but describing the experience accurately as he had perceived it, what was going on? Who was the swarthy-skinned stranger whom the witness seemed to know already? Although he sounds superficially very similar to the sinister Men In Black (see page 263), he did not arrive after the event to warn the witness off telling anyone about it, but actively encouraged Luciano Galli to have the experience. As in many other contactee stories, the choice of the witness seems almost arbitrary and pointless, for although they may be decent upright citizens, the people who are chosen are without real influence in the world.

The description of the mothership is interesting because it seems to predate the great evocative images of lesser UFOs entering and leaving such a gigantic vessel – as in the *Star Trek* and, more impressively, the *Star Wars*, movies – but of course they came much later.

This was just one of a wave of UFO sightings and apparent contact with their occupants that happened in 1957.

7 SEPTEMBER 1957, RUNCORN, CHESHIRE, ENGLAND

At 2.15 on the morning of 7 September 1957, "a quiet gentleman",[45] James Cook of Runcorn, Cheshire was out and about and staring in amazement at an unknown illuminated object in the sky, which changed colour from blue to white, back to blue, and finally to a deep red colour. Then it zoomed from its position in the sky, landing only a few feet from him. A ladder appeared from the UFO, while a disembodied voice said: "*Jump* onto the ladder. Do not step onto it. The ground is damp."

James Cook did as he was told, climbing into an empty chamber lit – dazzlingly – by some unseen light. The voice instructed him to remove his clothes and don the overalls, seemingly made of plastic, which he found nearby. Apparently he was now ready for an adventure.

The voice told him to leave that craft and go into another one that he found close by, where he discovered twenty very tall people. Then the UFO took off on a tour around the universe. The aliens told him that they came from a planet named Zomdic which was unknown to Earth's astronomers. They also said that, in the words of John A. Keel[46] "*the saucers were used only in the vicinity of the Earth and could not operate in outer space*" (his italics).

John Keel believes that this was an extraordinarily significant statement, which he takes as evidence to support his theory that the "aliens" are "ultraterrestrials" – beings from close to the Earth, not outer space (see the Explanations section). However, there may be a flaw in his argument, at least where this case is concerned: weren't they operating in outer space when they took Mr Cook on his extraterrestrial jaunt? Or did they stick to their outer limit, whatever that may be?

The aliens had a message for James Cook. They told him sternly that, "The inhabitants of your planet will upset the balance if they persist in using force instead of harmony. Warn them of the danger."

Not unnaturally, the earthling pointed out that nobody would listen to him, to which the alien riposted, somewhat sardonically, "Or anyone else, either."

After flying about in the UFO, the witness was dropped off exactly where he had been picked up, some hours later. He told the authorities – and later, Thelma Roberts of *Flying Saucer Review* – about his experience, but made no attempt to exploit his adventure in print. As far as is known, he never had another brush with the beings from Zomdic. He did have one thing to show for his adventure, however – a small burn mark on his hand when he had touched the rail of the ladder before earthing himself by putting his feet on the damp ground.

This is a fascinating story for several reasons. For a start, like many other contactee experiences of the 1950s, the aliens do not appear to have coerced him, although they did *instruct* him to climb aboard their UFO. It seems that he was more than willing to go with them. This was not by any means an abduction, at least as more recent UFOlogists have come to think of the phenomenon.

And what about the "live" superstructure, necessitating an earthing manoeuvre by putting both feet on the wet ground? Is this a clue that some form of electrical trigger was involved? Perhaps Mr Cook had suffered his one and only episode of Temporal Lobe Epilepsy (TLE, see page 464), which is a transient electrical storm in the frontal lobes, producing vivid otherworldly visions and mystical or bizarre sensations. The mind is extremely creative: it could easily have dramatized and externalised the results of the malfunctioning brain.

But supposing Mr Cook was *not* suffering from TLE or any other brain malfunction, however transitory. Supposing it happened literally as he described – beings from Zomdic arrived and took him off on a tour of the heavens and lectured him about the lack of harmony on Earth. Why choose him? He himself acknowledged that he had no influence over the ways of the world, something that they seemed to agree with. And why bother to go to such trouble just to impress upon one powerless man something so vague and predictable? While few would doubt that the world – possibly even the universe – needs harmony, why not give some kind of concrete information about how to achieve it?

10 JANUARY 1958, CURITIBA, BRAZIL

On 10 January 1958 Captain Chrysologo Rocha was having difficulty believing the evidence of his own eyes. He and his wife were sitting on the porch of their home overlooking the sea at Curitiba, Brazil, when what he later described[47] as a "new island" appeared. Seizing his binoculars, he focused on the anomalous hump among the waves and was astonished to see it expand before his eyes. Calling to the others in the house, a small group soon gathered to watch the sight.

Whatever it was, it had a lower part below the water line and an upper section that hovered above the waves. Suddenly both parts sank, as a ship sailed slowly by. After it had gone, the USO (Unidentified Submarine Object) rose into sight once more – and the eight witnesses on the shore noted that the two halves of the object were connected by several bright "shafts", by means of which small things "like beads in a necklace" moved up and down. Suddenly all this stopped and the USO submerged.

One of the group was an army officer's wife, and she alerted the Forte dos Andrades barracks at Guaranja, but although they responded as quickly as they could by scrambling a jet, nothing was found at the spot on the ocean where they had seen the strange USO disporting itself.

16 JANUARY 1958, TRINIDADE ISLAND, OFF THE COAST OF BRAZIL

Some 750 miles (1,200 km) off the coast of Brazil the *Almirante Saldanha* of the Brazilian Navy was docked at Trinidade Island. On board was professional photographer Almiro Barauna, who had been engaged in recording underwater scenes on that fateful 16 January 1958.

When, just after 12 noon, a strange object was spotted heading towards the island, Barauna was asked to record it on his camera, succeeding in taking six photographs in just 15 seconds. The ship's commanding officer, Captain Bacellar, demanded that the film be developed there and then, also insisting that the photographer strip to his swimming briefs before going in to the darkroom to prevent his smuggling in any fake pictures. When developed, the pictures revealed a hazy, bright, round object surrounded with a ring, like Saturn – but not, apparently, a structured craft.

All in all, there were fifty witnesses on board the ship, not to mention the evidence of the photographs themselves, which has led many – including Joscelino Kubitschek, the Brazilian president at the time – to declare them absolute proof of the UFO phenomenon. However, while the witnesses were genuine enough, and the photographs show something unusual, they cannot be considered proof of anything other than an unexplained event.

20 DECEMBER 1958,[48] DOMSTEN, MALMOHUS, SWEDEN

On the evening of 20 December 1958, Hans Gustavsson, twenty-five, and his thirty-year-old friend Stig Rydberg were returning from a dance to their home in Hälsingborg. Reaching the tiny rural village of Domsten, they encountered an unknown object standing on the ground, which they later described as being, "saucer shaped . . . about 16 feet (5 m) wide and 3 feet (1 m) high, with three legs". The UFO was lit from inside by a weak light that emitted no warmth. The witnesses remarked that there seemed to be a "dark core" at the heart of the light.

As the two men stood looking, astounded, at the craft, they suddenly came under attack from four 3 foot (1 m) tall entities, who although they appeared to be strangely limbless, nevertheless seemed to make a grab for the men, endeavouring to drag them towards the dimly-lit object. Rydberg and Gustavsson put up a fight, but discovered it was impossible to get a good grip on their attackers because their arms simply went through them as if they were made of jelly. As the assailants turned their fury on Gustavsson, his friend dashed to the car and sounded the horn furiously to try to attract help. It was at this that the creatures abandoned the fight, rushing into the craft and taking off out of sight.

In yet another out of the way spot we find a grounded craft with a variation of the alien theme – here the creatures are jelly-like, not hard and hairy as in the Venezuelan cases, but just as keen to attack humans who had stumbled upon them. But tantalizingly as ever, there are no clues as to the nature of the aliens' activities deep in the Swedish countryside, nor of their intentions. Neither can we deduce their attitude to humans from the skirmish: for although they appear to be fairly belligerent, it is possible they were only acting in self preservation, to prevent the men from getting too close to their secrets. Yet there is something animal-like about their reaction to Rydberg sounding the horn: most human attackers in the back of beyond would have taken absolutely no notice and simply carried on beating the living daylights out of their victims, but these creatures were scared off by it.

22 DECEMBER 1958, MUSZYN, POLAND

On 22 December 1958 an anonymous witness – later identified as Dr Stanislaw Kowalezewski – gave this account of taking a photograph of a UFO seen over the Polish town of Muszyn to reporters from the highly respected monthly science journal the *Dookola Swiata:*

"When I looked out of the window at about 3 o'clock on 22nd December I saw a strange shimmering light coming from the clouds. Although I had no filter for my camera I took the photo through the window, thinking that the light-sensitive material would record the orange glow."[49]

The witness had a clear view of the main road to Zegiestowa, besides the River Poprad and the local railway from the window. He went on:

The whole landscape was bathed in an orange light which was about 500 metres [1,600 ft] from my window. When I developed the film I saw a large grey lens-shaped object on the negative and at the same time one could make out the fairly clear outlines of trees and other details. The Sun is in the background, on the extreme right, and the mysterious object is in the foreground. A number of specialists have examined the negative and found nothing suspicious about it, whilst other experts declared that it must be a genuine UFO.[50]

In general the history of UFOs in Poland follows the western pattern – although it began ten years after the West's – but fledgling Polish UFOlogists were disapproved of as having "occupied themselves uncritically with Western sensation-mongering", largely because "in the first place they have not been able to distinguish the essential in these Western stories from the things of minor importance."[51]

Although UFOlogy was very quiet during the days of the Iron Curtain, it is now opening up and is no longer quite so frowned upon.

5 MARCH 1962, SHEFFIELD, YORKSHIRE, ENGLAND

On Sunday morning, 5 March 1962, fourteen-year-old Alex Birch and his twelve-year-old friend David Brownlow went out to play about with a Brownie 127 camera in the field close to their homes at Mosborough, near Sheffield, when an astonishing thing happened. As they took pictures of a friend, sixteen-year-old Stuart Dixon, who turned up unannounced, they were astounded to see five UFOs hovering above them.

Three months later Alex told the *Derbyshire Times*: "They were not moving and they made no sound. They were vivid, just hanging there. After a second or so, some big white blobs started to come out of them and they were sort of hazy and obscured. I got my camera up and took a shot of them. A second or two later, they disappeared at terrific speed towards Sheffield."[52]

His story was not believed by his parents and Alex's great sighting made no waves at first. He couldn't even have the photograph developed because he was short of funds, although his mother gave him the necessary money later. But it was Alex's English teacher, Colin Brook – something of a UFO buff – who was to begin the process of making Alex's story known. Certain of the photograph's authenticity, he said:

"There is no doubt that they mentioned the sighting before the photograph was developed. This seems to discount [the possibility] that they are making up a story after having seen what may have been caused by a fault in the lens or in the emulsion of the film."[53] He also said: "It is unlikely that they have indulged in trick photography as their equipment is simple and the line of trees clearly visible."

Perhaps tellingly, the boys let the adults – including Colin Brook and Alex's father – promote the "flying saucer" angle while they themselves maintained a low profile, in a way markedly reminiscent of the two young girls at the centre of the Cottingley Fairies controversy, who let the adults ensure that the story snowballed.[54] In fact, Alex told a reporter[55] that "the possibility that they might have been flying saucers did not cross my mind at the time", while Stuart Dixon claimed many years later that the UFOs were really a flock of starlings, although he added, ". . . a strange sighting of starlings, I will admit."

Then all hell broke loose. Alex's photograph appeared in newspapers and on television all over the world, apparently also triggering a massive wave of UFO sightings in his local area, including movie footage filmed by Walter Revill of a strange flying saucer over Walkley.[56] That UFO stayed around for over a week, appearing with such marked regularity that it suggests the misidentification of either some natural phenomenon such as a bright planet, or an aircraft.

Invited to address the inaugural meeting of the British UFO Research Association (BUFORA), Alex told the members "what they wanted to hear"[57] – namely that they had seen "flying saucers", while the two other boys declared a similar belief, Stuart Dixon saying that "I think they were space objects", while David Brownlow added "I think they was [sic] flying saucers." Clearly, the boys were in too deep to back down.

There appears to have been little or no real criticism of the boys' story – and of the photograph – among the BUFORA membership.

(After all it made sense that the founding members would have been more inclined to be enthusiastic about the ETH than sceptical of it.) Member and author of UFO books, Alan Watts, no doubt summed up the feelings of his colleagues when he wrote: "If we want the truth, I would say we couldn't do better than take these to be fairly normal Adamski-type 'saucers' and argue it out from there."[58]

Although Alex Birch's father wrote about the sighting and photograph to the Ministry of Defence, they were obviously in no hurry to investigate either. Finally, after some lobbying by the media, Alex and his father visited Department 56 at the Air Ministry in Whitehall, in central London, where Flight Lieutenant R.H. White and Flight Lieutenant A. Bardsley interviewed both of them for some time. (The Air Ministry said it was two hours, but Alex was later to claim it was nearer seven. He also alleged he had been "under duress" during the questioning.)[59]

Alex's memory of the momentous visit to London included an odd detail, if true. He said that "some men and a doctor" took both the camera and negative of the photograph apart, and ". . . told me they were not flying saucers but Russians."[60] What an astonishingly irresponsible thing to say to a child – especially one with so many contacts in the media! Did they hope he would pass it on? If so, why? Did they intend to start a Cold War scare – perhaps to hide something else? Or did they *know* the "UFOs" were Russian craft, perhaps taking part in some military exercise in conjunction with the Allies – a highly unlikely scenario, admittedly? But again, it is preposterous that MoD men would share such classified information with a young lad from the provinces with the world's media camped at his door.

The Air Ministry men were not – at least officially – impressed by Alex's photograph. Writing to Mr Birch on 25 September, Flight Lieutenant White remarked: "When you brought the negatives along on August 27 for us to have a look at them, two exposures on the film were missing and you explained that these had been spoiled. It is also a possibility that the photograph of the 'flying objects' is the result of an imperfect exposure . . . To sum up, the photograph can be explained in mundane terms and does not mean that the so-called 'unidentified flying objects' must have been over Sheffield at the time it was taken."

However, in a later memorandum to his colleagues, White wrote: ". . . the sequence of exposures on the two strips of negatives we saw do not exactly fit the boys' "[61] story." To be charitable, however, Alex had complained that, "They asked me . . . questions for so long I got muddled", which might explain the discrepancies. It must have been a daunting experience for a raw teenager to be confronted by professional interrogators for at least two hours, and he may have made some slips.

Afterwards, a print of Alex's UFO picture ended up in the files of the US Air Force's Project Blue Book, where it languishes with the official statement: "Insufficient data for evaluation."[62] Ten years later, when Alex went public with his hoaxing confession, he received a visit from two mysterious Air Force officers, who told him – cryptically – that: "We think we have sorted it out"[63]. A very odd statement, considering that Alex had just "sorted it out" for them by confessing!

The confession did not make nearly such an impact as the original picture (as is the way with such things), although Alex did appear on a BBC2 news programme, where he declared that he had simply painted the objects on a sheet of glass. But according to a statement he made to the *Sheffield Star*, his father knew nothing of the prank at the time. In fact, many years afterwards, Alex stated that his father, who only got to hear of the hoax the day before the television revelation, had implored him not to make the confession publicly. But it was out, and the damage was done.

When the Press descended on David Brownlow, he told them: "We just got a piece of glass and painted five saucer shapes it. Then we took a picture . . . It was just unbelievable that everybody was taken in by it."[64]

However, it was not quite such a neat story as it seems. As David Clarke points out: "None of the papers quoted the views of the third boy, Stuart Dixon. In 1999, he claimed he told journalists he still stuck by the original claim, and that was why he was not quoted as it did not fit the story that was emerging."[65] So was there some real doubt about the sighting? How could there be if Alex had simply painted the objects on glass? Had he and David Brownlow been in cahoots, keeping the secret from their friend? It doesn't seem likely.

Or was Dixon simply too stubborn, or too ashamed, to admit his part in the hoax?

In 1998 however, to add another layer of obfuscation to this already murky story, Dixon agreed that it was a fake, saying: ". . . We had a teacher at school who was a UFO freak. I said: 'Let's fake a photo of one,' and so we did. It was a perfect picture. This teacher fell for it straight away. Next thing it was in the papers and on TV all around the world. We just painted them [the objects] on a pane of glass. If you look at the original negative, you could see the British Oak pub in the background, the chimney stack is at a slant and you can actually see the edge of the pane of glass that we painted the UFOs on to. The more people believed in it, the more it took off and mushroomed. We all agreed to stick together and stay with the story, and that's what we did for ten years."[66]

(Of course this completely contradicts his earlier statement that alone of the three witnesses he had stuck by the original story.)

Yet, as Alex admitted later: "The hardest thing is getting people to believe us now we've admitted it's a hoax. But it's true." (This is a common syndrome: history has shown how many people react with disbelief to the unmasking of alleged miracles or paranormal experiences as fraudulent.)[67]

It might be assumed that this would have been the end of the argument about the Alex Birch photograph: it was a fake, therefore the story is dead. Indeed, there was no story in the first place. But there are good reasons to take it seriously and even delve further, putting the story in a wider context – and in the process perhaps shedding some light on similar cases and the whole question of hoaxing.

It has emerged that Alex and his parents claimed to have seen UFOs before the 1962 photograph was taken, and that as the witness has a long history of paranormal experiences, including poltergeist activity and hearing mysterious, banshee-like howling, for which he seems to act as a focus or magnet. Also, in 1972, he was struck by lightning, which – along with other "major electrical events" – researcher Albert Budden has demonstrated is associated with the perception of an array of UFO-related experiences, including being abducted by aliens. Alex also suffers terribly from ferocious migraines, especially in the period leading up to violent thunderstorms – another classic sign of the psychic personality. He has remarked: "My wife says I attract these phenomena which happen everywhere I go, and is ever present with me, sometimes with long breaks of no occurrences, then blocks of intense sporadic incidents. I always feel ill afterwards."[68]

Today Alex Birch is a successful businessman who has a website devoted to UFOs and related mystical subjects,[69] which carries his own musings on the unexplained. He writes: "Perhaps we are in the infancy of our species. We peer into the Dark, fearing it, yet seeking within it a reassurance that we are not alone. Perhaps in the black void are beings not unlike us, but maybe wiser, better, who will tell us secrets that will save Us from Ourselves." These words from one of UFOlogy's classic hoaxers would not sound out of place coming from any of the many "genuine" contactees, prompting the heretical thought that in the end, there is not much difference between the "real" experience and the one that hovers always at the edge of manifesting, but never quite does so.

As for the infamous box Brownie camera, it is now an exhibit in the Roswell Museum, where the Alex Birch UFO is represented as being authentic. Perhaps that says it all.

31 OCTOBER 1963, SANTOS, BRAZIL

Eight-year-old Rute de Souza was playing close to her family home at Iguape, near Santos, Brazil when she was distracted by a loud roaring sound. Then a silver craft descended from the clouds towards the Peropava River, but on its way the strange machine hit the top of a palm tree and began to wobble dangerously. Rute watched in fascination as the UFO splashed into the river, before running to tell her family.

Soon Rute's mother and uncle, Raul de Souza, stood rooted to the spot as the river seemed to be "boiling up" – first with water, then mud – where the UFO had splashed down. Not far away, several fishermen watched the spectacle with open mouths, one of whom, a Japanese man called Tetsuo Ioshigawa, later described what he had seen to investigators and newsmen. He said that the object, shaped like a "wash basin", and some 25 feet (7.5 m) across, had only been about 20 feet (6 m) above ground level when it collided with the palm tree.

Naturally assuming that they would find a wrecked flying saucer on the river bed, the authorities instigated a thorough search with divers and mine detectors, but found nothing at all. Charles Bowen reported that: "Speculating about the incident in the *Bulletin* [the journal] of the Aerial Phenomena Research Organisation (APRO), Jim and Coral Lorenzen wrote that the reported size of the UFO suggested that it could have carried a crew, and if so, then repairs may have been affected that would have enabled the craft to escape."[70]

But quite apart from the astonishing leap of logic involved in making such a speculation, there are a number of problems with it. If the UFOnauts were so bad at navigation that they hit a palm tree and crashed into the river, it seems unlikely they could have carried out the presumably sophisticated operation of retrieving it from the river in order to repair it. None of the witnesses saw the crew escape, so presumably they went down with their ship. Once again, none of this makes sense. It seems almost as if the whole thing was staged simply to mystify the witnesses with its inconsistencies and absurdities.

24 APRIL 1964, SOCORRO, NEW MEXICO

On the evening of 24 April 1964 at Socorro, New Mexico, an event took place that was to achieve a special place in the vexed annals of UFOlogy.

Policeman Lonnie Zamora set off after a speeding car, and after chasing it for a while, heard a loud explosion and saw a flame descending – "bluish and sort of orange too . . . three degrees or so in width" about a mile (1.6 km) away. There was no solid object in the vicinity of the mysterious flame.

Abandoning the speeding motorist, he drove over rough terrain towards the location of the "explosion", thinking that a dynamite shack had gone up. Suddenly he came upon "a shiny type object" squatting in a gully some distance away (later he realized it was about 800 feet – (240 m). At first Zamora thought it looked like "a car turned upside down" and became aware of two "people in white coveralls very close to the object". Estimating later that he only looked at them for about two seconds, he did not notice "any particular shape or possibly hats or head-gear", and although normally humanoid in shape, "possibly they were small adults or large kids".

Strangely not stopping to see if any help was needed at the site – he still thought a car had overturned – the police officer drove on, contacting the sheriff's office to report a possible accident. Going round a hill, he found he was overlooking the gully with the object in it. As he got out of his vehicle, he heard a few dull thuds, like a car door shutting hard. Now he had a clear view of the "overturned car", he realized it was actually an egg-shaped "aluminium-white" object, about 12 to 15 feet (3.5 to 4.5m) in length, sitting on short "legs". The figures were nowhere to be seen.

Almost as soon as he saw it, the thing started to emit a whining sound, spewing out orange and blue flames, its exhaust whipping up the dust around it. Panicking, the police officer ran, knocking off his glasses – which he needed, being very short-sighted – and when he looked back, the object was now in the air, flying off at a low altitude in a south-westerly direction. When it disappeared over a mountain range, the witness felt able to call the police station and urge them to look out and see if they could see anything unusual. He also asked Sergeant Sam Chavez to go on his own to the site of the encounter.

Later Zamora said he went down to the gully and noted that the brush was "burning in several places", and sketched the object from memory before his colleague arrived on the scene. When Chavez arrived a few minutes later they both found "four fresh indentations in the ground and several charred or burned bushes. Smoke appeared to come from the bush and [Chavez] assumed that it was burning, however no coals were visible and the charred portions of the bush were cold to the touch".

The FBI were contacted, who alerted the nearby White Sands test site, and despatched agent Byrnes and Army captain Richard Holder who was in charge of a tracking station outside Socorro. After interviewing Zamora, Holder measured the marks in the ground where the object's legs had stood, which became known as "pad-prints", and took samples of the soil and burnt vegetation. He noticed that the burning was haphazard and occasional, and some of the grass closest to the burned area was unburnt. Several shallow marks near the padprints were categorized as the "occupants" footprints.

Project Blue Book's staff investigator, T/Sergeant David N. Moody was assigned the case, together with the UFO Investigation Officer at Kirtland Air Force Base. They interviewed Zamora and Chavez and took measurements at the site of the encounter, also checking it for radiation – it tested negative. Moody was not too impressed with Zamora, whose account he found "vague". It appeared that the object, whatever it was, had not shown up on local radar.

Then, within four days of the event, the UFOlogists arrived: Coral and Jim Lorenzen of APRO[71] and slightly later, Dr J. Allen Hynek, and then Ray Stanford of NICAP showed up. The place was abuzz.

The presence of so many investigators seemed to galvanize the memory of the manager of a local service station, who told them that an unknown tourist had said he had seen a UFO fly over his car, as he headed north along US 85. Unfortunately, this key witness was neither identified nor found.

Unlike Sergeant Moody, Dr Hynek was impressed with police officer Zamora, concluding that: "Zamora, although not overly bright or articulate, is basically sincere, honest and reliable. He would not be capable of contriving a complex hoax, nor would his temperament indicate that he would have the slightest interest in such."

But what was the egg-shaped object? Zamora and Chavez seemed to want it to be declared a secret weapon, so they could stop worrying about it, but none could be found to match the description. When the soil samples and brushwood samples were analyzed, no chemical traces were found of propellant or anything else.

Of all the alleged landing/occupant cases listed in Project Blue Book's files, this is the only one that remained marked as "unidentified". However, it is not without its critics: Curtis Peebles in his *Watch the Skies!* (1995) points out that there was one dissenting voice in the Socorro area, that of Felix Philips who lived just 1,000 feet (300 m) away from the site of the encounter and heard no explosion, concluding that it was a hoax on Zamora's part.

Philip J. Klass – perhaps not surprisingly for a dedicated debunker – cited the lack of radar tracking of the UFO and failure to find any chemical evidence of the event in the soil samples, besides the haphazard burning of the brush, which seemed incompatible with Zamora's report of fierce heat and flame. Also, the "padprints" indicated a "nonsymmetrical landing gear that would have been unstable"[72] and which "could have been made by lifting rocks from the sand and using a shovel. It would be a simple matter to make four marks at right angles to each other, quite another to make the distance between them equal."[73] Other objections include the fact that apparently it would have been impossible for the unidentified tourist to have seen the UFO from US 85 and Zamora on the mesa simultaneously, although it has to be said that it seems unfair to blame an anonymous tourist's claims, bogus or not, on Zamora himself and allow them to taint the overall case, even in small measure.

Curtis Peebles adds: "The final factor was where the 'landing' took place. The site was between two main roads into town – US 60 and 85. The land itself was owned by Socorro's mayor. After the landing, the nearly impassable road was graded and plans were made to use the landing site as a tourist attraction. (The town lacked industry and was dependent on tourists.) The implication was the landing was a hoax, to bring in tourist dollars."[74]

Perhaps Peebles' quest for the ultimate rational answer has led him to be over-sceptical in this case, for although it is one thing to acknowledge that it is human nature to exploit the unexplained for hard cash, it is quite another to suggest that the whole drama was fabricated in order to do so – especially when there is absolutely no evidence to support the idea (although there are some valid objections to the witness's statement). Tourist attractions and even theme parks *do* spring up on sites of genuine historical interest – although the portrayal of the original events that took place there may not be particularly accurate. On the sceptics' logic, because of the tourist hype involved, neither the Alamo nor the life of Jane Austen could ever have happened!

If a *believer* had drawn their own conclusions on such slight evidence, Klass and Peebles would have poured scorn on it as badly researched and the investigators as partisan from the outset. It is a pity that in their haste to relegate all UFO reports to the wastebin, they leave such a bad taste in the mouth, especially considering the sterling work they do in presenting a much-needed rational approach to many other cases.

MAY 1965, YORK, NORTH YORKSHIRE, ENGLAND

One very warm Sunday afternoon Myra H. was alone in the house in York, trying to revise for her 'O' Level examinations – just a month away – but kept being distracted by the distant rumbles of thunder and the gathering gloom that heralded a spectacular storm. Feeling headachey, she stood just outside the kitchen door to get some air, and it was then she noticed a pungent smell "like burning cardboard".[75]

As the odour grew stronger, she became increasingly lethargic, as if something was drawing energy from her body, although her mind was just as alert as ever. "It was like being in a peculiar dream", she said later. Then, to her amazement and horror, a strange "machine" suddenly materialized at the end of the garden among the cabbages.

"I was rooted to the spot," said Myra. "It was a large globe – perhaps about six or eight feet in diameter and about four feet high, which shone like metal, although it seemed to be surrounded by beautiful pearlescent light that subtly changed colour. I remember there were soft shades of lilac and lemon in there somewhere, shifting and turning, with a popping, fizzing noise."

Beautiful it might have been in retrospect, but at the time Myra was terrified, even though she discovered her feet refused to move. "I was literally rooted to the spot," she says. "I felt as if the globe would come and roll over me and I'd be dead."

As soon as she thought that, the globe *did* begin to roll towards her, very slowly. As it rolled over the cabbages they flared up as if a match had been held to them, but the flames went out immediately, as soon as the globe had gone. When the thing got to within about six feet (1.8 m) of her, it stopped. "It just sat there, fizzing to itself," says Myra, "and I know this sounds fanciful, but I could have sworn it was *thinking*. There was something about the way it sort of huffed and puffed right in front of me, as if deciding what to do."

Suddenly Myra snapped out of her strange lethargy and stood up for herself. "I got angry. I said to the thing, 'Oh just go away and frighten someone else, can't you!' " – at which it changed colour rapidly several times and made an even louder fizzing sound, then it shrank to almost nothing – a small sphere about the size of my fist – and exploded, leaving a horrible smell of rotten eggs.

"I've never experienced anything like it before or since," says Myra, now a fifty-three-year-old teacher in north Yorkshire. "At the time I went to the library and looked for some answers, and the only thing that sounded anything like my experience was the UFO phenomenon, although it wasn't really that close. For ages afterwards I wondered if I'd been visited by little men from Mars or somewhere, but deep down inside I didn't really believe that. I didn't know what to believe. The thing that still puzzles me after all these years is that the phenomenon seemed to be *intelligent*. Maybe I was imagining it, but it seemed to react to my thoughts and words. When I stood up to it, it seemed to lose heart – I know that sounds ridiculous – and exploded."

It seems certain that what Myra experienced was ball lightning, a globe of electrical energy generated by the brewing storm. But, as in many other similar cases, this natural phenomenon seemed to exhibit a certain amount of crude intelligence, which cannot, at our present stage of knowledge, be readily explained.

JULY 1965, VALENSOLE, FRANCE

At 6 o'clock on the morning of 1 July 1965 Maurice Masse, a farmer of Valensole, France, heard a peculiar noise coming from his lavender field, which he saw was emitted by some form of craft sitting among his plants. Striding up to it in order to tell the pilot to go and park elsewhere, he soon realized he had made a big mistake.

The craft, no larger than a family car, was egg-shaped with a circular cockpit and was supported on a pivot and thin legs. To the front of the grounded UFO were two little people – less than four feet tall – wearing one-piece greenish-grey suits. Their heads were huge, an estimated three times the size of human skulls, with a residual mouth, no lips and normal, if small, hands. On suddenly becoming aware of Masse's approach, one of them pointed a small tube at him, at which he became paralyzed.

For approximately a minute, the beings stared at the helpless lavender farmer, communicating with each other with a sort of throaty gargling sound, although their little mouths made no movement. Later, Masse was to admit that something about them conveyed more "friendly curiosity than hostility", and that he was not particularly scared of them, despite what they had done to him.

Then the entities entered the craft, the door closing "like the front part of a wooden cabinet", although Masse could still see them through the cockpit window. The craft rose into the air, hovered, then when it was about 60 yards (54 m) away from him, it either flew off so fast that Masse's eyes could not follow it, or literally vanished. As the witness said, "one moment, the thing was there, and the next moment, it was not there anymore".

After it had gone, reality struck – hard. Masse was still paralyzed and alone in the middle of a lavender field. He was terrified that he would simply die there. However, after a period of about 20 minutes, he began to regain the use of his limbs. The after-effects, which lasted several weeks, included lethargy and sleepiness, to such an extent that he found it hard to remain alert for just four hours at a stretch, which seriously impeded his work as a farmer. The scene of the encounter was also plagued with tourists for some time after the event.

Masse himself has claimed that he knew intuitively that the entities meant him no harm, although they manifested a distinct indifference to him, as can be seen from the cool manner in which they paralyzed and abandoned him to his fate.

3 AUGUST 1965, LOS ANGELES, CALIFORNIA

At about 11.30 on the morning of 3 August 1965, police patrolman Rex Heflin was sitting in his truck at the side of the Santa Ana Freeway attempting to make contact on his radio with the road maintenance superintendent to report that trees were blocking the view of a railroad crossing sign when, to his annoyance, the radio went dead. Immediately after this he noticed what he took to be an aircraft approaching high in the sky, but then he realized it was a classic disc-shaped UFO with a domed top (similar to those reported, and photographed, by George Adamski). Fortuitously, as part of his equipment as an official of the Orange County Road Department, he had his Polaroid 101 with him, so he grabbed it and snapped his first photograph.

Officer Heflin alleged that as the UFO arced over the road he took his second picture, and a third one shortly after the object wobbled on its axis then shot vertically upwards and accelerated beyond the road in a north-westerly direction. The only trace the departed UFO left behind was a trail of vapour (or smoke) in the air – which Heflin also managed to photograph. On returning to his vehicle, he discovered that the radio was back in working order.

Once off duty that afternoon, he showed his colleagues his Polaroids of the UFO, and allowed them to make copies (presumably not an easy task with Polaroids, especially at that time). Heflin said later: "time passed and apparently more copies of the pictures were made and handed out to various friends of friends, until most of Santa Ana was saturated with the UFO pictures."[76] One friend, with the witness's permission, even sent one off to the prestigious *Life* magazine, where although considered "the best that *Life* had seen so far", they were rejected as being "too controversial".[77]

Then the *Santa Ana Register*, the local newspaper, took the story up, printing an interview with Heflin together with cropped versions of the photographs on 20 September 1965. Almost immediately, however, a mystery – which would not go away – developed. As the demand grew for Heflin to supply more and more copies, it transpired that he had no original prints. And although the *Register* said he had given them prints from copies of the Polaroid, Heflin claimed they were originals. Then the episode became rather murky.

The witness changed his story, saying that he had given the original prints to a man who claimed he was from the North American Air Defense (NORAD), but never returned them. (Heflin had trustingly allowed him to walk off with them without issuing a receipt.) But some months afterwards, NORAD denied any involvement with the case, while Colonel George P. Freeman, the Pentagon spokesman for Project Blue Book made a statement to the effect that other "mystery men" were putting pressure on many UFO witnesses all over the United States. Was this an epidemic of MIB phenomena?

Doubts circulated about the Heflin images, but then in April 1969 aerospace engineer (a former NASA scientist) John R. Gray wrote an article for *Flying Saucer Review* in which he analyzed the photographs in minute detail,[78] concluding that they were probably genuine.

However, this is not a common view. Even at the time the American UFO organization Ground Saucer Watch (GSW) computer-analyzed the images, and although they could not be certain, erring on the side of their being fakes – a position shared by Dr W. Hartman of the Condon Commission, who concluded that the Heflin case was "of little probative value" and that the images contained "no geometric or physical data that permit a determination of distance or size independent of the witness' testimony".[79] Perhaps it is significant that Dr Hartman found it easy to reproduce similar images by "suspending a model by a thread attached to a rod resting on the roof of a truck and photographing it . . ."[80]

LATE AUGUST 1965, SEATTLE, WASHINGTON

An anonymous woman suddenly woke in the early hours, one morning in late August, 1965, in her Seattle, Washington, home. She discovered she was completely paralyzed. Because of the warm summer weather, her window was open, and she watched helplessly as a tiny, dull grey, "football-sized"[81] sphere floated in and hovered over the floor near her startled eyes. Later she told Seattle investigator J. Russell Jenkins that she had no wish to cry out, even though the weird object lowered three tripod legs onto the carpet.

It put down a tiny ramp, on which six minute people, all wearing tight-fitting clothing, climbed out and began to repair the mini-UFO right then and there. Apparently completing the task to their satisfaction, they piled back into the UFO, took off and flew out of the window. As soon as they had gone, the paralysis left the witness's body. She remained convinced that she had been awake throughout the encounter.

Sceptics are convinced that such experiences can be explained by the physical phenomenon of sleep paralysis, which is a natural process that happens to everyone to prevent them acting out their dreams. Sometimes an associated phenomenon takes place where the individual is asleep and dreaming, but believes themself to be awake and having a weird experience, typically of encountering strange entities in their bedrooms. Yet once again, this is one of those explanations that Charles Fort (see Explanations) suggested may itself require an explanation. For while sleep paralysis undoubtedly exists as a function of the brain, why do the entities conform to a pattern that does not rely on the prevailing cultural expectations? In this case the witness saw very tiny beings, not spacemen in shiny suits and "goldfish bowl" helmets, as might be expected in the 1960s during the Space Race.

How could she know that her "vision" corresponded not only with other cases from many periods in history, but also with the activities of the Otherworld, or fairy realm? Could it be that the brain, as a sort of filtering mechanism, or "receiver" (see page 267), *used* the phenomenon of sleep paralysis, in other words an alternative state of consciousness, in order to reveal the existence of the mini people – or did the mini people use the phenomenon of sleep paralysis in order to make themselves known to the witness?

As John A. Keel remarks wryly, "You can see why very few witnesses to this type of event would be anxious to tell anyone about their experiences. And you can see why almost none of these stories ever appears in print, except in occult-oriented literature. Nevertheless, if we hope to assess the true UFO situation, we must examine all of these stories."[82]

23 MARCH 1966, TEMPLE, OKLAHOMA

At about 5.30 on the afternoon of 23 March 1966, fifty-six-year-old electronics engineer William "Eddie" Laxton was driving alone on an isolated section of Highway 70 near the Oklahoma-Texas border to his place of work, the Sheppard Air Force Base near Witchita Falls, Texas, where he taught electronics. Suddenly a massive – 75-foot (22-m) long and 8-foot (25-m) high – "fish-shaped" object appeared directly in front of him. Jamming on his brakes, his car finally came to a stop about 50 yards (45 m) from the object, which was now lying across the road at a 45-degree angle.

Eddie Laxton said later that, "There were four very brilliant lights on my side, bright enough so that a man could read a newspaper by the light a mile away." He also recalled that it appeared to be illuminated inside and had "a plastic bubble in front which was about three feet in diameter, and you could see light through it." With his technician's eye – and what his friends call a "phenomenal memory",[83] he noted that the object had a tail with stabilizers that were about 2 feet (750 cm) long.

The UFO, Eddie noted, bore recognizably terrestrial black letters and numbers in a vertical line on its side, which read either: T L 4 7 6 8 or T L 4 1 6 8. It also had a single porthole about 2 feet (600 cm) wide, which was split into four, and a small open doorway underneath about 4 foot (1.3 m) high and 2 foot (750 cm) wide, which was emitting white light. The occupant, however, was busy *outside* the craft, using something like a flashlight to examine the underside. When he became aware of Eddie getting out of his car, the entity turned and retreated up a metal ladder into the craft. Eddie said later, "I'm sure it was aluminum. When the door snapped shut, it sounded like when a door closes." (This is reminiscent of officer Lonnie Zamora's observation of the sound like a "car door closing".)

The being was, apparently, human-like, being about 5 foot 9 inches (1.7 m) tall, weighing an estimated 180 (81 kg) pounds, with a "light complexion". Eddie recalled that he "got the impression due to his stooped shoulders he was about thirty to thirty-five years old. He wore either coveralls or a two-piece suit that looked like green-colored fatigues. I got the idea that he had three stripes above and three below [on his sleeve]. The above stripes were in an arch and the below stripes were in a wide V shape." Eddie also reported that the "man" wore a "mechanic's cap" with the peak turned up.

Almost immediately "the craft started up . . . it sounded like a high speed drill. It lifted off the ground about fifty feet [15 m] high and headed toward the Red River. In about five seconds it was a mile [1.6 km] away."

As the UFO rose into the air, Eddie admitted, "The hair on the back of my hands and neck stood up" – although whether this was through fear, awe or some physiological reaction to an electrostatic field, is unknown.

Wound up by the experience, the witness climbed back in his car and drove on until he came to a massive tanker truck parked by the roadside and talked to the driver, C. W. Anderson, who said he, too, had seen something apparently keeping pace with him that had eventually flown off towards the Red River. But Mr Anderson was not the only witness to the UFO that night. Later, after the story came out in the local newspapers, several other truckers went on record as having seen similar UFOs along that particular stretch of road – but a few months previously.

A man of integrity, Eddie reported his experience to the Air Force, and as a result, as he said, "A colonel and other officers wanted to see the spot where the object had been. I went out with them and showed them the place. They asked me a lot of questions while their men searched the place with all kinds of instruments. They seemed to know just what they were doing."

Later Eddie said of the "alien" he had encountered: "He looked just like you or me. If I met him tomorrow in a bar, I would know him instantly." Clearly, this was no towering Venusian or spindly Grey, but someone wearing prosaic overalls and a workman's cap – almost disappointingly human-like. But was he human? Was he tinkering with some kind of secret terrestrial craft? If so, why did the Air Force personnel not swear Eddie to secrecy or put pressure on him not to go public with his story? Interestingly, it seems that the authorities were almost as perplexed as he was.

Conspiracy theorists may become excited by the Air Force's interest in the UFO – does it reveal a covert agenda of some sort? – but considering that an unknown craft had apparently come and gone as it pleased within United States air space, surely it would have been curious, not to say alarming, if they *hadn't* shown an interest.

4 APRIL 1966, BENDIGO, VICTORIA, AUSTRALIA

Ronald Sullivan, thirty-eight, was driving along a road between Bendigo and St Arnaud, just east of Bealiba, (a small town about 130 miles – 210 km – north-west of Melbourne), late on the night of 4 April 1966 when,[84] without warning, the beam of light from his headlights bent to the right. After driving to the nearby town of Wycheproof to have his headlights checked at a late-night garage, he dropped in at the police station at Maryborough near Melbourne where he told the duty officer that he had managed to regain control of the vehicle with difficulty. Apparently he had then come to a halt as he observed multicoloured "gaseous lights" in the adjacent field. Then an unknown object appeared, which zoomed upwards and vanished.

Obviously it was not an auspicious place. When Ronald revisited the scene of his sighting a few days afterwards, he discovered that another driver[85] had been killed there the day before.

The police found a "circular depression" about 5 feet (1.5 m) across and about 5 inches (13 cm) deep in a newly ploughed field close by. (This sounds very like one of the early type of crop circles – a plain circle with no embellishments.) Although this report contained few other details, it did elicit a professional view from Stephen L. Smith, a scientist who worked closely with the Cambridge University Investigation Group about the unexplained bending of the headlights beam. He suggested[86] that an illusion could have occurred by "the sudden extinguishing of the left-hand component of the headlamp beam, which 'through its divergent character, would seem to have been bent to the right . . . [due] to a freak of reflection caused by the absence of dust particles by which headlight beams are normally seen.' "

Pondering on this case, Charles Bowen mused: "If hallucination were the cause of the phenomenon, then was it spontaneously generated in the witness's brain, or was it caused by some outside agency – perhaps just a force field emanating from the object he had observed?"[87]

MAY 1967, MALAGASY REPUBLIC

In May 1967 a curious incident befell a French Legionnaire named Wolff and twenty-seven of his colleagues serving in the Malagasy Republic. On a training exercise, they halted for lunch in the bush, when they were astonished to see a solid object like a "shining egg" fall swiftly from the air with a motion like a falling leaf, all the while emitting a piercing whistle. Immediately after it thudded onto the ground, the soldiers discovered they were paralyzed, but regained the use of their limbs when the object took off again – as if "sucked up into the sky" – apparently within seconds. Yet when they checked the time, they realized that *three hours* had passed . . .

Legionnaire Wolff described the object as being about 23 feet (7 m) high and 10 to 13 feet (3 to 4 m) in diameter. There were traces of the UFO: three holes in the ground where the "feet" had stood, beside a 10 foot (3 m) crater with a layer of vitrified crystals at the bottom.

The soldiers had no idea what had happened to them during the missing time, but felt the after-effects of their experience for some time. They all suffered intense, pulsing headaches and had a buzzing sound in their ears for days afterwards. Unfortunately they took no photographs of the craft or the landing site, but M. Wolff did tell his story to the French research group *Lumiéres dans la nuit (Lights in the night)* some time afterwards.

13 AUGUST 1967, PILAR DE GOAIAS, BRAZIL

On 13 August 1967 in the late afternoon, forty-one-year-old Brazilian ranch worker Inacio de Souza and his wife Luiza were on their way back to the ranch from shopping in the local village, Pilar de Goaias, which is some 150 miles (240 km) from the capital of Brazil, Brasilia.

On their approach to the ranch, they saw three beings – whom they described as "people" – apparently "playing" on the landing strip. (The owner of the ranch, a very prosperous man, flew several planes.) According to Luiza, the entities were wearing skin-tight yellow clothing, although her husband thought they were naked (she was probably right – women have a better eye for such things).

Just as the beings noticed the de Souzas and began to approach them, an unknown craft appeared on the scene – with the appearance of "an upturned basin" – either on or hovering just above the ground at the end of the landing strip. Terrified, De Souza reached for his .44 carbine and fired at one of the figures. This appeared to cause the craft to emit a beam of green light that knocked the ranch-hand to the ground. Luiza rushed to his aid: as she did so, the beings entered the "basin" which then took off at high speed, emitting a sound like the mass humming of bees.

However, that was by no means the end of the story for the de Souzas. In the following days, the ranch-hand suffered tremors in his head and hands, besides extreme nausea and violent headaches.

Informed of the incident (although whether all the details were included is unclear), the ranch owner immediately flew his worker 180 miles (300 km) to Goiania to see a doctor.

It was discovered that de Souza had suffered what appeared to be burns on his head and upper body, which took the form of 6-inch (15-cm) wide circles, which the doctor initially ascribed to an anti-reaction to a poisonous plant – but when he heard about the incident he revised his diagnosis. Now he was convinced de Souza was in the grip of a disease that had caused him to hallucinate, making it clear that the ranch-hand should keep quiet about his story, which he perceived as utter nonsense.

However, events took a tragic turn. Blood tests revealed that de Souza was in the last stages of leukaemia, with a very poor prognosis: he was not expected to live beyond sixty days. He died on 11 October 1967, his body covered in ugly yellowish-white patches.

Was de Souza a victim of alien malice – or, given the circumstances, self defence? (After all, he shot at them first.) Or is the very act of being around UFOs and their occupants a hazard to human health?

Clearly, the description of the aliens given later by the couple is too brief to be of much use, but they seem to have more in common with the Greys – they were reported as "playing" on the runway, which may be an association with their child-like proportions – than the tall blond Venusian types reported by the likes of George Adamski. But once again, there are many more questions than answers. What were they really doing on the runway – assuming that they had not parked their craft simply in order to play? Was this another example of the aliens' apparently endless fascination with human technology? (However, it is hard to imagine what they would have found quite so exciting about a long stretch of tarmac.) On the other hand, if they *had* stopped in the back of beyond just to have a game, it may say a great deal about their true nature.

30 OCTOBER 1967, BOYUP BROOK, WESTERN AUSTRALIA

At about 9 p.m. on 30 October 1967 A.R. Spargo, "an employee of a large labour force",[89] was driving on his own close to Boyup Brook in Western Australia when without warning, his car stalled. The radio and all the electrics were completely dead. Suddenly a strong beam of light illuminated his car, coming from the underside of a "mushroom-shaped craft, 30 feet [9 m] or more in diameter, hovering above the treetops at an estimated 100 feet [30 m] above the ground." The UFO emitted a vivid blue glow.

Spargo later told Dr Paul Zeck, a UFOlogist and psychiatrist – and reporters from the *Western Australian* – that he "seemed to be surrounded by the beam. It was two to three feet in diameter, and brilliant on the outside. Yet I could see up it, and there was no glare or anything inside the tube . . . I had the most extraordinary feeling that I was being observed through the tube. I couldn't see anyone – I could just make out the shape of the glowing craft. I felt compelled to look up the tube. But I didn't feel any fear, and I don't remember thinking of anything in particular.

"After about five minutes it was switched off – just like someone switching off an ordinary electric light. The colour of the craft seemed to darken, then it accelerated very swiftly and disappeared toward the west at terrific speed."

Then he was moving at speed along the road, with no memory of the car starting or of driving off: it was as if his mind had been wiped clear for just a few moments. In fact, he knew exactly how long he had been operating on autopilot – his Omega chronometer watch was inexplicably five minutes slow.

This was in the 1960s. If it had happened less than twenty years later, it might automatically have been assumed that Mr Spargo had been abducted, albeit very briefly. Who knows what incredible tales he would have had to tell if subjected to hypnotic regression?

12 DECEMBER 1967, ITHACA, NEW YORK

Rita Malley, a mother of two, was driving home on the evening of 12 December 1967 along Route 34 to Ithaca, New York, with her five-year-old son Dana in the back seat, when she had a shocking experience that was so weird it was reminiscent of science fiction rather than what we have come to expect as "real life".

When she first spotted the red light that appeared to be following her, she guiltily thought it was the police – she was driving just above the speed limit – so she glanced apprehensively out of the window. This was no police car. It was a surreal lighted object "shaped like a disk about the size of a boxcar, with a domed top and square red and green windows," Rita remembered later.[90] The UFO was flying just above the power lines on the left side of the car.

Then, to her horror, Mrs Malley realized she could not control the car, and shouted to her son in the back to brace himself for what she believed would be the inevitable crash. But Dana made no sign or sound. "It was as if he was in some kind of trance," his mother said. Then the nightmare took another turn. "The car pulled over to the shoulder of the road by itself," she explained, "ran over an embankment into an alfalfa field and stopped."

The bizarre experience continued. Mrs Malley said: "A white swirling beam of light flashed down from the object . . . and I heard a humming sound. Then I began to hear voices. They didn't sound like male or female voices but were weird, the words broken and jerky, like the way a translator sounds when he is repeating a speech at the United Nations. But it was like a weird chorus of several voices.

"I became hysterical. My son would not respond to my cries. I knew the radio wasn't on. The voices named someone I knew and said that at that moment my friend was involved in a terrible accident miles away. They said my son would not remember any of this. Then the car began to move again, although still not under my control. We came up out of that field and over the ditch as if it were nothing, and then back onto the road."[91]

Hysterical and horrified, Rita Malley found she could control the car once more, and drove at speed, not stopping, until she reached her home. Her husband, John, later told journalists, "I knew something was wrong the moment she walked into the house. I thought maybe she had had an accident with the car or something."

The next day, still traumatized, she discovered that her friend had been in a serious car crash, just as the mechanical-sounding voices had said.

Mrs Malley told her adventure to local UFO investigators and to the *Herald-Journal* of Syracuse, New York, who published a brief story on 21 December 1967, although she was not mentioned by name. She told the reporters that whenever "memories of the episode would flood her mind, she would break down sobbing all over again."

John A. Keel points out that, like many other less well-known encounters with UFOs, a child was present – who then went into a trance – and the fact that the witness could no longer control the car is also a frequent occurrence. Then the occupants of the UFO proved to Mrs Malley that they knew not only the name of her friend, but also the immediate future of that friend. But were they really predicting the friend's car crash – or did they cause it?

1968, SCUNTHORPE, LINCOLNSHIRE, ENGLAND

This particularly interesting story of what appeared to be an attempt by jet planes to intercept a UFO was sent anonymously to the office of *The Unexplained* in 1981:[92]

> What I am about to tell you is the absolute truth. It happened eighteen years ago in a village called Susworth, near Scunthorpe in Lincolnshire. It was a summer evening, between six and half past, and I was on my way to the village shop. What made me look up I don't know – perhaps I was alerted by the unnatural stillness: there was no sound from birds or insects. At any rate, what I saw in front of me, about half a mile [800 m] away, was a big black object, something like the spinning tops we used to play with when we were children. There was no sound coming from it, no lights of any kind, and I couldn't tell whether or not it was spinning. It seemed just to stay still, there in the sky, until after a few minutes I heard the sound of jet engines over to my right. When I looked, two jets were approaching. They passed right in front of me, then turned towards the object – but in the space of just a few seconds it had shrunk into nothing. By the time the jets had circled back, it was nowhere to be seen. I've never been able to forget it.

What made the birds and insects so quiet at the approach of the UFO? Was it some kind of repressive electromagnetic force field? Other witnesses to both UFOs and ball lightning have reported a similar phenomenon of unnatural stillness in the air – as have witnesses to paranormal phenomena, such as the materialization of ghosts or the onset of "time slips". In such cases, it has been speculated that the phenomenon itself needs energy with which to manifest, seeking to draw it from the atmosphere – and often even from the witnesses, who frequently report sensations of lethargy. Yet, as always, there is a paradox. If this spinning-top UFO was a psychical manifestation of some sort, how could it be real enough to cause the scrambling of jet planes to intercept it?

JUNE 1968, CORDOBA, ARGENTINA

Shortly after midnight of 13 June 1968, thirty-nine-year-old motel owner Pedro Pretzel was walking home near the town of Villa Carlos Paz, Cordoba, Argentina, when he saw a strange object, apparently on the nearby highway. The "machine",[93] which was beaming an extremely intense red light at the motel, some 55 yards (30 m) away, could only be seen for a matter of seconds. Shocked and disturbed, Pretzel rushed back to the motel and found his daughter, nineteen-year-old Maria Eladia, lying in a dead faint behind the kitchen door.

After her father revived her, she told him how, after seeing some guests to the door, she noticed that the motel lobby was flooded with a strange light. As she approached it, she nearly jumped out of her skin. Facing her was a 6-foot (2-m) tall, blond "man" wearing a sort of "diver's suit" apparently made out of vivid blue overlapping scales, like a tropical fish. In one hand he was holding up a blue sphere, which revolved. On one finger of his other hand was a massive ring, which he waved in front of her eyes until she felt completely drained of energy. The entity's hands and feet emitted an odd light, which seemed to add to Maria's sensation of losing control.

The witness later recalled that the being – who smiled constantly – impressed her with a sense of "goodness and kindness", although his attempts to communicate with her failed. She thought his words, which she "heard" telepathically, for his lips did not move, sounded "like Chinese".

Maria stood unable to move for some minutes while the humanoid entity tried to communicate with her, then he walked slowly and deliberately to the open door and out of it, after which it closed by itself. At that moment, the witness collapsed in a dead faint. She was very weepy and agitated for some time after the encounter. Senor Pretzel reported the incident to the police, who "promised to investigate it".[94]

What on earth happened to poor Maria that night? Charles Bowen asks: "Did [she] witness a projected image . . . that was emitted from the UFO her father had seen on the nearby highway? If she had been witness to such a phenomenon, then it is possible that her father came on the scene just as the image was about to be withdrawn. Could the 'humanoid' have been a hologram transmitted by laser beams and projected against, say, the glass of the lobby windows?"[95]

Bizarre though the idea of a projected image may seem, there are other occasions in the history of both UFOlogy and the paranormal that hint strongly at such an explanation. However, Charles Bowen asks sagely: "But however the strange and alarming effects were produced the questions remain. Why? And by whom?"[96]

18 OCTOBER 1968, MEDULLA, FLORIDA

At 7.30 on the evening of 18 October 1968 in Medulla Florida the McMullens' family dog began to bark furiously and howl. When they looked out, they were astonished to see a transparent reddish-purple object about 10 feet (3 m) above the ground, in which two ordinary-looking men were visible, apparently "pumping a horizontal bar up and down".[97] There was a strong smell of ammonia.

Then the saucer-shaped craft, which they estimated to be about 30 feet (9 m) in diameter, lifted off and flew away. Slightly earlier, two other people witnessed a strange light ascending into the sky from the area around the school, which was close to the McMullens' house. Apparently anomalous explosions were also reported at the same time.

As John Keel asks: "Was this transparent sphere a spaceship from another planet? Not very likely. The witnesses saw nothing inside it except the men and the bar. No machinery. No wonderful apparatus."[98]

There have been many other sightings of strange craft with little apparent technology, and others where a strong odour of ammonia – sometimes sulphur or the "smell of bad eggs" – hung in the air. But what does this mean? Are the UFOnauts really from other star systems, or is there another, perhaps even more disturbing, explanation that fits the facts much more closely?

4 JULY 1969, ANOLAIMA, COLOMBIA

When Arcesio Bermudez witnessed a shining sphere moving just above a field near his home at Anolaima, Colombia, his first instinct was to try to communicate with it using his flashlight, in which he was joined by his companions. But despite his friendly intentions, such proximity was to have a deadly result for him personally, which may not only serve as a warning, but also suggest the nature of certain UFO characteristics.

APRO, which studied this case, concluded their report in their *Bulletin* with these sombre words: "Within two days of the observation, the principal witness, Mr. Arcesio Bermudez, was taken very ill; his temperature dropped to 95 [degrees] F, and he had a 'cold touch', although he claimed he did not feel cold. Within a few days his condition became far more serious; he had 'black vomit' and diarrohea with blood flow. He was taken to Bogota and attended by Dr. Luis Borda at 10. a.m. on July 12 and br Dr. Cesar Esmeral at 7:30 p.m. At 11:45 pm., local time, Mr. Bermudez died."[99] The witness's symptoms were strikingly similar to those of gamma ray poisoning.

LEARY, GEORGIA, USA, OCTOBER 1969

The autumn of 1973 saw a flurry of UFO sightings, many of them reported to the National Investigations Committee on Aerial Phenomena (NICAP). Among them were reports from Governors Ronald Reagan and Jimmy Carter, both of whom were to become presidents. In this atmosphere of heightened interest in the UFO phenomena, Governor Carter responded to NICAP's inquiries about an alleged sighting of his own, a few years before, by filling in their usual form in this way:[100]

> NAME: Jimmy Carter
> OCCUPATION: Governor
> ADDRESS: State Capitol, Atlanta
> PHONE: (404) 656 1776
> EDUCATION: Graduate in Nuclear Physics
> MILITARY SERVICE: Navy
> Carter and ten members of the Leary, Georgia, Lions Club witnessed a UFO shortly after dark, 30 degrees above the western horizon, in October of 1969. The group of persons observed the object for ten to twelve minutes, starting at 7:15 EST. The object was at one time as bright as the moon. The object changed size, color, and brightness. The object was sharply outlined and self-luminous. The object came close, moved away, came close, then moved away. It was about the same size as the moon, maybe a little smaller, varied from brighter/larger than planet to apparent size of the moon. The object to distance, then disappeared. Estimated distance difficult to determine, maybe 300-1,000 yards [270-900m], about 30 degrees above the horizon.
> [signed] 9/18/73 Jimmy Carter.

When, three years later, in the middle of the 1976 presidential campaign, Carter – then Democratic candidate – announced that he had seen a UFO in 1969 the UFO community exploded with delight. Was the ever-elusive "proof" finally at hand? After Carter's election, on 18 April 1977 the influential American magazine *U.S. News & World Report* carried this historic announcement in their 'Washington Whispers' column:

> OFFICIAL WORD COMING ON UFO'S. Before the year is out, the Government – perhaps the President – is expected to make what are described as "unsettling disclosures" about UFO's – unidentified flying objects. Such revelations, based on information from the CIA, would be a reversal of official policy that in the past has downgraded UFO incidents.[101]

Predictably, the White House was overwhelmed with floods of mail from UFO enthusiasts, especially "True Believers", and NASA was roped in to help answer it, although the agency was extremely reluctant to go further and undertake any kind of UFO study, on scientific grounds. Although it agreed to test any UFO-related organic or inorganic material, as NASA Administrator Robert A. French wrote in a formal letter to the Press on 21 December 1977:

> There is an absence of tangible or physical evidence available for thorough laboratory analysis. And because of the absence of such evidence, we have not been able to devise a sound scientific procedure for investigating these phenomena. To proceed on a research task without a disciplinary framework and an exploratory technique in mind would be wasteful and probably unproductive. I do not feel that we could mount a research effort without a better starting point than we have been able to identify thus far. I would therefore propose that NASA take no steps to establish a research activity in this area or to convene a symposium on the subject.[102]

The high-profile Carter sighting stirred up public interest in the subject, but UFOlogy suffered a setback when it was revealed that what he and his friends had seen on that night in 1969 was no UFO but actually the planet Venus.[103] If nothing else, this serves as a cautionary tale to all UFO witnesses – after all, Jimmy Carter, as the details on his witness report reveal, was a graduate in Nuclear Physics and therefore hardly an ignorant hick. If someone like him can mistake the planet Venus for some kind of unknown craft – and by doing so prompt nationwide UFO mania – what chance do less educated people have in correctly identifying anomalous lights in the sky?

27 June 1970, RIO DE JANEIRO, BRAZIL

On 27 June 1970 Aristeu Machado and his family of five daughters were sitting on the verandah of their Rio de Janeiro home, from where they could see the road below, which led ultimately to the South Atlantic Ocean. Together with their house guest Joao Aguiar, of the Brazilian Federal Police, they were all engaged in playing a game while lunch was being prepared by Maria Nazare, who called to them for a time check: it was 11.38 a.m.

A couple of minutes after that Aguiar commented on the activity of a "motor boat striking the water",[104] which seemed to be causing rather too much spray as it zoomed along. Soon everyone was looking out to sea, noting what they took to be "two bathers" onboard, apparently signalling to them with their arms. Aguiar later said[105] that these figures were wearing "shining clothing, and something on their heads". The "motor boat" could now be seen to be a craft made of a greyish metal of some description, with a see-through dome on top. About 20 feet (6m) long, it did not bob up and down in the water like a normal boat.

Thinking the signallers were in trouble, Aguiar dashed off to a nearby hotel to telephone the Harbour Police, who said they would come quickly. After having been away for about half an hour, Aguiar then returned to the Machado residence and took up his position on the verandah once more, his eyes still anxiously fixed on the ocean.

A few moments later he – and the others – were stunned to see the "boat" suddenly rise out of the water, hurtling over the surface for about 300 yards (280 m) before ascending into the air in a south-easterly direction. The huddle of people on the verandah realized that the object was in fact, disc-shaped. As they watched, startled, a six-sided protruberance – with flashing green, then yellow and red, lights – withdrew into the belly of the UFO.

The housekeeper, Maria Nazare, who had joined the group on the balcony, said later that she could see two beings inside the craft as it flew off. The others noted that the UFO's appearance changed from dull metallic look to a translucence once it was in the air. The manoeuvres took place in complete silence.

A "hoop-shaped" object "about the size of a trunk or chest" appeared on the sea where the UFO had first been spotted, which sank briefly before splitting into two. One of the objects set out towards the shore, but abruptly made a right-angled turn and headed away again – directly against the prevailing current.

About 20 minutes after Aguiar had telephoned for help, the police launch from Fort Copacabana arrived, which suggests that they must have also witnessed the UFO's movements as they made their way to the spot. They retrieved a mysterious red cylinder from the sea, and dashed off back to base without saying a word to the bemused witnesses (even though one of them was a Federal Police Officer).

Apart from a brief news item in the local newspaper *Diaria de Noticias* the next day,[106] the story simply died. As far as can be ascertained, no other witnesses came forward.

Was the UFO (or more accurately, USO – Unidentified Submarine Object) some kind of secret weapon that was known to the local authorities? Or was it a genuinely unknown craft? If so, does this story imply that the UFOnauts have bases under the sea? And what happened to the mysterious red cylinder that was taken away from the scene?

1970 (?), SOUTHEND-ON-SEA, ESSEX, ENGLAND

Mrs Shelagh Padmore and her husband were being driven home in the early hours, one morning in 1970[107] by her cousin, who suddenly pointed out "a very bright light descending vertically on his side of the car."

Intrigued, they all stared out at the sight. Shelagh says: "We thought at first that it was a falling star, but when it suddenly stopped and with a burst of red flame shot across the front of the car to my side, stopped for a while and then dropped vertically again, we realized it must be something much stranger."

They drove on, but ". . . we noticed that the object was travelling with us and I wound down the window to make sure it was not a reflection. We noticed that there was no sound from it, which ruled out the possibility that it was a helicopter or an aircraft."

Then things became even more surreal. Shelagh goes on: "As we travelled along, we noticed that the object was following our every movement. When we stopped at traffic lights, it stopped with us, when we slowed down, it did too. Several times we tried to lose it by turning quickly into small side streets, but whenever we came out it would be waiting for us on the main road, so in the end we stopped trying."

Tagging along behind the car like a faithful dog, the UFO was still there when they arrived home. Not surprisingly, Shelagh admits that, "By this time I was feeling just a little nervous and it was a great relief that, when my cousin drove off, I saw the object move off again, following the car as before."

The poor cousin, however, telephoned the Padmores with the not entirely unexpected news that the "UFO" was stationary above his house. Shelagh recalls that: "My husband and I went into our back garden and, looking in the direction of my cousin's house, could see the object, just hanging there in the sky."

In the end, both the Padmores and the cousin gave up on the phenomenon and went to bed, and when they looked out the next morning the "thing" had gone. Although the local newspaper evinced absolutely no interest in the UFO, two years afterwards they carried an account of a woman who had had *exactly* the same experience in the early hours of one morning in the same district . . .

The Padmores' UFO had all the characteristics of that little-known (and even less understood) natural phenomenon, ball lightning (see also the case of Myra H. on page 82). The spheres of electrical energy appear to attach themselves to human individuals – and sometimes even exhibit signs of a rudimentary intelligence – before dissipating into the air. But why one should form that night, under those circumstances and favour that area, must remain a matter of speculation.

30 AUGUST 1970, RIO DE JANEIRO, BRAZIL

For some reason, the Brazilian experience of UFOs and their occupants tends to err on the deeply unpleasant side. Another example of this terrifying phenomenon was the case of Almiro Martins de Freitas, a security guard who worked for the Special Internal Security Patrol Service on the Funil Dam at Itatiana, in the state of Rio de Janeiro.

On the evening of 30 August 1970, de Freitas was patrolling the outside of the facility after a heavy rainfall. Nearing the end of his round, he encountered a strange humped-shaped object on a hillock, flashing a row of red, blue and orange lights.

Forcing himself not to turn and flee – as was his instinct – he crept towards it, managing to get within approximately 50 feet (15 m) of it before anything further happened. At that point, however, a terrible noise abruptly rent the air, like the sound of a jet engine far too close for comfort, deafening him. Drawing his revolver, the security guard fired at the object, resulting in its emitting a bright flash of light that knocked him to the ground. Blinded, de Freitas fired wildly in the direction of the lights, but this seemed to provoke a wave of intense heat. He found he couldn't move at all.

A passing motorist and another security patrolman later found de Freitas standing strangely erect by the mound waving his revolver and, shouting: "Don't look! Beware the flash! It's blinded me!" Trying to calm him, they carried the traumatized security man to the car. Although de Freitas recovered his mobility, he remained blind.

Later it was noted that on top of the mound there was a circular area that had remained dry during the downpour, which seemed to corroborate de Freitas's apparently wild story about what must be designated as a UGO – an Unidentified Grounded Object – being seen there.

De Freitas was taken to a hospital in the city of Guanbara, where he was subjected to a battery of physical and psychological tests by psychiatrists and ophthalmologists. He was found to be perfectly normal in every way except for his blindness, which they ascribed to shock – in other words, "hysterical" blindness. The doctors noticed that he became extremely agitated whenever he was asked about his experience.

By 3 September the newspapers had got hold of the story, but almost immediately the Brazilian Government made it abundantly clear that such publicity was neither welcome nor advisable, sending their own investigators to research the matter further. After that it became impossible for independent researchers to find out any more about the case.

It is interesting that in both this case and that of the ranch hand de Souza, the aliens only responded negatively when provoked by being shot at (if, indeed, there were entities involved in the de Freitas affair: it may be that it was the craft itself that took action against him).

26 JUNE 1972, FORT BEAUFORT, CAPE PROVINCE, SOUTH AFRICA

On the morning of 26 June 1972, South African farmer Bernardus Smit, was busy in a field on his property, Braeside Farm, near Fort

Beaufort in eastern Cape Province, when one of his workers – Boer ("Farmer") de Klerk – raced towards him, greatly agitated. Apparently he had been inspecting the irrigation system when he caught sight of what looked like smoke rising from a small copse. Going closer, he was astonished to see a shiny object, "with a star on top" rise above the trees, then remain stationary, hovering. At that point de Klerk had rushed to warn the boss that something very strange was happening on his land.

Going to investigate, the farmer saw the UFO glowing bright red, then green and yellow: whatever it was, he decided he did not like the look of it, so he raced home to telephone the police – and collect his .303 rifle. Without waiting for the police to arrive, he fired a round at the unknown object, but – unlike the unfortunate South American witnesses described above – seemed to get away scot-free. Even when the police arrived and shot at it, no retaliation was taken.

The UFO, having changed its colour to a dull grey, ceased its chameleon-like behaviour when Smit's shot hit the star on its top, but began instead to emit a humming sound and disappeared into the thick bushland. Later, the ground where the UFO had been was found to bear marks, as witness to the incident. A local newspaper commented wryly:

"I suppose the good men at Fort Beaufort were only behaving in the traditional South African way of life – what we don't understand, we shoot."

However, although that was the limit of Smit and de Klerk's brush with UFOs, the event proved to herald a major wave of South African UFO sightings, and later, some alleged abductions.

JULY 1973, TOMAKOMAI, JAPAN

In July 1973, twenty-year-old university student Masaaki Kudou was doing his holiday job of security guard at a timber yard at Tomakomai, on the south coast of Hokkaido, a northern island of Japan, when a profoundly disturbing scenario unfolded.

After patrolling the yard by car, he stopped to listen to the radio and keep an eye on the bay below. Suddenly he observed a spectacular "shooting star" flash across the sky, which then behaved unlike any shooting star he had ever seen. It stopped abruptly, disappeared from sight, then reappeared, expanding and contracting rapidly, until it reached "the apparent size of a baseball held at arm's length"[108] – when it zoomed about the sky in all directions, demonstrating a high degree of manoeuvrability. Kudou watched, dizzied by the display. Then when it began to drop towards the sea – with a "falling leaf, spiralling motion" – it turned a greenish beam of light upon the sea, sweeping around until it had come disturbingly close to the witness. When it had reached a height of 70 feet (20 m) from the surface of the waves, it disgorged a transparent tube which seemed to make a gentle *min-min-min-min* sound, which grew deeper as the tube descended. The tube touched the waves and seemed to glow, before sucking water up into the body of the hovering UFO.

Wondering if he was awake or dreaming, Masaaki Kudou looked briefly away, and when he looked back the object had withdrawn the tube from the sea – and was beginning to move towards him with "what seemed to be infinite menace".[109] The UFO was now about 160 feet (50 m) above his parked car, and Kudou could see it clearly. It appeared to be white, gave off a self-generated glow, and was "as smooth as a pong-pong ball". There was a row of portholes, through which a "shadowy human-shaped figure" could be seen, besides two smaller creatures outlined in another window. This sight terrified the witness, who sat moaning with terror in his car, while observing three or four new UFOs – very like the first – arrive on the scene, together with a massive dark brown one that resembled "three gasoline drums connected together lengthwise", that hovered, making no sound.

Then the shining globes abruptly entered into one end of the large dark brown object, which zoomed away. The student witness felt drained and numb and had a splitting headache, and his car radio was emitting strange noises. The phenomenon had lasted an estimated twelve minutes.

It is interesting that the UFO took water from the sea: there are many similar incidents reported throughout the world. In some cases the sucking up of water is connected with the making of crop circles, especially in rice-growing areas. And the headache and radio interference suggest, once more, a link with an electromagnetic phenomenon.

23 JANUARY 1974, THE BERWYN MOUNTAINS, NORTH WALES

Known as "the British Roswell", this remains one of the most hotly-debated cases in the history of British UFOs.

Immediately before the event under discussion, there was a wave of mysterious helicopter sightings across the north of England, that equally mysteriously ceased immediately after the Berwyn Incident. Many people thought "it seemed to be looking for something."[110]

Perhaps it was looking for something that fell from the skies for a few terrible seconds just after 8.30 on the evening of 23 January somewhere around the villages of Llandrillo, Llanderfel and Corwen, after one, or possibly two, explosions shook the neighbourhood, ending with a loud rumbling. Some reported seeing a light flash across the sky, while others said there was an eerie afterglow – and a few claimed to have seen light beamed into the sky from the ground.

Everyone believed that there had been a disaster – perhaps a plane had crashed – and alerted the emergency services, while a local nurse, Pat Evans, drove off by herself to offer assistance. The story goes that as she reached the remote high ground she encountered a huge glowing ball of light that pulsated, changing from red to yellow then white, which she called "fairy lights". Realizing that the lights could not be reached by foot she drove off, only to be stopped and ordered off the mountain by a posse of soldiers and police.[111] More turned up rapidly to cordon off the mountain.

In the aftermath, the area swarmed with the military – even the farmers were not allowed near to tend their livestock – and military aircraft and helicopters flew repeatedly over the mountain, as if searching for something. Had a UFO crashed on the Berwyn Mountains? Mysterious strangers also arrived in the villages, whose fascination with the event was only equalled by their reluctance to speak about themselves and their reasons for being there.

The Press loved it, suggesting that an aircraft may have crashed – which would explain the noise and the lights – while others plumped for a meteorite falling to Earth, or even, apparently ludicrously, the lights of poachers' torches. However, more suggestive and sinister elements began to creep into the story: one local newspaper claimed that, "There is a report that an Army vehicle was seen coming down the mountain near Bala Lake with a large square box on the back of it and accompanied by outriders."[112]

Although the Press soon lost interest, certain UFOlogists in the north of England received a series of official-seeming documents from a shadowy organization called the Aerial Phenomena Enquiry Network (APEN), which sensationally alleged that *they* had retrieved an alien craft from the Berwyn area after the incident, having been quicker off the mark than the authorities. They also claimed that they were arranging for a central witness to the crash to undergo hypnotic regression to try to elicit more details – something that was virtually unknown at the time. APEN also made similar claims about other UFO incidents in Britain, especially the Rendlesham Forest case. (See page 128.)

The matter rested in relative obscurity until leading British researchers Jenny Randles and Nicholas Redfern separately picked up the story in the 1990s, lecturing and writing about it to large and appreciative audiences. Had Britain found its own Roswell? It certainly seemed so when veteran British UFOlogist Tony Dodd told of an anonymous informant who claimed to have known of "two large, oblong boxes" that were taken from the crash site and sent to Porton Down, a government research establishment. The informant said, "We were shocked to see two creatures which had been placed inside contamination suits. When the suits were fully opened, it was obvious the creatures were clearly not of this world and when examined were found to be dead. What I saw in the boxes that day changed my whole concept of life."[113]

The mystery man described the beings thus: "The bodies were about five to six feet tall, humanoid in shape, but so thin they looked almost skeletal with covered skin." He added, "Some time later, we joined up with the other elements of our unit, who informed us that they had also transported bodies of 'alien beings' to Porton Down, but said their cargo was still alive."

Researcher Andy Roberts found that, despite claims to the contrary, there *was* abundant documentary evidence of the incident to be found. He notes[114] that Nicholas Redfern's emphasis on the helicopters seen in the area around the time of the "crash" was a "red herring" – the truly significant aspect of the event being the mysterious *lights* seen in the sky. Realizing this was the key to the controversy, Andy Roberts checked with the Astronomy Department of Leicester University, and discovered that "a number of outstanding bolide meteors were seen that night" coinciding with "the approximate times given by witnesses in north Wales". Roberts notes: "Bolide meteors are considerably brighter and longer lived than ordinary 'shooting stars'. They can appear to be very low, depending on the position of the witness, and often trail 'sparks' of blue and green across the sky." He adds: "Such meteors are responsible for many misperceptions of UFOs and even fool the emergency services, who are often called out to 'plane crashes', only to discover the witnesses had seen a bright bolide meteor."

But what about the huge explosion, followed by the terrible rumbling sound? Roberts dug further, finding that seismologists had found that they were due to an *earthquake* of a magnitude of 4-5 on the Richter scale, the epicentre being in the area around Bala. As he notes dryly, "To cause a reading of that magnitude, a solid object – meteorite or UFO – would have weighed several hundred tons and left a massive crater."

Andy Roberts also managed to track down the nurse, Pat Evans, to double-check her story. He discovered that she had driven to the mountain and had seen a circular ball of light that changed colours like "fairy lights", but she had not been stopped by the police, the military or anyone else. In fact, she made it clear that she saw "not a living soul" there that night, and is furious that her testimony had been embroidered in this way. But who were the strange "officials" who descended on the villages after the Incident? Nothing more sinister, as Roberts found out, than a team from the British Geological Survey based in Edinburgh . . .

And the lights that were seen to beam up into the sky at the time of the event were shown to have probably come from the torches of a team of poachers who were being chased by a group of police that night – who also, of course, carried torches.

But what about the mysterious APEN documents and Tony Dodd's anonymous informant? Roberts points out that both of them only surfaced after the story had been published in *UFO Magazine* in 1996. Redfern's informant's telephone line is now "dead", while Tony Dodd has refused to elaborate on his story, or even endorse the truthfulness of his witness. Roberts asks, "If the military had obtained aliens, dead or alive, would they really ferry them by truck? Surely a helicopter would have been the fastest, most efficient and secret form of transport. Porton Down, the research establishment to which they were taken, would hardly compromise security or contamination by opening the boxes in the presence of what were essentially delivery boys."[115]

As for APEN, most open-minded UFOlogists believe them to be a hoax, perhaps something similar to the more famous – and enduringly controversial – MJ12 papers (see Conspiracies, page 419).

1974, GREECE, AND SOME YEARS LATER, SOUTH AFRICA

South African Basil Papadimoulis reported a UFO sighting[116] dating back to 1974, which happened when he was on holiday in Greece. He remembered feeling that something peculiar was happening. Looking up, he saw "an oval-shaped object, at least three times bigger than the brightest star we can see".[117]

Standing transfixed, he noted that "it was bathed in a strange, glowing light that was yellow and incredibly deep, and as the object travelled along it moved up and down very smoothly; it seemed to me that it was dancing between the stars."

The UFO was noiseless, and left no mark such as a vapour trail in its wake. Then when it reached the end of the stretch of land, "it suddenly stopped, then dived down at tremendous speed. It made a sharp turn and went up again at double its previous speed. Then it simply disappeared."

But that was not the end of Basil Papadimoulis's UFO experience. Some years later, after he had moved to South Africa, he saw another, although perhaps this was not too unexpected as he freely admits that he had "been waiting and hoping to see one ever since the first time, and constantly searched the sky." He explains:

> One morning I became very excited, certain that I would see a UFO that day, and every 10 minutes I left the shop where I worked to look at the sky. But nothing happened. Then, at about 7 p.m., I just turned my head – and there it was: an enormous oval object, with the same glowing light as before, moving between the clouds.

This case illustrates the difficulty of analyzing such personal UFO experiences, especially as it seems there were no other witnesses on both occasions. Was the young man so determined to see what he wanted to see that he unconsciously misinterpreted quite ordinary events – such as an aircraft flying high overhead – or a natural phenomenon like a strange formation of clouds, turning them into wonderfully exciting UFOs? Did he *create* the UFOs by the latent powers of his mind, as magi are known to have created *visible* thoughtforms? Or did his utter belief and expectancy – the two proven requisites for pro-actively inducing phenomena – somehow invite the UFOs into his reality? Were they in some way actually answering his call?

22 SEPTEMBER 1974, LAUNCESTON, TASMANIA, AUSTRALIA

It was raining and the hills were covered in mist as Mrs W.[118] parked her car close to the junction of the Didleum and Tayene Plains roads about 30 miles (50 km) north-east of Launceston, Tasmania, Australia, late in the afternoon of 22 September 1974. She was waiting for a relative, whom she had arranged to meet at that spot.

Listening to the car radio at 5.20 p.m. Mrs W. was surprised when it suddenly began to emit a high-pitched whining sound. At the same moment the surrounding area was abruptly lit up, as was the interior of her car. About to turn the radio off, she looked up and saw a luminous silver and orange UFO – about the size of a big car – coming down towards her, travelling slowly but apparently purposefully.

Panicking, the witness started to reverse up the road as fast as she could, but the UFO continued its remorseless approach until it was fence-high in the centre of the road about 30 to 35 yards (25 to 30 m) away. Although profoundly shaken, Mrs W. managed to take in a detailed description of the object, noting that it was domed on top and silvery-grey underneath, with a band of what appeared to be portholes with approximately six bands below it and a small revolving disc at the base, below which was a tube that protruded slightly. All other details were obscured by the extremely bright orange light it gave off.

Still reversing wildly, the car accidentally ended over the edge of the road, its wheels stuck. At that point the UFO hovered in front of Mrs W.'s startled gaze, before flying off in a south-westerly direction, and abruptly rising vertically at some velocity.

Fortunately Mrs W. was only about a mile (1.6 km) away from her house, so she leapt out of the car and ran home as fast as she could – all the way having the uncanny feeling that she was being watched. Having gasped out her story to her husband and son, they inspected the car and the site of the encounter but found nothing noteworthy. However, when the car was retrieved the next day the bonnet was completely clean – which was very odd, as it had had muddy cat-paw marks all over it when she had set out on her journey. The car radio had also been affected by the encounter: although in perfect working order before, afterwards it suffered badly from sound distortion.

For some time afterwards Mrs W. herself was traumatized: shaky and nervous. Less serious but irritating was the fact that her newly permed hair had gone completely straight again!

Although Mrs W. reported the incident to the Royal Australian Air Force, they found nothing that could have explained the incident.

OCTOBER 1972, WAKEFIELD, YORKSHIRE, ENGLAND

Barney Baines (a pseudonym),[119] a fifty-three-year-old bus driver from Wakefield in Yorkshire in the north of England, was returning to his home on the outskirts of the town at around 8 o'clock on a chilly and windswept October night in 1972 when he heard an unusual noise coming from some distance above his head. Looking up, at first he saw nothing, but after a while his eyes adjusted to the angle and he saw, to his surprise, "what looked like an old-fashioned child's top, quite small, revolving slowly in mid-air". Looking around Barney was surprised to see the street was empty, so there was no one with whom to compare notes. Meanwhile, the "top" was moving "at a stately pace" in the air, emitting little flashes of electricity. As it passed by, blue lights crackled off the telegraph wires, and Barney began to feel nauseous, weak-kneed and headachey.

The "top" moved slowly just above the telegraph wires, which began to make a loud humming noise. Feeling as if he might collapse, Barney somehow staggered to his house at the end of the street, where he lived alone (his wife having died the previous year and his daughter was yet to move in). By this time he was experiencing bad pains in his chest, and sweating heavily – classic symptoms of a heart attack – so he flung himself towards the hall telephone with the intention of phoning for an ambulance, but was immediately thrown off his feet by an electric shock. The phone, which had always behaved normally, was now "live". As Barney pointed out, the shock could have killed him quite literally – especially as he appeared to be already in the throes of a heart attack – but strangely it had the opposite effect. Lying gasping on the floor in the hall he gradually realized the chest pains had gone completely, and he no longer felt sick. Sitting up gingerly, he noticed that the street lights – which he could see through the frosted glass of the front door – seemed to be flickering on and off wildly.

By now not knowing what to think, he opened the door and saw "a large object, exactly like a flying saucer you see in films" hovering over the street, although the "top" was nowhere to be seen. Barney stared, feeling completely disoriented by the ongoing weird experience, "not knowing whether I was dreaming or what". Absolutely silently, the UFO floated "majestically" over the street, so huge it blotted out the sky, with Barney "standing there with my mouth wide open". Then it passed out of sight, but when Barney stuck his head out of the door to check on its progress there was nothing there – both the flying saucer and the "top" had totally disappeared.

Feeling slightly shaky, Barney wondered whether to ask the neighbours if they had seen anything unusual, but was afraid they would ridicule him, so he settled instead on asking their opinion about the safety of his telephone handset. Two of his neighbours came to examine it, but although they found nothing odd – it was no longer "live" – they volunteered the information that all their electrics had gone berserk earlier in the evening, with televisions, lights, cookers and refrigerators behaving strangely, making loud humming noises and flickering on and off several times.

Trying to put it out of his mind, Barney slept soundly, having a particularly vivid dream in which his dead wife came to visit him, explaining that "souls are light-stuff and can be made use of time and time again". When he woke, he was so disoriented that he was convinced she was sleeping next to him, as she had for over twenty-five years. It seemed to him that the experience of the night before had somehow triggered the dream, although he admitted shyly that "in my heart of hearts I don't believe it was just a dream. I think she really came to me".

Barney Baines added that he had never had any such experience before and had no interest in, or particular knowledge of, either UFOs or the paranormal, nor was he a believer in Spiritualism.

It is interesting that Mr Barnes' experience was so clearly bound up with electricity – but was a freak surge in electricity the cause of some kind of hallucination, and later, his physical distress, or had a genuine UFO caused the electrical phenomenon? And had the UFO experience somehow created his vivid dream of his wife – or was the dream a genuine paranormal encounter with her? Perhaps the electrical conditions had temporarily opened a spiritual or magical portal, enabling the late Mrs Baines to "come through". Whatever the cause, a lasting result was Barney's great sense of personal comfort and closeness to his wife, and an abiding, but not uncritical, interest in the world of the unexplained. The UFO changed his life for the better.

1975 (or 1976), WIRRAL, MERSEYSIDE, ENGLAND

The following UFO report was filed informally five years after the event,[120] and is the story of one S. Copestake from Wirral, Merseyside, who related:

> One night, at about 7 p.m., my brother and I had just finished watching a TV programme when I just rose from my chair and walked outside. It was December and so it should have been pretty cold, but I didn't feel cold. I had gone out in my slippers and without putting on a coat. I walked along the front of the house and stopped at a dividing fence and turned left to face the

> field where, passing low over the flats [opposite], was an object rather like a cottage loaf [a large flattened sphere topped with a smaller one], with the lower half being red and the top being white and having the appearance of rotating.

Mr Copestake watched, entranced, still not feeling the cold, then "the object passed behind some trees and immediately I went back inside and sat down as if nothing had happened.

"Later, I thought about my sighting and came to the conclusion that the object was not of any conventional type and was certainly too big to be aircraft landing lights [sic]."

The sighting, not surprisingly, had a great – and lasting – effect on the witness, who added: "Since then I have been interested in UFOs, and have also seen another UFO in the same place."

Did the UFO somehow call to the witness, coaxing him out into the cold night – or did the witness somehow create the sighting from the unknown depths of his own psyche? Perhaps there is a simpler explanation: he was simply in the right place at the right time to see an extraterrestrial – or, at least, unknown, craft fly over head.

29 AUGUST 1975, NOË, HAUT-GARONNE, FRANCE

Shortly before 11 p.m. on 29 August 1975 forty-eight-year-old businessman M. Cyrus drove from Longages to pick up the Route Nationale just south of Noë, a deeply rural area in south-western France. As he drove along, under a bright moon, he noticed a strange "aluminium-covered" machine in a field to his right. As he drew level with the object it began to glow phosphorescently from underneath and then floated up to meet his startled gaze.

As Monsieur Cyrus jammed on the brakes, the machine tilted backwards, showing its underneath, while intensifying the light, which became so unbearable that he covered his eyes with his arm, ending up in the ditch. At that point the UFO shot completely silently up into the sky where it hovered immediately above the stricken car.

A passing motorist stopped to give aid, saying "I thought your car was exploding" (presumably because it was emitting some kind of light or smoke). Dazed and shaken, Cyrus checked "to see if I was still alive", before muttering "God heavens – is this it?"

The UFO was still there, high in the sky, but now it had taken on a reddish tinge. It stayed there for about 15 minutes, shining a beam of light on the car. By this time a small crowd had gathered, the consensus being that Cyrus should report the incident to the gendarmerie, but he replied, puzzingly: "You all know me; I'll go to the gendarmerie tomorrow. Now I'm off home!" Once there, his wife was shocked by his distressed state.

In the aftermath of the incident, Cyrus began to suffer inexplicable physiological symptoms, including abrupt phases of drowsiness even when driving, and black spots in front of his eyes, especially first thing in the morning. However, unlike many other UFO contactees, he had no radiation-like burn marks – nor was his car damaged in any way. Indeed, uncharacteristically for a close encounter vehicle, its lights and engine had continued to function perfectly immediately before, during and after the experience.

UFO investigators – from the UFO organization *Lumiéres dans la nuit* – attempted to find trace marks of the UFO on the road, but found nothing. Aerial photographs also failed to find evidence of the encounter. Monsieur Cattiau, one of the UFO investigators, located the first man on the scene, but unfortunately, although he had been happy to talk to the gendarmes, he refused to give a statement to the UFOlogists. But M. Cattiau did discovered something particularly interesting about M. Cyrus – he had a long history of encountering UFOs.

Back in 1957 he had been at a vineyard in Quillan, in the Aude district of south-western France (not far from the mysterious village of Rennes-le-Chateau)[121] when, at about 8.30 in the evening, he saw two orange cigar-shaped UFOs hovering over the vines 200 yards (180 m) away. Cyrus called to the grape-pickers who, on seeing the object, ran towards it, causing it to fly away without making a sound.

Then one night in autumn 1974 – just the year before the encounter that left him in the ditch – he and his wife were driving on the road to Muret when they saw some bizarre flashes of light to their left. As they looked, a massive orange globe appeared, lighting up the countryside all around them, and, as they moved off, keeping pace with the car for a distance of about 5 miles (8 km).

Arriving at the village of Ox, the apparently very composed M. Cyrus and his wife compared the size of the sphere to the church there, and found that the UFO was indeed as huge as they had estimated (although, annoyingly, they give no estimated size). As their strange little convoy – a car and a UFO – passed by an electrical transformer, the latter exploded. Next day they found out that the circuit breaker had been tripped inexplicably.

Just weeks before M. Cyrus's experience on the road to Noë he heard guttural voices speaking in an unknown language on his car radio, although on both occasions the radio was not switched on, reminiscent of the bizarre voices that came through the radio in Jean Cocteau's surreal film *Orphée.*[122] As Charles Bowen commented:[123] "While this is not strictly within the UFO realm, one is forced to wonder whether or not M. Cyrus is a deep-trance subject, or perhaps possesses a degree of clairvoyance – in which case something could well have "beamed in" on him, setting him up for the big encounter of 29 August. It would answer many a question if we knew *why.*"

This is a very pertinent observation. Although researchers like to compartmentalize the unexplained into various categories – UFOs on the one side and the paranormal on the other – it may well be, as the evidence frequently suggests, that the two are inextricably interlinked. Yet having said that, we are no nearer to an explanation or solution. Even if, as Charles Bowen shrewdly comments, M. Cyrus is a "deep-trance subject" – meaning he is capable of temporary dissociation of personality, effectively opening the way for other entities to "come through" – does that mean he is particularly attractive to UFOs? Or that there is something about him that enables them to manifest with unusual ease around him?

The exploding of the electrical transformer would excite those who agree with Albert Budden's "electrical trigger" theory (see the Explanations chapter), which indeed may prove crucial in understanding this story. Is M. Cyrus particularly sensitive to electrical discharges, to the point that they induce visual hallucinations in him – or conversely, are the electrical disturbances a *result* of UFO activity, as so often appears to be the case? Is the apparent explosion in UFO sightings in the twentieth and twenty-first centuries due simply to the recent preponderance of electrical and electromagnetic systems around the world? We are surrounded by telecommunications masts, radio waves, satellite systems and microwaves as – literally – never before. Perhaps there is a race of beings that feed on such emissions, which finally has enabled them to materialize in our visible spectrum . . .

JANUARY 1977, DROMORE, CO. TYRONE, IRELAND

One of the few known cases of UFOs from Ireland comes from the early 1980s.[124] Mr P. Hughes reported that:

"It was the middle of January, 1977. That particular winter was unusually cold with plentiful snow and frost. On the day of the sighting at approximately 2.30 p.m. there was a heavy snow shower that lasted about 20 minutes. After the snow stopped falling it shone brightly and the sky was dark blue. My sister and I, aged seven and eleven respectively, decided to go out." That was the beginning of their own magical "winter's tale".

Not long after leaving their house, the little boy noticed "a peculiar object descend rapidly to about 100 feet [30 m] above the ground (a distance of about 330 yards [300 m] from where I was standing). I pointed it out to my sister, and was relieved that she, too, had seen it. We both watched it make two small but perfect circles. Then it rose vertically very swiftly until it was no longer visible."

Unlike any plane the children were familiar with, "the object had some peculiar characteristics. It was an oval, silvery object, about 30 feet [10 m] long. It had neither wings nor propellers. The object did not appear to have an engine as it made no sound."

Mr Hughes added that he "was not close enough to it to notice any windows or symbols, although of course it is possible that it had both."

The most interesting fact about this sighting is that it took place in Ireland, one of the most resolutely UFO-free countries on Earth. Yet Ireland still has a thriving belief in the "Little People", and reports of leprechauns and the like remain a feature of that ancient land. Perhaps that is the secret: if, as writers such as John A. Keel and Hilary Evans suggest, there is only one phenomenon that takes on different guises depending on the prevailing collective unconscious of the culture, it is happy to retain its fairy face in Ireland, while it dresses up as extra-terrestrials in the space-age United States and the more demonic goatsuckers in the voodoo lands of South America.

1977 (?),[125] GLASGOW, SCOTLAND

Some time in the 1970s, Joe McGeough was lying awake one night in the Glasgow Infirmary (where he was being treated for a skin condition) when he noticed a bright light in the distant sky. At first he dismissed it as a plane, but as it came nearer he saw it had no navigation lights. He says:

> By this time I was beginning to make out the object's shape. It continued to come closer. I sat up in bed and put on my glasses so I could see it better. I found I could make out its shape quite clearly: it looked like two saucers – one upside down on top of the other, with a slightly higher section on top.

Greatly intrigued, Joe climbed out of bed, and went across to stand by a window to get a better view. Deciding he ought to have some corroboration of the sighting, he roused another patient who soon joined him at the window, exclaiming: "It's a flying saucer!" The two patients watched it for a further minute, when it disappeared from sight behind a building. Joe discovered afterwards that one of the nurses, and several other local residents, had seen the same UFO that night.

One of the more interesting things about post-1947 "saucer" sightings is the fact that, contrary to popular belief, Kenneth Arnold did not describe his UFOs as "flying saucers" at all. He said that they moved like a saucer would if tripped across water: but even so, the term "flying saucer" became synonymous with unexplained craft seen in the skies, and strangely, *saucer*-shaped UFOs were suddenly seen everywhere. It is as if the mis-quote hit a nerve in the public, and the sound-bite took on a life of its own.

SPRING/SUMMER 1977, YORK, YORKSHIRE, ENGLAND

This story concerns Cyril Picknett,[126] a retired Civil Servant and former regimental sergeant major, and some of his neighbours who also had anomalous experiences – all slightly different – at roughly the same time in the Burnholme district of York, in the north-east of England.

One unusually warm night – around 10.30 p.m. – in late spring Cyril went out into the front garden to enjoy a last cigarette under the stars, his wife having gone to bed early. This was uncharacteristic: although a keen gardener, once night fell he usually stayed indoors. A quiet side road, it was empty of either people or traffic. Enjoying the night, he saw "out of the corner of my eye" a bright spark of light that suddenly burst into his field of vision from right to left. It travelled – completely silently – at enormous speed, quite low, before disappearing out of sight roughly in the direction of Selby, to the north. The whole thing was over in a flash.

Cyril said nothing to anyone about this at the time, but felt strangely compelled to return to the front gate a couple of nights later. "It was as if I was waiting for something," he said. "As if I knew something else would happen out of the ordinary."

It did. As he stood there, something extraordinary happened. "There was this coach – an old fashioned railway coach, Victorian, Edwardian, something like that," he said. "It went very slowly over my head in a giant arch so low that I could see the people inside. There was a man in a tall hat reading a newspaper and a woman in a bonnet smiling at me. Then they'd gone out of sight and everything went back to normal. I just went inside and had a cup of cocoa. There didn't seem any point in mentioning it. People would have thought I'd gone bonkers. But I did see it. It was the strangest – and most wonderful – thing I'd ever seen in my life and probably will ever see. I'm very glad I did, though. It was marvellous, really."

It transpired that two neighbours had also had a peculiar experience at roughly the same time that Cyril saw his anomalous flying train. Both about forty-five years old, Brenda was a widow and Shirley's husband worked on the oil rigs, so as next door neighbours they had naturally become friends.

They described how one evening in May 1977 they were eating a late supper together – at around 11 p.m. – when Shirley, whose house they were in, popped into the garden to call the cat in. Looking up, she screamed for Brenda to "come outside *now!*" There was a huge globe of bright white light hovering over the tree-line at the bottom of the garden, just beyond the "beck" (stream). The ball of light moved rapidly over the trees, abruptly turned at right angles, then came straight at them. The two women stood rooted to the spot, paralyzed with fear, but just as it seemed to be upon them – roughly 3 yards (2.7 m) away – the globe suddenly rose into the air and bobbed around over their heads for a few seconds before zooming off into the dark. They stood there shaking, then finally went inside and had a couple of large brandies. The cat went missing for four days, which was unusual for a creature so dependent on its creature comforts.

Afterwards the women said that although they felt confused and scared, there was a certain element of feeling "special" for having had it happen to them. At first they thought that the "thing" might be from another planet come to harm them – an ET invader – but afterwards thought it unlikely. They had no idea what it might have been, but added "Whatever it was, it seemed to know what it was doing. The way it changed direction was not how you think of robots or anything like that. It seemed to be alive, somehow."

What was the huge globe of white light? Was it that most recent of additions to the list of accepted natural phenomena, ball lightning? Balls of lightning (BOLS) often appear to react intelligently to their surroundings, although whether this is an illusion on the part of witnesses, or evidence of real intelligence – or intelligent control – is unknown.

Were the two anomalies – the vision of the train in the sky and the mysterious globe in the garden – connected? It seems so: it would be very strange indeed if Cyril's only paranormal event of a lifetime more or less coincided with the weirdness just up the road in his neighbour's back garden. Maybe there was something odd in the air at that time, or, as the occultists say, the "veil was thin" between this world and another dimension.

SPRING/SUMMER 1977, WEST BROMWICH, MIDLANDS, ENGLAND

On fine evenings in the spring and summer of 1977, Ken Wintle of West Bromwich in the Midlands, liked to enjoy a quiet drink in the garden of his local pub with his wife, sometimes accompanied by their small daughter Julie. On one particularly cold evening, however, everyone else was huddled inside at the bar, while he stood outside admiring the stars "and thinking how sharply they stood out in the heavens",[127] when suddenly he "received the shock of my life" . . .

He goes on: "Right in front of me, and headed straight towards me, was a formation of UFOs. I gazed with trepidation and walked forward a few steps to make sure I wasn't dreaming, then stopped dead in my tracks. An indescribable feeling came over me as I realized that what so many people had reported was, for me, now true."

Checking that the objects were still visible, Ken dashed into the bar to secure another witness, beckoning to an acquaintance called Dennis to come outside. Ken says: "As Dennis reached the door I pointed upwards saying 'Pigeons': I did not want to implant any suggestion of UFOs in his mind. His eyes widened in surprise and his mouth gaped. 'No, they're not pigeons', he said quietly, looking as though he had just seen a ghost."

At that moment little Julie looked out of the door, spotted the UFOs and said, awestruck, "Coo!" Then Ken, Dennis and Julie went outside and watched the objects, which were by then almost directly overhead. Dashing round to the street side of the pub, they found that the glare from the street lights made it impossible to see the UFOs, so they returned to the bar, shaken, and discussed what they had seen.

Ken recalls: "I asked Julie what she thought the objects were and she replied 'Swans'. I didn't disillusion her. A few years later I mentioned the fact that she had seen some UFOs, but she had forgotten all about the incident."

Having discussion their sighting at length, Ken and Dennis were agreed on these details: "Each UFO was shaped roughly like an elongated diamond, with the top and bottom squared instead of pointed. The UFOs were flying low, at about the same height as light aeroplanes, and not very fast. They were all more or less the same size and slightly longer than a light aeroplane.

"About a dozen UFOs flew in a triangular formation, adjoining a similar group consisting of about eight craft. At the rear two more UFOs moved from side to side while maintaining their position in the formation; they wobbled slightly, rather like a child's spinning top. The entire squadron was orange coloured and silent."

The next day Ken gave the details of the sighting to a reporter from the local newspaper, who obviously thought it a huge joke – until he realized that the witness was entirely serious about it. After the article came out, several other people contacted Ken with their own stories of UFOs. The incident was to trigger for him a life-long fascination with anomalous objects seen in the sky – as he wrote: "Since then I have spent a great deal of time sky watching and have seen several different types of UFO: the last sighting I made was in daylight, and the craft was of the classic 'flying saucer' type."

But was little Julie nearer the truth than either of the adult witnesses that night, when she called the "UFOs" "swans"? Now we know that even Kenneth Arnold's classic "flying saucers" were probably white pelicans flying in formation, this sighting may have come into the same category of understandable misidentifications. Yet there are questions that remain to be answered: do swans – or indeed any such large birds – fly at night? Even if the objects were birds of some description, what light source would be strong enough at that time of day to illuminate them high in the sky and cause the optical illusion? And even if Ken's first UFO sighting was a misidentification of some sort, what about the subsequent sightings? Perhaps there was an element of "artefact induction" (see page 470), where a mistake actually ushers in genuine phenomena.

11 JANUARY 1978, GLOUCESTER, ENGLAND

In the early hours of 11 January 1978, Joan Mandeville-Dell, who lived just outside Gloucester, woke up and went downstairs to get a drink of water, but when she returned she found her bedroom "was glowing pink."[128] Looking out of the window, Mrs Mandeville-Dell saw "an enormous red object like a huge tower with six yellow lights, one beneath the other."

Hardly able to believe her eyes, she looked around for other witnesses, but – due to the early hour – no one was up and about, so she telephoned the police and immediately a young constable visited her house. She goes on: "He watched the object with me, but could not explain what it was. He suggested it might be the television mast at Much Marcle, [a village] just under 20 miles [32 km] away, but I could see the mast on the horizon. It appeared as tiny red dots with no shape at all."

The UFO remained in the sky for another hour or so, before disappearing from sight. Mrs Mandeville-Dell reported the sighting to the local newspaper, who carried a request for other witnesses, but no one else came forward.

This is an interesting case, because – thanks to the witness's quick thinking, and the unusually swift police response – there was an "official" observer of the UFO. Even so, no one was any the wiser about the nature of the object.

Perhaps it is significant that Mrs Mandeville-Dell admits that she had "several experiences of UFOs", which suggests that she attracts them, or is particularly open to them in some way. But does this mean that once they have been "summoned", they can be seen by others? Or is she just lucky in her frequent UFO sightings?

28 AUGUST 1978, STOWMARKET, SUFFOLK, ENGLAND

On 28 August 1978, Audrey Harvey of Stowmarket saw that she had noted in her diary that there was an air attack (RAF) exercise taking place at nearby Wattisham, but that was not all that happened that day. As she later wrote:[129]

". . . at 10 p.m. I saw [a] UFO. There was no engine noise, just a square yellow light preceded by a white intermittent glow like a lighthouse's. It went down in the fields behind some bushes, in the direction of Battisford. I thought of running to a neighbour, but wondered how to explain what had happened – especially if they had seen nothing."

On a separate occasion, Audrey "was out walking with the kids and the dog. Phantoms [jet planes] were flying in to land above us.

One came in low and after it had gone over us there was an almighty crackling blue glow and echo."

But there was more. Audrey says: "I instinctively turned as I felt something else was coming in the wake of the Phantom. There was nothing there, but I froze on the spot. So did my seven-year-old, who sensed something was happening – but my nine-year-old, who was just a few yards away, saw, heard and felt nothing."

Puzzled and disturbed by the experience, Audrey spoke to a pilot about it later, "who said the plane had probably flown through a storm and electrical current was crackling off its body". But Audrey was not convinced, adding, "I felt it was more". However, it seems as if the pilot may have been right up to a point: from Audrey's description it seems that the weird attendant phenomenon may have been ball lightning.

The first sighting, though, does seem more like a genuine UFO, although whether it appeared to investigate the RAF exercise, or was somehow part of it – perhaps a secret aircraft of some kind – will never be known. Audrey, clearly an intelligent witness, said cynically, "I know better than to question the RAF, as they are covered by the Official Secrets Act", but added, "Even so I do wonder if anything showed up on their radar screens."

Because no one else in the neighbourhood mentioned any similar sightings to the witness, she was left wondering: "Can one create a sighting from the imagination?" which is an excellent question, to which the answer is almost certainly *yes*, but how easy it is to succeed in doing so is another matter.

JULY 1979, MARYLAND, USA

In July 1979 Liverpudlian, Amantha Jones, aged twelve and her cousin Lisa were on holiday in Maryland, when they noticed a strange silvery object high above them in the sky. As Amantha later wrote:[130]

> . . . it seemed fairly unremarkable and, not paying much attention, we continued to play ball. As the sky gradually grew darker, the object became clearer, and lit up. Dusk had fallen by now, so

my cousin Jason decided to fetch a spotlight and make signals to the object.

But this innocent gesture had an unsought result. Amantha says: "At an incredible speed, the object came at us. There it was, huge, domed, with brightly coloured revolving lights. We stood, stunned, for some seconds, and then ran into the house. After a while we ventured out again, and this time were caught in the beams of two giant spotlights. This time when we ran inside it was for good!"

It transpired that the two young cousins were not the only witnesses to the passing of the giant UFO that night. Amantha noted that ". . . next morning, there was an article in the *Washington Times* saying that there had been a number of reports of radiation burns from an area not far from the Capitol."

Despite this apparent evidence of hostility – either deliberate or accidental – on the part of the UFOs, Amantha was deeply impressed by "the alien craft – it was beautiful". Indeed, she waxed lyrical about the (presumed) aliens, adding: "What intelligence those creatures must have – probably more than we shall ever know!"

But she wondered "why . . . didn't the occupants [of the UFO] beam us up into their spaceship? I'm sure with their technology, they could easily have done so." Amantha seems to be one of the few who regret not being abducted by aliens.

AUGUST 1979, SUDBURY, SUFFOLK, ENGLAND

On a pleasant August evening in 1979 Sally Plumb and her two nieces were enjoying a snack in her garden when, as she said:[131] "Something made us look up. What we saw was a silent, silvery Zeppelin-shaped object hovering just above the treetops. It was so frightening that we just sat and looked at it."

As it moved away slowly, the witnesses dashed to tell the rest of the family about it, but when they returned, there was no sign of the dirigible. The others, who had not seen it, "said it was probably a plane or something" but Sally had her doubts.

What was the "Zeppelin"? The answer must be probably an advertising airship, such as the more recent Goodyear balloon, or a dirigible from an air show. Yet of course the possibility remains that it was a *genuine* UFO, one that will always remain unidentified, moving majestically through Sally's memories like a magical galleon.

NOVEMBER 1979, LIVINGSTON, SCOTLAND

On the morning of 9 November 1979 forestry worker Robert Taylor set off in his truck with his dog to check on certain trees close to the Edinburgh-Glasgow motorway (the M8). Parking at the end of a path, he walked the rest of the way to inspect the plantation, but as he came to a clearing he was stunned to see a huge globe squatting on the ground. It was dark grey, roughly 12 feet (3.6 m) tall and 20 feet (6 m) wide, and seemed to partially dematerialize from time to time, as if – unsuccessfully – trying to camouflage itself or become invisible. Approximately half way down the outside of the object were what might have been windows and a ledge from which protruded things like "bow ties".

As Robert Taylor stood there, not knowing what to think, two mine-like spheres with spikes around them hurtled towards him from the direction of the object, each just 2 feet (600 m) in width. Their spikes grabbed Taylor by the trouser legs and dragged him towards the UFO, but he was assailed by a "choking" smell and became unconscious.

When he regained his senses he seemed to hear a swishing noise, but the object was no longer there. He had great difficulty in walking, and his dog seemed to go wild, running around in circles, barking madly. Somehow Taylor managed to reach his truck but in his enfeebled state only succeeded in getting it bogged down in soft ground, so he had to abandon it and travel home on foot – a mile or so away. Once there he felt ill for days, with a raging thirst and bad headache.

There were physical traces of the encounter: besides badly torn trousers, investigation of the site revealed holes that may have been caused by the spiked objects, but it was a puzzle how the UFO could have landed there without being seen from the road. Some have speculated that the UFO may have arrived in the clearing by another route, or even simply materialized there, behind the trees.[132]

Since then, despite the efforts of sceptics such as Steuart Campbell to discredit his experience, Robert Taylor has become convinced that the object was a spaceship, and that his attackers were robots used to keep people away. Once again, however, the logic fails to hold up. If the space people were so advanced that they could materialise where they liked, surely they could also become invisible at will to prevent humans from stumbling upon them? If they have no wish to be seen, why do they allow so many witnesses to see them?

AUGUST 1980, WALNEY, CUMBRIA, ENGLAND

One warm evening in August 1981, L. Corkill of Walney, Cumbria, was relaxing listening to music in his bedroom with his brother Roy. Suddenly they were jerked out of their pleasant reverie by the sight of "a jumbo jet flying past and behind it, at a slightly higher altitude, was a small UFO, apparently flying at the same speed as the plane."[133]

The witness kept his eye on the object, while going to the window for a better view. After about 30 seconds he fetched his binoculars, saying that "as it happened, they were already properly focused so I got an excellent view of the object immediately. It was silver, its wings were set further back than those of a normal plane. It had a rounded 'nose' and there were no engines visible." The two brothers watched it for about five minutes before it "just disappeared" – although whether this means it vanished from sight instantaneously or flew away out of sight is unknown.

The description of the object, with its rounded "nose" and swept-back wings, sounds very like the prototype of an experimental aircraft rather than the more usual type of reported UFO – although there is always a question mark over such cases. Who knows? – it could have been an alien craft deliberately mimicking the appearance of secret planes. Whoever or whatever they are, the UFOnauts are certainly masters of disguise.

27 DECEMBER 1980, RENDLESHAM FOREST, SUFFOLK, ENGLAND

The former Cold War base of joint RAF/USAF operations at Woodbridge, Suffolk is surrounded by the somewhat scruffy swathe of

Rendlesham Forest, scene of one of the most enduringly controversial encounters in the history of UFOlogy. Now, in the days after glasnost and perestroika, the base, together with neighbouring Bentwaters, has been sold off, but still sees the occasional re-enactment of the events of December 1980 by dogged UFOlogists and television crews.

In the early hours of 27 December two security men were patrolling the perimeter of the camp when they spotted some unknown lights deep in Rendlesham Forest. Fearing that they might belong to a stricken aircraft – although they heard nothing to suggest a nosediving or crashlanding plane – they sought permission to leave the camp and enter the forest to investigate. Permission granted, they crashed through the trees, only to be confronted by a large triangular craft. Unable to make anything of it, they had to leave the scene, and the UFO seemed to disappear.

However, two nights later, it returned to the same area of forest, and this time a team led by the deputy base commander, Lieutenant-Colonel Charles Halt, went out to reconnoitre. This is his official report, dated 13 January 1981, the subject given as "Unexplained Lights" and copied to RAF/CC:

1. Early in the morning of 27 Dec 80 (approximately 0300L), two USAF security police patrolmen saw unusual lights outside the back gate at RAF Woodbridge. Thinking an aircraft might have crashed or been forced down, they called for permission to go outside the gate to investigate. The on-duty flight chief responded and allowed three patrolmen to proceed on foot. The individuals reported seeing a strange glowing object in the forest. The object was described as being metalic [sic] in appearance and triangular in shape, approximately two to three meters [2.18 to 3.2 yd] across the base and approximately two meters [2.28 yd] high. It illuminated the entire forest with a white light. The object itself had a pulsing red light on top and a bank (s) of blue lights underneath. The object was hovering or on legs. As the patrolmen approached the object, it maneuvered through the trees and disappeared. At this time the animals on a nearby farm went into a frenzy.

The object was briefly sighted approximately an hour later near the back gate.

2. The next day, three depressions 1" [3.75 cm] deep and 7" [17.5 cm] in diameter were found where the object had been sighted on the ground. The following night (29 Dec 80) the area was checked for radiation. Beta/gamma readings of 0.1 milliroentgens were recorded with peak readings in the three depressions and near the center of the triangle formed by the depressions. A nearby tree had moderate (.05-.07) readings on the side of the tree toward the depression.
3. Later in the night a red sun-like light was seen through the trees. It moved about and pulsed. At one point it appeared to throw off glowing particles and then broke into five separate white objects and then disappeared. Immediately thereafter, three star-like objects were noticed in the sky, two objects to the north and one to the south, all of which were about 10 [degrees] off to the horizon. Two objects moved rapidly in sharp angular movements and displayed red, green and blue lights. The objects to the north appeared to be elliptical through an 8-12 power lens. They then turned to full circles. The objects to the north remained in the sky for an hour or more. The object to the south was visible for two or three hours and beamed down a stream of light from time to time. Numerous individuals, including the undersigned, witnessed the activities in paragraphs 2 and 3.

[signed] Charles I. Halt, Lt Col, USAF
Deputy Base Commander

As the man who ran Ministry of Defence's UFO desk, Nick Pope, points out,[134] certain important elements in the story are missing, so the full truth about the Rendlesham Forest incident is unlikely ever to be known. However, Charles Halt has spoken to the media many times since 1980 – for example, saying on ITV's *Strange But True?* (1994), "I knew there was something there, but I was also convinced there was a logical explanation." Back on the night in question, leading the men into the depths of the forest must have been a nerve-wracking experience, even for trained soldiers. Enemy personnel and bombs are one thing, but possible alien technology encountered at the dead of night in a forest must have been quite another – indeed, the tape recording of the men's journey into the unknown reflect the attempts to keep their nerves under control as they forced their way through the vegetation towards the UFO.[135] In 1994 Halt said to the *Strange But True?* team: "It pulsated as though it were an eye winking at you and around the edges it appeared to have molten metal dripping off it." He added, apparently from the heart, "Here I am, a senior official who routinely denies this sort of thing and diligently works to debunk them, and I'm involved in the middle of something I can't explain." (It is interesting to note that he is seemingly admitting to being part of a scheme to cover up the truth about UFOs, routinely denying "this sort of thing" and "diligently works to debunk them". Why was he now admitting to such a stance in public? Was his new-found openness to the phenomenon also part of an official plot? He may have said, "I was really in awe", but anyone can learn lines.)

However, there were other witnesses besides the immediate team led by Halt. Airman John Burroughs was mending some light-alls when a vivid blue light shot past him just above his head, apparently making the broken lights come on, then go off again. Another witness was nineteen-year-old Airman Larry Warren, who was with a second investigative team that left from the East Gate of the camp. He saw everything around him vividly lit up by an anomalous light and a ball of red light – which he initially took to be an A-10 aircraft about to land. There was a strange circle of mist in a nearby field, over which the red light hovered before exploding, incredibly without either heat or sound, "into a galaxy of coloured lights". Inexplicably, these lights then took on the shape of a solid craft, about 20 foot (6 m) high and 30 foot (9 m) in diameter, with blue lights underneath. The UFO seemed to be illuminated – so brightly that it was difficult to look at – with a mother-of-pearl or rainbow hue. While others dived for safety (having first recorded the object on video and cine film), Larry Warren was glued to the spot.

Next day, in the aftermath of all the excitement, depressions were found in the soil at the scene of the encounter, which tested positive for radiation (see Halt's report, above) – twenty-five times the usual level for the forest, or so it was claimed. Then came the secrecy, covering the incident like an impenetrable fog. As Nick Pope records: "Personnel at the base, such as Sergeant Jim Penniston, noticed unusual activity in the days that followed. Unscheduled flights came and went. Penniston was told to keep quiet and forget them. Even Halt was left in the dark. His superiors told him to submit his report to the Ministry of Defence, which he did." Nick admits that the MoD behaved less than openly about the case, adding: "My predecessors do not seem to have acknowledged this [Halt's report] and Halt, quite rightly, finds this incomprehensible. The report was forwarded to the ministry by Squadron Leader Donald Moreland, specifically to the forerunner of my department, Defence Secretariat 8. Here the trail goes cold, and it is not clear from the files what happened next. Lieutenant-Colonel Halt, bewildered and angry, is still awaiting his reply from us fifteen years later."[136]

Neither did Halt receive any elucidation from his own government, although it seems that the mysterious aircraft noted by Penniston, were – according to former US Army Intelligence officer Clifford Stone – carrying high-ranking USAF officials whose task it was to collect all the evidence of the sighting, such as the video and cine film footage, together with various samples from the forest itself. This would have been sent to Washington where it would be kept, in classified isolation, by the Pentagon, like the boxed-up evidence from an X-File. (There is a rumour that the film was quickly airlifted to the USAF base at Ramstein, then in West Germany, but no one knows for sure.) Whatever happened to the material from the Rendlesham Incident, Charles Halt never saw any of it again. He was promoted and has since retired.

Back in Britain, however, his report caused a slight stir when the routine answer from the MoD – that whatever had gone on at Rendlesham it had "no defence significance" – prompted an exasperated Lord Hill-Norton, former chief of defence staff and admiral of the fleet, to comment: "Either the Americans, and indeed the deputy base commander, were hallucinating, or they believed that something had landed there and they had taken photographs and records of it. In either event, it must be of interest to the defence of the United Kingdom."

No one is more unequivocal on the subject of the potential UFO threat to British (and world) security than Nick Pope. He told leading British UFOlogist Tim Good (author of *Alien Bases*, 1998): "If, as the evidence suggests, structured craft of unknown origin routinely penetrate the UK Air Defence Region, then it seems to me that, at the very least, this must constitute a potential threat. How can we say there's no threat when we do not know what these objects are, where they come from, or what they want?"[137]

Yet although clearly the MoD had no desire – at least officially – of pursuing the Rendlesham mystery, others had. Indefatigable UFO researcher Jenny Randles, together with BUFORA colleagues Brenda Butler and Dot Street, undertook to comb the area around Rendlesham to find other witnesses who might be able to complete the frustratingly fragmented picture of what happened in December 1980. They found that some local people had seen lights zigzagging through the trees on the nights in question, but had been told in no uncertain terms to keep their mouths shut. One man even had his property encircled with barbed wire and MoD "Keep Out" notices.

One of the key witnesses the women found – given the pseudonym "James Archer" – claimed to have seen the triangular craft sitting on the forest floor on three legs, and although careful not to mention aliens, he did describe seeing non-human shapes moving about inside the craft. According to this witness, the UFO zoomed off at an incredible speed from a stationary position.

However, Airman Larry Warren, nineteen at the time of the encounter, has since gone public with his story, in *Left At East Gate*, co-authored with fellow American Peter Robbins, a colleague of Budd Hopkins, the alien abduction researcher. Warren recalls, sensationally, that his orders were to guard the glowing object in the field, but as he did so, three creatures with the by-now classic description of a frail, bug-eyed, Grey alien emerged from the craft *and had a brief meeting with the base commander, Colonel Gordon Williams*. Then a smaller craft, also triangular, separated from the mothership and appeared in a nearby clearing. Warren and his fellow airmen were warned by "intelligence types" not to discuss – and certainly not to make public – what they had witnessed in the forest.

After the publication of the Butler-Randles-Street book, *Sky Crash: A Cosmic Conspiracy*, in 1984, Sergeant Adrian Bustinza came forward with his version of events, which tallied in most details with that of Larry Warren, describing the craft as massive and saucer-shaped and detailing how many photographs had been taken of it.

The events at Rendlesham Forest are still the subject of heated debate. What really happened there back in December 1980? Did alien craft land in the trees not far from a USAF base, and did its occupants really meet with the base commander, as if by prearrangement? The story is so sensational it seems like science fiction – and that is precisely what many commentators think it is. Yet if the voices on the tape in the "Halt Package" are those of actors simulating terror and awe, surely they deserve to be nominated for the top thespian awards, and it corroborates some of the detail in Halt's official report – for example, one of the men can be heard saying, "All the barnyard animals have gotten quiet now," implying that they had previously been frenzied, just as Halt detailed.

Nick Pope says, ". . . fact and fiction intertwined until Rendlesham Forest became a myth . . . [it] is the British Roswell – we know that something happened, but we don't know what." He points out that the trail has long since gone cold: even the trees have been felled, perhaps, as some rumours have it, as a result of the dangerous levels of radiation in them. Certainly Nick is impressed by the evidence of the radiation, contacting the Defence Radiological Protection Service, and although he discovered that the radiation level at Rendlesham was ten times the norm, not twenty-five times as previously alleged, it still suggests that something unusual had happened there.

One apparently astonishing explanation for the sightings in the forest was put forward by British science writer Ian Ridpath on *Strange But True?*. He claimed that the pulsing lights were nothing more than the revolving beams from the Orford Ness Lighthouse, some six miles away from the scene of the encounter. When this was put to him on the television programme, Charles Halt said (rather mildly, considering): "A lighthouse doesn't move through a forest, doesn't explode, doesn't change shape, doesn't send down beams of light." And of course the USAF personnel were considerably more familiar with the lighthouse than Ian Ridpath. Others suggested the sightings were due to the re-entry of a Russian rocket, or the depressions in the ground came from hopping bunnies (no doubt demented by the radiation). Perhaps more believable is the idea that the sensational UFO story was concocted to hide an explosion of a nuclear reactor on board a crashed aircraft. Although Nick Pope doubts that such a drama could have been kept from the media, surely the UFO story very successfully *took its place,* as, in this hypothesis, it was designed to do. UFOs have a way of providing excellent cover: once an installation is "tainted" by stories of Greys and UFOs, no serious investigative journalist would be seen dead within miles of it – perhaps as intended . . .

An interesting twist to the Rendlesham story is the sea-change in Jenny Randles' beliefs. Once a determined conspiracy theorist, as can be seen from *Sky Crash,* even she now evokes the lighthouse theory to explain the lights, while other associated phenomena are hypothesized as being the result of electromagnetic forces. While there may be some truth in these ideas – which are detailed in her more recent book *The UFOs That Never Were* (2000), co-authored with Dr David Clarke and Andy Roberts – it comes as a shock to many that such a dramatic shift is possible, although her honesty in admitting she was mistaken can only be admired. This change of heart serves to underline the fluid nature of this research, and the necessity of keeping an open mind.

A final thought: witness Clifford Stone believes that a craft of superior technology landed in the Rendlesham Forest in 1980, controlled by an intelligence "that did not originate on Earth". While that may one day turn out to be no more than the unvarnished truth, conspiracy theorists may like to note that he is an ex-intelligence officer, presumably with the same loyalties and agenda as he had when he was still in that job. (Intelligence officers, ex or otherwise, have a habit of cropping up in the background of major UFO stories.) Whether that has in any way coloured his pronouncement about Rendlesham is unknown, but that simple fact will no doubt cause further speculation.

29 DECEMBER 1980, HUFFMAN, TEXAS

If ever one is tempted to think that encountering UFOs is just a bit of fun – an exciting novelty with which to impress the media – the Cash-Landrum case should be the ultimate cautionary tale.

On the evening of 29 December 1980, Betty Cash, Vickie Landrum and her young grandson Colby were driving to their homes in Dayton, Ohio along a lonely stretch of road when they saw a huge diamond-shaped UFO in the sky. Getting out of the car about 55 yards (50 m) away, they seemed to be caught up in a blast of heat that made them all break out in an instant lather of sweat, and their skin felt as if it was burning. Emitting shrill beeping sounds, the UFO gave off such an intense light that it hurt their eyes, flames coming from it sporadically. Then it flew off, leaving the trio distressed and shaken, although they had no idea then how traumatically the experience would affect them.

Arriving home in Dayton, Vickie announced that her head hurt and she felt very ill, but by midnight she was a good deal worse. Both she and Colby were in a terrible state: they had both suffered bad burning of the skin – as if sunburnt – had a fever, vomiting and uncontrollable diarrhoea, besides excruciating headaches. The pain from the burns was intense, taking several days to bring under control, but the headaches continued without responding to any form of treatment for about three weeks. After that, the symptom lessened, but recurred regularly for a year.

Their brush with the UFO had other effects, as if it somehow damaged the witnesses's immune system: they suffered severe skin problems and other infections, besides the recurring headaches, sickness and diarrhoea, but their eyes were severely affected by their experience. Vickie's eyesight began to deteriorate rapidly after the encounter, so much so that she had to invest in a series of glasses – each stronger than the last – and have treatment for bad eye infections. To a lesser extent, her grandson Colby has suffered similar eye problems.

Perhaps almost as bad was the loss of hair that Vickie experienced after the incident: she lost over 30 per cent of her hair in large patches – a terrible psychological trauma for any woman – although thankfully it grew back, uncharacteristically frizzy. Colby lost a small area of hair but this also grew back quite quickly.

Meanwhile Betty, too, was suffering acutely. She said: "The blinding headache that developed within an hour or so made me feel like I was going to die".[138] Huge water blisters, "the size of golf balls" erupted on her face, neck and scalp, one of which obscured her right eye. She also had an aversion to any heat source, even warm water or sunlight. This sudden sickness and debilitation was devastating for Betty, previously a very active person, who had run a restaurant and was planning to open another in the near future. All those plans were ruined: she spent five periods in hospital, including one in intensive care. She also lost her hair, which regrew but was never the same, and suffered permanent scarring from the almost constant skin eruptions. Her doctors pronounced themselves baffled by her condition, which, they speculated, could have been caused by exposure to "some kind of electromagnetic radiation".

As for young Colby, apart from the physical reactions to the encounter, he suffered such terrifying nightmares for some weeks afterwards, that during a "re-enactment of the event set up by investigators", Vickie said, "He was so terrified I though he would die of fright", besides developing a fever.

Investigators discovered that others had also seen a giant diamond-shaped UFO around that time, including deputy sheriff Frank Chinn of Echols, Kentucky, who spotted "an upside-down diamond with flashing lights in the middle" on 28 December, and Jan Moffat who saw a strange bright light in the sky north of Houston, although only for a short time. Two witnesses, however – six-year-old twins Jason and Jesse Williams, of Ohio country, saw four silver UFOs flying low over the houses, which they described as "triangle things", thus predating the "flying triangle" flaps by about ten years.

But what were these UFOs? What produced such horrific side effects in the witnesses? If not craft from outer space (and there is always the possibility that they might be), could they have been secret weapons, terrible machines of destruction developed in great secrecy by the US military, presumably part of the legendary "Black Ops"? One objection to this theory is that such a craft has not been deployed since the sighting in 1980, although there have been opportunities to do so – during the Gulf War, for example. Although sophisticated weaponry did terrible damage to the vehicles and "soft targets" – people – on the road to Bazra, there is no evidence that the survivors suffered similar chronic side effects from simply *witnessing* the USAF's best flying overhead. The three witnesses in this case had no direct contact with the UFO other than simply standing in the road below it.

Whatever the nature of the craft that caused the damage, one very like it seems to have been busy for several years before the Cash-Landrum case. In December 1967 Maryellen Kelley of Mohomet, Illinois saw a huge orange UFO from a distance of about 40 yards (36 m), which was flying roughly 50 feet (15 m) above the ground. Immediately she fell to the ground as a massive electric shock shot through her,[139] her face reddening as if badly sunburnt, her legs and hands were burnt, and her eyesight was damaged. She also "developed earache in her left ear, nosebleeds, pains in her chest and excessive thirst." Others suffered a similar fate.

Also in 1967, Steve Michalak came across a "cigar-shaped object emitting a brilliant, purple light" near Falcon Lake, east of Winnipeg, Canada, receiving facial and chest burns, and later, sickness and diarrhoea, loss of power in his limbs, weight loss and blackouts.

There was more to come. *The Unexplained* noted: "In Finland in November 1976, nineteen-year-old Eero Lammi was knocked to the ground by a luminous ray from a UFO and suffered burns on his chest. His injuries were similar to those of a twenty-year-old man from Tyler, Texas, who was hit in the chest by a luminous ray from a UFO in January 1979. His chest was marked by a large diamond-shaped burn for many months".[140]

Over in Italy, the *Data Net Report* told the story of nineteen-year-old Osvaldo d'Annunzio, who had been paralyzed by witnessing a hovering UFO, "so that he was unable to run away. His face was severely burned, and he suffered afterwards from violent headaches. He commented: 'The cows in the surrounding meadows changed colour and did not resume their true colour until after the UFO's departure'."

And on 10 September 1981 a young woman from Plymouth, England, received a burn on her hand from simply witnessing a UFO in the sky . . . The tally goes on. But there is a difference between the cases just cited and that of the Cash-Landrum trio. Since the incident neither of the women has been able to work: the constant skin problems mean that they would be barred from pursuing their careers in the food industry, and in any case their recurring illnesses were too unpredictable for them to be considered reliable workers. In many respects, their UFO sighting was the end of their lives. In all the other cases, the witnesses made a complete recovery.

27 AUGUST 1981, KING'S SUTTON, OXFORDSHIRE, ENGLAND

At 11 a.m. on 27 August 1981 Beryl Hayward and her ten-year-old son were enjoying the beautiful sunny day as they walked to the post office in the picturesque village of King's Sutton in Oxfordshire. They were nearly there when her son suddenly said, as she later recalled: "'What's up there, Mum?'" Mrs Hayward followed his gaze and there "gliding away across the sky almost directly above us was a round disc shape."[141]

She went on: "I don't know whether it was the Sun reflecting off it or whether it was flashing white lights, but it was shining very brightly – just like a mirror. I couldn't see any details, only the round shape, but one edge looked quite sharp against the sky as if it were the underneath of something.

"It moved very slowly then seemed to stop and hover. By this time I was quite excited. Suddenly it started to change to an orange-red colour and began to pulsate. Then it turned completely red and, as we watched, I wondered why it seemed to be getting smaller – then realized it was in fact going straight up into the air at terrific speed."

Stunned and excited, mother and son continued to crane their necks as they looked skyward. Mrs Hayward said, "I watched until it was just a speck of red, then it disappeared. But high in the sky, as far as I could see, I'm sure I could just make it out again as a small white disc."

Obviously discussing her strange sighting with others, later that day the witness discovered that other people had also seen the UFO, including a man thatching a roof. Mrs Hayward went to see him, and "he said he had never seen anything like it and that in his opinion it was definitely not an aeroplane or balloon."

The orange-red pulsating light and sudden spurts of speed are reminiscent of a great many other UFO sightings, suggesting a common origin.

1981, RUNCORN, CHESHIRE, ENGLAND

One evening in 1981, John Walsh, his wife Bell and his father went for a late night walk along the Bridgewater canal close to their home in Runcorn, Cheshire. But as they approached one of the bridges, all of them "had a strange feeling."[142]

John and Bell said later: "We had just finished comparing how we felt when a bright white light sped across the sky towards us. It was travelling very fast and we ran onto the bridge to get a better view. The light came from the direction of some laboratories on our left and went over us towards an old water tower. As it did so, it changed back to a brilliant white."

They watched the light for about ten minutes, during which it made no sound. Then "when it disappeared from our view a small light plane came from the same direction that the light had. That also passed over us and the noise of it was very loud – unlike the light. It was so quiet you could hear the water lapping at the sides of the canal."

The Walshes added: "Where we live we see many types of aircraft including helicopters. We always hear them before we see them. What we saw that night definitely made no sound."

9 JUNE 1982, CLAPTON, SOMERSET, ENGLAND

On 9 June 1982 Andrew Matthews has just met up with three friends after a hard day's haymaking when they all spotted something very unusual in the sky, heading towards them from the north-east.[143] It was a "bright pinky-red ball, which appeared to be about a mile [1.6 km] away and looked as if it was about to come straight up through the valley."

The four friends ran to get a closer look, noting that it "seemed to be about 200 feet [60 m] off the ground and glowed uniformly with a pinky-red colour." After a headlong dash of about 100 yards (90 m) the group stopped, just in time to see "the ball [start] to drop towards the ground behind a tree." They estimated that the UFO must have been about 5 miles [8 km] away, considerably more distant than they had first believed. Andrew said:

"After a few seconds the object reappeared, rising above the tree, then dipping down behind it again. That was the last we saw of the object itself, although as we stood watching we saw a strange light flashing from behind the tree for about ten minutes or so. All this took place around 9.15 p.m. when dusk was falling."

On reaching the farmhouse they shared their experience with the other haymakers, who ridiculed them, but the farmer and his wife said they had seen the UFO too.

1983, CALIFORNIA

The following story, if true, is a warning to those who are too quick to pronounce all UFOs extraterrestrial craft.[144]

Trying to achieve a deeper tan, a Californian sun-worshipper – having heard that a more effective kind of ray operated above the city smog – tied forty-two helium-filled balloons to a deckchair that was in turn firmly fixed to the ground by a length of rope. The idea was to rise to 6,000 feet (1,800 m) to do his spot of sunbathing, but in the event he rose higher than his wildest dreams when the rope snapped "and the deckchair rose, untrammelled, to a height of 15,000 feet (4,500 m), where a passing airline pilot reported him as a UFO sighting.

"Prepared for all possibilities, he pulled out an air pistol and shot the balloons one by one. His deckchair demolished a power cable, blacking out the whole area, and he arrived back on Earth much paler than when he left."[145]

While it is undoubtedly true that, as Hamlet said, "There are more things in heaven and earth . . . than are [usually] dreamt of . . ." surely it is not often that they are sunbathers aping Icarus.[146]

JUNE 1983, ST PETER PORT, GUERNSEY, THE CHANNEL ISLANDS

In early June 1983, thirty-one-year-old Londoner Tracy Wellman (a pseudonym), a busy freelance television researcher, had taken the opportunity to have a short break in the picturesque seaside town of St Peter Port on Guernsey. Booking into a pleasant little Bed and Breakfast establishment on the outskirts of the town, she spent her first evening dining with a French friend in a local fish restaurant before walking back to the B & B alone in the moonlight at around midnight. Just 50 yards (45 m) away she saw a light bobbing up and down against the darkness of a field or park. Taking it to be someone walking with a torch, she thought nothing more of it, until the light suddenly appeared immediately in her path, about 10 yards (9 m) away, but by now it was huge – she estimated it as being "about 20 foot (6 m) tall, with a sort of cap on the top, like a metal dustbin lid. It was curious because it was just a shapeless light underneath, but at the top it appeared to be structured, like a machine of some kind, made out of hard metal."[147]

After standing stock-still for what seemed like hours, but was probably – on her later reckoning – only about eight minutes, Tracy seemed to wake up. "It was as if I'd been paralyzed", she said. She took a step back, and the "thing" started to "fizz, like an indoor firework". At this she turned and ran up the path, but even though she was very scared, curiosity got the better of her and she slowed down, sneaking a glance behind her. Expecting to see it on the path where it had been when she fled, she saw nothing, but a slight fizzing sound from above made her look up. To her horror, the object was now only feet above her head, and was emitting a strange greenish glow. Tracy was utterly terrified, believing it meant her serious harm, perhaps even intending to kill her. Muttering half-formed prayers under her breath – as a very lapsed Catholic – she fell on her knees on the path.

"Although when I look back I wonder how I could have been such a coward – normally I'm very feisty – I seemed to have no choice but collapse on the ground. I felt as if all the strength had been drawn from me," she recalls. "Perhaps it was just fear or shock. But at the time I felt that the fizzing globe was out to get me, and that it was somehow attacking my nervous system. I admit to feeling absolutely abject fear. I felt I couldn't escape no matter what I did – as if something evil was attacking me just by being so physically close. I just prayed and shut my eyes."

Behind her closed eyelids the terrified young woman perceived an increasingly bright light, which she assumed was the globe descending on her. Just when it became almost unbearable, there was a loud "pop!" and the light vanished.

"There was a terrible smell, like bad eggs," Tracy shuddered, "but nothing else was left to show for the experience. There was no sign of the globe and no sound except for distant traffic. I was very shaken, and stumbled off to the B & B where I confess I burst into tears all over my landlady. I started to gabble the story but her face was such a picture of incredulity – mixed with some compassion, because she was a nice person and I had obviously had some kind of nasty experience – that I gave up. She got me a hot drink and tucked me up in bed. I slept like a log without any bad dreams."

The next morning, after sleeping late, Tracy plucked up courage and retraced her steps along the path where she had encountered the strange globe. There was no physical trace, although there was a slight whiff of the same bad egg smell, and she found a couple of small personal belongings that had fallen out of her handbag as she panicked.

After moving to St Malo for two more nights, her holiday was over. From that day to this Tracy has never had any other kind of UFOlogical or paranormal experience, and actively steers clear of the subjects (even turning down a lucrative contract to work on a major television series about the unexplained). Telling me the story was, she later claimed, "a great unburdening, a sort of catharsis", and although she was quite happy to go into details, she has no intention of ever telling her experience again.

However, she was eager to make the point that she *had* been drinking on the night of the experience, something that is – if we are to believe the vast majority of UFO witnesses, who all seem to be exceptionally upright citizens – extremely rare in such cases. Most reports emphasize their utter sobriety, and indeed, there is usually no reason to doubt it. But here Tracy admits quite cheerfully to having had "at least a bottle of white wine with my friend over dinner" although she added that this was her usual quota for a night out, and was in her experience by no means enough to create hallucinations. In any case it would be a unique kind of white wine that conjured up such a bizarre vision, complete with sulphurous smell. (Tracy said firmly, "I don't do drugs".)

But what was the strange fizzing globe? Why did it block her path and then hover over her? Was it a machine of some sort, and if so, who did it belong to? Who or what was directing it?

It is possible that it was some form of natural electrical phenomenon, such as a BOL or ball lightning, about which very little is known (and until recently was deemed officially non-existent by scientists), although it is hard to imagine how an electrical discharge could appear to have a "dustbin lid" at its apex. However, it may be that a purely natural phenomenon had somehow triggered a partial hallucination in Tracey's brain, adding the detail of the lid to an otherwise undecorated ball of electricity or light. (Perhaps the alcohol aided the process of misperception, although there are enough cases on record of extraordinary visions of one sort or another without the benefit of "at least" six glasses of chilled Chardonnay.)

The smell of bad eggs – sulphur – is, of course, traditionally associated with the demons of Hell, presumably because witnesses in the past experienced the smell and the demons together. Perhaps our ancestors knew more than we do about the true nature of certain strange experiences. Perhaps they were right: the unknown creatures and objects that appear in our paths and traumatize us, giving off a pungent odour, are simply up to no good, never were and never will be. It is a possibility that many of the most intelligent and questing UFOlogists have seriously considered – as can be seen in the Explanations chapter.

1987, SWITZERLAND

One of the most enduringly controversial cases – which came at the watershed between the Contactee Era and that of the abuctees – was that of the Swiss, Eduard (Billy) Meier, who claimed many very close encounters with ravishing female aliens. What makes his story particularly interesting, though, is the fact that he took dozens of what appear to be very unambiguous photographs of their visiting spacecraft.

However, computer analyst William Spaulding has demonstrated convincingly that the photographs are fakes. It seems that Meier took them by pointing the camera directly at the Sun – in order to obscure the giveaway wires from which he had suspended model spacecraft. Another investigator discovered what appeared to be the models themselves in Meier's garage, although the latter claimed that they were made as reconstructions after the event.

Yet, despite the claims of sceptics that the Meier case is one of clear-cut fraud, there are still those who believe it to be genuine, largely thanks to Gary Kinder's book on the subject, *Light Years.*

NOVEMBER 1987, GULF BREEZE, FLORIDA

In November 1987, Gulf Breeze resident Ed Walters saw a UFO, snapping five photographs of it with his Polaroid camera – and so beginning an extraordinary saga of multiple witnesses of things seen in the sky, intensive investigation, massive publicity, and alleged proof of hoaxing.

During that first incident Walters claimed he was physically lifted into the air and paralyzed by a blue beam. Meanwhile, elsewhere in Gulf Breeze, the former editor of the local newspaper, Charles Somerby, and his wife Doris also saw an unknown object in the sky at the same time. When Walters presented his five UFO photographs to Somerby's paper, there was no going back, although at first he protected his identity under the name of as 'Mr Ed'. The newspaper article encouraged another witness, one Mrs Zammit, to come forward with the startling news that she had seen the blue beam that had levitated Ed Walters.

For six months UFOs seemed to be attracted to Walters as if by a magnet. He photographed dozens of them, being alerted to their presence by a strange buzzing noise in his head – which conspiracy theorists believe may be an implant that had been put in place during an early, unremembered, abduction. (Buzzing noises herald a whole host of UFOlogical and paranormal experiences – including the vision at Fatima, Portugal – and are also associated with the onset of some forms of Temporal Lobe Epilepsy, which may or may not be significant.)

Gulf Breeze became inundated with tourists and investigators: mass skywatches were rewarded with the sight of multiple UFOs dancing about the skies, some of which were caught on both video and stills cameras. Various tests were set up, including a lie detector test, which convinced the examiner that Walters "truly believes that the photographs and personal sightings he has described are true and factual to the best of his ability."

Walters was also given a sophisticated stereo camera by image analyst and respected optical physicist Dr Bruce Maccabbee, which produced more photographs that were, however, dissimilar to the original Polaroids in that they showed only a small object, brightly lit, and not a recognizable UFO as in the first set of images.

While the crowds gathering around Bay Bridge did appear to see strange things in the sky, they may have been the victims of at least one hoax, possibly more: servicemen from the nearby Pensacola Naval Base have admitted to playing up to UFO fever by launching eerily-lit hot air balloons made of plastic bags and candles into the night sky. Indeed, some of the video footage does appear to show melted plastic "droppings" falling down, which some UFO hard-liners eagerly interpreted as "satellites" from the "mothership".

Ed Walters himself came under attack when it was alleged that models of the "UFOs" in his photographs had been found in his house, but such claims have been vigorously rebutted by others. This is one of the more complex and tantalizing cases, for while some of the strange lights over Gulf Breeze may have originated in the men's quarters at the Naval base, it seems unlikely that all of them were pranks.

Perhaps there was an element of trickery involved, which then triggered *genuine* sightings – a form of mass "artefact induction" (see the chapter on Explanations), as if the possibility of seeing such things became acceptable to the masses, opening their eyes to strange phenomena that may always be present in the skies.

Then in May 1988 Walters alleges he suffered from missing time, which some – notably investigators from the Mutual UFO Network, MUFON – speculate may have been an indicator of an abduction. They go further: perhaps the fact that he was about to be examined for signs of implants made the aliens remove them . . . In any case, after the episode of missing time, Walters no longer experienced the buzzing noise, and no longer saw or photographed UFOs.

29 NOVEMBER AND 2 DECEMBER 1989, AND 30-31 MARCH 1990, BELGIUM

An extraordinary wave of UFO sightings was reported in Belgium in the late 1980s and early 1990s – so extraordinary in fact that military jets were scrambled to intercept the mysterious intruders in their sovereign national air space. On 29 November reports of sightings flooded in from all over that tiny country, from ordinary people, the military and the police. But it was the events of 2 December that came to be seen as sensational evidence of the ETH by UFOlogists worldwide. The gendarmerie of Liège, the largest Flemish-speaking city in Belgium, reported many sightings of a UFO to the radar controllers attached to the country's Quick Reaction Alert – a small number of jets kept on standby in case of foreign aggression[148] – who saw evidence of an intruder for themselves on their screens. Acting with commendable swiftness, they scrambled two F-16s within five minutes. But although they arrived at the location of the sighting very fast, the jets saw nothing strange in the Belgian skies, and the radar blip simply disappeared as they grew close to it.

Later, another sighting was reported near Maastricht in Limbourg, 40 miles (64 km) from Liège, and although this one had no radar blip, the air-traffic controllers immediately dispatched the F-16s to check out the situation. This time, they did find something – but it was only a very terrestrial laser show . . . From then on it was decided not to scramble expensive and valuable aircraft on what could be a wild goose chase unless there were at least confirmatory radar blips.

All this was to pale into insignificance compared to the events of the night of 30-31 March 1990, when an estimated 2,600 UFO sightings[149] – or perhaps even 2,600 individual UFOs – buzzed Belgian and north German airspace. (And even that may simply be the tip of a very large and worrying iceberg: presumably not every sighting was reported.)

Most of the reports – from the Wavre area, to the south of Brussels – merely described coloured lights, although many were seen to be attached, in a triangular pattern, to the underside of massive low-flying craft, cruising the night sky at a leisurely 30 or so miles per hour (48 km/h). While some cruised so slowly as to appear almost arrogant, others contented themselves with hovering, then abruptly shooting off at incredible speeds – certainly much faster than the familiar 1,100 mph (1,760 km/h) of the Belgian F-16s.

The police alerted the NATO radar stations at Glons and Semmerzake, which confirmed that they had observed radar blips of anomalous – and therefore illicit – craft in their airspace. Once again, two F-16s from the Quick Reaction Alert were scrambled and within minutes could be seen on the radar screens chasing in on the mystery targets. But just when they seemed to be closing in, the objects disappeared from the screen – the ultimate evasive action – but not before displaying sensational aerial capabilities. During the 75-minute chase, the UFO was observed to drop an incredible 1,300 metres (4,250 ft) in just a second, and when it reached a minimal 200 metres (650 ft) from the ground, it simply vanished.

In his *Open Skies, Closed Minds* (1998), the Ministry of Defence's answer to *X-Files'* Fox Mulder, Nick Pope, points out that, "Actual interception of the object was probably impossible", adding dryly: "F-16 radar screens tend to share one affinity with orthodox scientists: they filter out and ignore slow-moving or stationary targets on the assumption that they will not be jets. The pilots have no way, therefore, of using their own radar to close in on the object once it has slowed down or stopped."[150]

The UFO seemed to be under intelligent control – or perhaps even had an intelligence of its own, like a very advanced robot – with its descent into invisibility at the arrival of the jet fighters. As General (then Colonel) Wilfried De Brouwer, Belgian Air Force Chief of Operations, said, "There was a logic in the movements of the UFO."[151] (It is refreshing to note that such a distinguished military man could bring himself to utter that often anathematized acronym. It is hard to imagine a high-ranking RAF or USAF officer talking so freely, and so publicly, about "UFOs".)

As Nick Pope says, "The implications of the Belgian case are frightening. A structured craft flew over Belgian airspace, clearly observed from the ground and on radar, but it evaded the fastest, most sophisticated aircraft they possessed. They had, in effect, been powerless. What if that triangle, whatever it was, had been hostile? Whether extraterrestrial or very much of the Earth, what if it had started dropping bombs?"[152]

1992, CENTRAL LONDON, ENGLAND

A case from Nick Pope's files[153] tells how a man phoned him one day in 1992 to describe a UFO sighting that had been seen by hundreds of homeward bound commuters over the River Thames in central London. The witness said he had been crossing Waterloo Bridge when he saw something that stopped him in his tracks. There, just yards away, was a vividly coloured UFO hovering over the river, then as he watched, fascinated, it suddenly zigzagged away at an incredible speed.

Nick Pope comments on the bizarre nature of this report: if true, then hundreds of people must have seen the UFO, yet, "It was as though it never happened". Perhaps this extraordinary story underlines the essentially quixotic nature of the phenomenon: some may see the craft while many more are totally oblivious to them. Perhaps the phenomenon is almost self-regulating, deciding for itself the number (and identity?) of the witnesses. Perhaps a kind of natural screening out of unsuitable witnesses takes place: only those with open minds or in certain types of conscious states – in a reverie or daydreaming, for example – or natural psychics, can see them. There is also the disturbing possibility that our skies are always crammed with UFOs, but that we can only see them under rare and freakish circumstances, just as in the old fairy stories the entities became furious when humans could see them, yet they had existed prolifically all around mankind for countless ages.

FORT WAYNE, INDIANA, MARCH 1993

This quaint little story was reported in *Strange* magazine,[154] and is worth quoting in its entirety, if only for its novelty value:

"Emily Eck, a 7th grader at Memorial Park Middle School in Fort Wayne, Indiana, reported an unusual experience. She was babysitting a woman's son when, without warning, all the house lights went out. This left her in the dark with a crying child. A bright light shone in the front window, almost blinding her, and rain poured down outside.

Hearing a knock at the door, she said: " 'I saw the door knob turn. The door opened and a little purple man entered.' She described him as being four feet tall, with a 'curvy nose' and 'big circular ears.'
" 'I could not move while he walked into the kitchen and got a Dr Pepper,'she said. 'And as quickly as he walked in, he walked out. I

ran to the window, the little boy following me, and there was – an unidentified flying object. A UFO!'

"According to Miss Eck, it took off towards space, and the house lights turned back on. The kid remained quiet."

Assuming that Miss Eck was not merely seeking her fifteen minutes of fame, this charming – but apparently off-the-wall – story nevertheless has several of the elements of the classic encounter, with the failure of the electricity supply, bright light and hovering UFO. Yet the "alien" is uniquely purple – more like a cartoon character come to life than the classic "Grey" or blond Venusian type, suggesting that some kind of trickster was behind it, or that some electrical phenomenon had kick-started a strange little hallucination in Emily Eck's mind – which her small charge also seemed to share.

30 MARCH 1993, MULTIPLE SIGHTINGS OVER BRITAIN

Nick Pope recalls in his *Open Skies, Closed Minds* that it was three years to the day – 31 March 1993 – after the Belgian wave (see above) that he received a flood of UFO reports from all over Britain. The night before, a great many people, including trained observers from the military and police, had seen unknown objects in the skies, and the reports rapidly began to surface on Nick's MoD desk.

He notes[155] that the first call came from a policeman from Devon who, together with a colleague in a patrol car, had seen two bright lights and a dimmer one flying in formation, apparently leaving vapour trails. Nick said: "I probed carefully over the phone. From what he told me about their colour and movements, they were clearly not aircraft lights. Similarly, they were not fireballs. They were simply unidentifiable."

It soon transpired that many other people had seen the same – or similar – mysterious lights the night before, which made Nick think back to the Belgian wave. Although acknowledging that, "Colourful experiences like this produce excitement, and excited people don't always make good witnesses", he began to see a distinct pattern, asking: "Could we be looking at the same astonishing chameleon craft?" Thinking of the "uncanny anniversary" of the Belgian wave, Nick was in danger of becoming just as excited as the

British UFO witnesses, but knew he had to keep calm and think as objectively and logically as possible. He began to shift methodically through the mass of data that was accumulating on his desk, trying to make sense of the emerging pattern.

Most of the reported UFO sightings had occurred between 12.30 a.m. and 1.10 p.m., peaking at the latter time, and although the locations of the sightings did not produce a neatly straight line on the map (as Nick had hoped), it was obvious that they were concentrated on certain areas – Cornwall, Devon, Somerset and Wales, the south-western and most western part of the southern half of mainland Britain (Scotland and Northern Ireland seemed to be unaffected by the sightings), although there were a few scattered reports from further north, in Yorkshire. If this was the flight path of one single UFO, then it was, as Nick noted, "zigzagging in a random pattern through our skies at incredibly high speed".

It soon became obvious that several hundred people had seen the UFO, which is especially impressive for a sighting in the early hours of the morning, when most people were in bed and asleep. Later, Nick received reports from Ireland, France and – once again – Belgium. The UFO community began to seek official confirmation of their own data from Nick at the MoD UFO desk, although he was the first to admit that even he did not have nearly as much information as he would have liked.

One report, which Nick describes as "dynamite", came from a military patrol at RAF Cosford near Wolverhampton in the West Midlands: as he says, "An unidentified craft in any British airspace was threatening enough, but over a high security military establishment?" A report came in from an expert in aviation and mathematics who had timed the flight of the UFO between two points near Haverfordwest in Pembrokeshire, calculating that its speed was roughly the same as the contemporary F-16 – 1,100 mph (1,760 km/h) – and clearly, with his expertise in aviation, he was highly unlikely to have misidentified an aircraft.

Five members of a family in Rugeley, Staffordshire, had driven after a 200 metre- (600 ft-) wide UFO flying at just 300 metres (980 ft) overhead despite the unpleasant vibrating hum that it was emitting. Thinking the low-flying craft was about to land in a field, they made an emergency stop and leapt out, but the enormous UFO had simply vanished.

However, almost certainly the most sensational sighting of the night came from the meteorological officer at RAF Shawbury near Shrewsbury in Shropshire. For fully five minutes, he observed an object – which he estimated to be the size of a Jumbo jet – zigzag towards him in the sky at great speed, sweeping the countryside as it went with a powerful beam of light emitted from its underside, as if searching for something in the countryside below. Like the family in Staffordshire, the witness heard a strange vibratory hum.

Nick Pope says: "What could I say to this man? He was a trained observer, considerably more familiar with the night sky than I was. A patronizing lecture on aircraft lights seen from unusual angles seemed wholly out of place." But it was the description of the craft's probing beam of light, a searchlight scouring the countryside that disturbed Nick the most, with its implication of intelligent control and a distinct purpose. Was it looking for "what is usually in the fields on a mild, spring night – cattle?" Had the meteorological officer seen the prelude to what could have been a rare case of British cattle mutilation, but which was aborted at the last minute for some reason?

Needing to have all the facts at his fingertips, Nick checked with the Royal Observatory at Greenwich, but there had been no unusual planetary activity, no anomalous atmospheric conditions. Similarly, no military or civilian aircraft had been in the skies over Britain behaving in such an inexplicable, but distinctly sinister, manner. However, RAF Fylingdales, in the north of England, announced that debris from the Russian rocket Cosmos 2238 had re-entered the atmosphere the night before and had *possibly* been visible from some parts of Britain – an "explanation" that was immediately seized upon by determined debunkers (who, in its absence, would no doubt have been content to condemn all the witnesses as liars or fantasists). However, space debris burns up on re-entry, and does not fly slowly emitting weird hums and beaming white lights down on the countryside. That, however, was a mere detail to the sceptics.

Nick "took an unprecedented step and ordered a number of radar tapes to be impounded and sent to me. As these tapes are usually wiped for reuse, it was important to work fast."[156] He goes on: "There were a few returns which fitted the times and locations when sightings were made and after several hours of scouring the standard VHS videos I could isolate and identify these." However, it soon became obvious that the crop of the night's radar blips were a wash-out, experts dismissing the apparent anomalous ones conventionally enough, such as the returns from tall trees. Whatever had buzzed Britain that night, it had not been picked up on radar. Deeply disturbing to the MoD man was the fact that without radar no aircraft would have been scrambled to defend the realm . . .

There had long been rumours of the covert production of a "black op" craft, the legendary American stealth plane "Aurora", which is said to be designed specifically to evade radar. Were the night-time UFO sightings of such a craft, perhaps being test-flown over Britain? Perhaps not unexpectedly, the Americans declared themselves equally baffled by the UFO – as indeed they would if it *had* been the Aurora or something like it. (After all, the whole point of a covert operation is that as few people as possible know about it, to establish "plausible deniability".)

Nick's official report finally read: "Type of craft – unknown; origin of craft – unknown; motive of occupants – unknown. Conclusion – unsatisfactory."[157]

1993-97, BONNYBRIDGE, SCOTLAND

Over the course of about four years in the 1990s, the area close to Bonnybridge in central Scotland became the focus for hundreds of UFO sightings, becoming known as the "Bonnybridge triangle", although other villages and towns – such as Denny, Falkirk, Grangemouth and Shieldhill – were also UFO hotspots. Almost every shape and size of object ever reported were spotted in this place, from small balls of light to Adamski-type saucers, cigar-shaped "mother-ships" and the more modern black flying triangles. Although intense media attention threatened to turn the area into a circus, and almost certainly contaminated the "true" reports with entirely mendacious accounts (perhaps a few pounds for a quick quote did not come amiss now and then), Bonnybridge remains a major enigma of recent times.

One day in the mid-1990s when the Slogget family were walking on the moorland above Bonnybridge the road was suddenly flooded with a vivid blue light and then they saw a huge UFO emerge from behind some trees, looking – as Carole, the twenty-six-year-old daughter, said – just like a giant "Tonka toy".[158] It emitted a loud whirring sound, which became a sort of "howling" when they heard a door opening. Then there was a brilliant flash like a flashgun going off as the UFO vanished behind the trees once more. Later it transpired that a local man and his twin daughters also witnessed a glowing object above the village at roughly the same time.

On another occasion when "Robert Muir" (a pseudonym) was taking some photographs of the brightly-lit BP AMOCO complex from a nearby hill he sensed that something was hovering above him. Looking up, he was horrified to see a "glowing disc"[159] zooming rapidly towards him. With great presence of mind, he managed to manoeuvre himself so he could take a photograph of the UFO – for all he knew it could have been the last picture he ever took – but the object immediately flew off.

Robert Muir's photograph shows a rather blurred light grey disc shape, somewhat darker grey at the edges. There is a small dome in the centre, with what looks like a brightly-lit window, on either side of which is an illuminated tube. The perimeter is surrounded by a suggestion of a "corona", as if it is "backlit". Investigators had the film examined professionally and could find no evidence of a hoax.

Dozens of other UFOs in the Bonnybridge triangle have been captured on film – both stills and movie – by both amateurs and professionals, as excitement about the sightings reached fever pitch. Television crews from as far away as Japan seemed almost permanently camped in the village, and "skywatches" on the surrounding hills became commonplace. Then matters seemed to take a rather sinister turn.

During one of the skywatches a local researcher stopped a passing car to ask the occupants if they had seen a particular light in the sky: they responded in the negative and drove off. It was a trivial enough incident and should have been instantly forgotten, but the next day the police visited the researcher and interrogated him about his activities. But why? What had he done wrong? Even if the people in the car had taken umbrage and reported him, how had the police got his name and address – and more to the point, why had they bothered? As Brian Allan of the Scottish Strange Phenomena Investigations organization says, "In spite of repeated requests for some sort of government assistance in attempting to find an explanation for the happenings, there has been continual 'stonewalling' by the authorities. The official answer most often received is, 'There is nothing here to investigate'. This response begs the question, 'If that's the case, then why are the security services here?' "[160]

Not surprisingly, many of the more extreme theories have been suggested to explain the Bonnybridge Flap, including intrusions from parallel universes and activities connected with "alien breeding programmes" (although, interestingly, there are no reports of abductions in this area at the same time). Brian Allan points out that: ". . . there is a geological fault beneath the village, a belt of an ultra hard quartz material. This fault created major problems with mining operations in the area and caused tunnels to be re-routed to avoid this layer of almost impenetrable material. It is feasible that this substance is prone to the 'piezo crystal' effect and generates electrical charges when subject to the mechanical stress of tectonic movement. This phenomenon helps lend credence to a current theory that many UFO sightings have their origin in highly unusual, natural, geo-magnetic anomalies creating an electromagnetic field, which in turn affects the temporal lobe in sensitive people [see page 464]."

He recalls that, "On a sky watch held one clear, cold night in October 1998, I witnessed a brief flash of bright blue light that seemed to emanate from the ground. The light appeared as a single ball, then flashed outwards from the ball in two opposing arms. While this sighting only lasted for a fraction of a second, it was seen by eight or nine people . . ."[161]

Since 1997 the area around Bonnybridge has seen considerably fewer UFOs, and it seems the flap is over.

BAKEWELL, ENGLAND, SEPTEMBER 1993

No fewer than thirty people saw a mysterious black triangular object fly over Bakewell in Derbyshire, England at 9.30 p.m. on 26

September 1993. It had a bright white light at each corner with red lights along the sides, and travelled at about 40 mph (64 k/mh). It made no noise.[162]

30 NOVEMBER 1994, SOUTHERN CHINA

One of the cases to emerge recently from China occurred on 30 November 1994 when guards at a factory in the southern region described a UFO flying over a nearby tree farm as "two spotlights in the air, a very brilliant ball of light changing colours from yellow to green and red, which passed above very noisily like a locomotive."[163]

Shi Li of CURO (the China UFO Research Organization) discounts the idea that the UFO could have been due to freak atmospheric conditions, explaining that the damage it caused was selective, cutting trees down but not telecommunications lines. He said, "A worker was even thrown into the air several metres away, but he was not hurt. There were no casualties with people or animals, but the force of the phenomenon was very strong. Nearby there is a train wagon factory and the roof of some of the wagons was thrown away, some wagons were even moved dozens of metres and steel pillars were cut. The guard at the factory saw something very noisy passing [through the air] like a train with lights". (Is this a version of the aerial train observed by Cyril Picknett in York in the 1970s?)

A similar phenomenon happened at another tree farm – this time in Guizhou – some weeks afterwards, alarming the authorities. Shi Li says, "This event had a very large impact on a national level, but they didn't reach a final conclusion, they say it's pending, that it seems to be unexplained." After due investigation and deliberation, Shi Li declares, "all of us Chinese UFOlogists reached the conclusion that it was an extraterrestrial spacecraft, a UFO. When it half-landed, it attempted to land but hit all the trees and cut them."[164]

This is one of the first UFO sightings to emerge from China, after the ban on UFO research was lifted in 1978.

6 JANUARY 1995, THE PENNINES, ENGLAND

Just seventeen minutes before the scheduled landing of their flight from Milan at Manchester Ringway Airport, Captain Roger Wills and Flight Officer Mark Stuart suddenly ducked when "[a] brightly lit wedge-shaped craft appeared only yards in front of them at 13,000 feet [3,900 m] over the Pennines. Air traffic control at Ringway told them that theirs was the only plane showing on radar."[165]

It seemed to the cowering crew members that a collision was imminent, but at the last moment the UFO flashed along to the right of the plane and vanished from view. The British Boeing 737 landed safely, with the passengers totally oblivious of their close-call with disaster. The pilots sent a sketch and other details of their encounter to the Joint Air Miss Working Group, which is part of the Civil Aviation Authority, but did not talk to the Press. One of their colleagues described them as "high-grade, sensible guys".[166]

9 FEBRUARY 1995, CHINESE AIRSPACE

According to Chinese UFOlogist Shi Li, a UFO was detected close to a Chinese airliner. He says, "On the radar screen they detected an oval object, which later changed to a round shape, about 2 miles (3.2 km) from a commercial airliner. The pilot didn't see a thing, but the control tower told him a UFO was flying parallel to them. At that moment, the anti-collision automatic system on the Boeing 737 turned on and the Control Tower gave instructions to the plane to climb over the cloud layer to avoid a collision."[167] Presumably, the UFO then disappeared.

JUNE 1995, REDDITCH, WORCESTERSHIRE, ENGLAND

Retired police officer John Hanson, formerly a police constable in the West Midlands police force, looked out of his bedroom window at his home near Redditch, Worcestershire at around 10.35 p.m. one night in June, 1995, when he saw something rather surprising. Later he recalled:

"I saw a silver, pear-shaped ball of light hovering over a tree. It was about 40 feet (12 m) off the ground, and about 20 feet (6 m) long and 5 feet (1.5 m) wide. Immediately opposite was a red, cigar-shaped object, about 30 feet (9 m) tall.

"The pear ball suddenly moved and there was a piece cut out of it like a wedge of cheese. It then changed into the shape of a jellybean and jumped onto the cigar-shaped object. The two objects then fused together, produced a rippling light and were gone."[168]

He added that the whole experience, which took just five minutes, was very similar to another incident that took place about 20 miles (32 km) away at roughly the same time.

Mr Hanson, now fifty-three, was adamant, however, that he was not "talking about flying saucers or spacemen, but an unidentified object – some form of energy source", which the UFO, with its shapeshifting qualities, may well have been.

As a policeman – and therefore a trained observer – John Hanson's testimony is particularly valuable:[169] it is interesting that, unlike many witnesses, his mind did not try to make the object into a spaceship complete with portholes. As far as can be ascertained, he merely reported what he saw, but the question remains as to its true nature.

JANUARY 1996, VARGINHA, SOUTH CENTRAL BRAZIL: "THE BRAZILIAN ROSWELL"

After terrified locals witnessed several UFOs over the small industrial town of Varginha, in south central Brazil, the area was already in a heightened state of hysteria when the most disturbing part of the story began to unfold. On the night of Friday, 19 January 1996, an American spy satellite picked up an unidentified object heading for the town. The local military were put on full alert, and – not unnaturally – there was frenzied activity behind the scenes, with the Americans, as one witness put it "in it up to their necks". Whatever the UFO was, both the Brazilian and American authorities were very interested in it.

Despite an official total media blackout, video cameras were much in evidence as the locals rushed to record the extraordinary drama being played out in their skies. Footage shows white objects gliding among the houses, sometimes just a few feet from the ground, at other times high in the air, moving fast. According to some witnesses, one UFO was long and cigar shaped, with smoke trailing out from behind as it moved, although many of the videos show an indeterminate whiteish shape with no vapour or smoke.

Some locals described one UFO as being "as big as a minibus", light-coloured, and "almost transparent".

Then at 7 a.m. on Saturday 20 January, the Fire Brigade received reports that strange creatures were running loose on the outskirts of the town, but when they got to the place they found the military already there. The Army and firemen then collaborated in order to capture the creature, which they finally located a few hours later hiding in the tall grass, apparently hurt. Seemingly "fragile", it proved very easy to capture. (According to American UFOlogist Stanton Friedman, this was "not a sophisticated alien".)

Witnesses heard three shots, then the soldiers emerged from the long grass carrying a sack with something moving in it. By early afternoon the live "alien" (if that's what it was) was taken to a high security military base, but Major Calza, of the Army unit, claimed it was not an extraterrestrial. He said the story was just a garbled version of a mentally handicapped dwarf taking refuge during a violent storm (in a sack carried by soldiers?). The firemen, while now denying they had captured an alien, refused to be questioned further, saying the matter was "classified".

Nick Pope, when asked his views for the television programme *Alien Encounter*[170] for the ITV series *Beyond the Truth,* said that as that particular area in Brazil is known for its strange animals, perhaps the alien was simply a misidentified creature from the neighbouring forest. Strange creature indeed: in the afternoon three local girls spotted a bizarre and terrifying entity crouched by a wall. When one of the girls screamed, it looked up, turning its big red eyes in their direction. They later described it as having "soft brown skin . . . and three horns on its head, visible veins in its arms", adding that it looked as if it was suffering, in pain. Panic stricken, the girls fled hysterically, telling their parents that they had seen "the Devil".

At 10 o'clock on the Saturday night a military patrol caught another unknown beast and – treating it gingerly, being afraid they might catch some kind of disease – took it to the local hospital, where it was found to be dead on arrival. Chaos ensued as the military hastened to seal off the area.

Some doctors and nurses believed the being was "the Devil's Son", while others favoured the more scientific, if barely less disturbing, explanation that it was the result of some secret genetic experiment. According to witnesses, of the fifteen doctors in attendance wearing surgical masks, one of them pulled out the creature's tongue, which was long (4.5 in, 12 cm) and flat, recoiling to leave just a slit for a mouth. Its eyes had no pupils; it had short skinny legs and arms. There was "an unbearable smell of ammonia".

The body of the wretched little thing was taken under tight security to an Army facility, before being transported to a restricted access laboratory at Sao Paulo University under the guard of intelligence officers, where Dr Badan Palhares, head of Forensic Medicine, was put in charge. However, in response to a later question from a student, he is reputed to have denied any involvement with alien creatures, but added cryptically: "Ask me this question in ten or fifteen years from now".

According to *Alien Encounter*, the creature was then flown to the United States in an unmarked transit plane. Meanwhile, back in Varginha, one of the soldiers died suddenly just weeks after being involved in the incident. There was no autopsy, but a blood test had revealed that his blood contained an unusually high level of toxins.

Many witnesses to the incident were traced and warned off by the authorities, and the military held their own Press conference at which they robustly denied their involvement, saying that they did not intend to explain their night manoeuvres to western UFO researchers, which seems reasonable enough – as far as it goes.

In some respects, however, that was only the start of the weirdness that attacked the Varginha area in the early months of 1996. An old lady saw another creature, which she described as being "very ugly", at the local zoo (outside, not inside, the cages). No one knows what happened to it. Perhaps this was connected in some way with the unexplained deaths of five previously healthy animals in the zoo. The Director said she had never known anything like it.

Then, curiouser and curiouser, the Arecibo Space Observatory in Puerto Rico – which has attempted to communicate with deep space for thirty years – had its own visitation from one of these creatures, which because of their preferred form of taking nourishment, rapidly became known as "Goatsuckers" (*Chupacabras*). Dismissed as the carcass of a wild cat by some, the dead creature found near the observatory was not the only bizarre member of this unknown species to plague the area. Others – horribly alive – were described as being one metre (2.5 ft) tall with wide wings, vestigial features, and, according to one terrified witness, "neon green eyes that turned red". It sported a fin on its back and walked erect on two legs, changing colour like a chameleon and vibrating very fast, producing a humming sound. On one occasion a Goatsucker was observed to levitate into a waiting UFO. On another, a man lashed out at one creature with his machete: when it made contact the being "sounded hollow like a drum". And although nicknamed Goatsuckers, these creatures also have a taste for other animals, tearing rabbits apart with extraordinary violence, but leaving them completely exsanguinated. One man found they had left "holes in veins, as in operations", and others noted that there were perfectly round holes in the base of the dead animals' necks or jaws. Sometimes the rectal area showed evidence of having been probed. In some cases three quarters of the blood was removed, resulting in such limp carcasses that it was impossible for *rigor mortis* to set in. The similarity with the phenomenon of cattle mutilations (see page 253) is compelling.

As thousands of witnesses reported sightings of Goatsuckers – although it seems that the term covered a variety of bizarre beasts – some had closer encounters than others. One man, hearing loud noises in his yard, found such an entity fighting his dog: when he shot at it, the thing rolled into a ball "like a worm", hit a wall and disappeared within seconds – "unbelievably fast".

Thousands of the local people lost livestock to these sinister unknown predators during that particular Goatsucker flap, although there have been many others in several Central and South American countries. Sceptics dismiss the phenomenon as merely misidentified local animals, contaminated with the heightened fear of mass hysteria. But it is strange that the local people seem at other times perfectly capable of recognizing, and reacting in appropriate ways to the local fauna, and that the military should take such an interest in a feral cat or unusually large bat, for example.

The possible explanations are, however, complex. Leaving aside the sceptics' old chestnut of "they must have imagined it" – imaginations that can rip farm animals apart and leave no blood behind must be worth investigating for themselves, if nothing else – there are several options, although to take them all equally seriously one has to adopt the often difficult Fortean stance of having no opinion, or at least suspending disbelief, if only temporarily.

If the creatures were directly connected with the UFO sightings around Varginha – and although it may seem a reasonable supposition that they were, there is no evidence for it – the question arises as to whether the UFO crashed or landed somewhere nearby, allowing the entities on board to escape into the countryside. Of course this opens up the familiar debate about the true identity and origins of UFOs: are they from other star systems? Were the "fragile" and "unsophisticated" creatures caught by the Brazilian authorities extraterrestrials? They certainly did not appear to be indigenous species, but being unknown does not necessarily imply they came from deep space.

If they did *not* hail from elsewhere in the universe, what are the other options? It may be instructive to recall the opinions of the doctors and nurses who dealt with the dead alien (using the word in its most basic sense): while some believed it to be the spawn of the Devil, others saw it as the result of some kind of top secret genetic experiment. Let us examine *both* of those ideas, in reverse order.

Did the fragile little alien originate in the laboratory of some sinister Dr Mengele-like figure, clinically manipulating human, animal and – who knows? – possibly even plant, DNA in the ultimate Luciferan experiment to challenge God? Is this cowering "demon", in reality a new species, born of Frankenstein biogenetics in some sterile chamber deep below the Earth or inside the bowels of a remote mountain? The stuff of nightmares, such a scenario is all too possible, for although similar things have been successfully fictionalised in *The X-Files*, the evil doctors of the Third Reich (and, indeed, elsewhere) came close to such a concept with their deeply shocking experiments on prisoners, some of which involved draining their blood and injecting animal or vegetable matter into their bodies. The notorious Dr Josef Mengele himself, creator of this living hell for many thousands of those who found themselves in his "care", is known to have fled from Nazi Germany to South America, although there are conflicting rumours as to what eventually became of him. It seems likely that if Mengele did make his home in South America – possibly

Brazil or Argentina – he would have set up another hellish laboratory deep in the rainforest far away from prying eyes, and who knows what creatures he spawned. But just how unknown flying craft fit into this scenario must remain a matter for speculation. Certainly, there is a widely-held view that UFOs – at least the classic disc shapes – are secret weapons developed from old Nazi designs, if not developed with neo-Nazi connivance,[171] being flown from hidden bases scattered across the globe, from Central America to Czechoslovakia. Perhaps an underground race, essentially the Fourth Reich, has developed such highly advanced technology that they can buzz our air space with impunity, although given their track record they seem uncharacteristically tentative in making their presence felt. Why are they not holding the nations of the Earth to ransom – an easy enough task, one might imagine, given their technological superiority? They could blast us from the air at any time and we would have little means of defending ourselves. Such apparent reticence is odd. Perhaps the Nazis have learnt the value of patience and a measure of subtlety over the last sixty years, although somehow one doubts it.

Leaving a putative Fourth Reich out of the equation, there is still the possibility that some other modern Dr Moreaus[172] lurk in the Brazilian forests, perpetrating their horrifying science on helpless individuals. It is interesting that several of the doctors and nurses who dealt with the Varginha alien immediately thought it reminded them of some such genetic experiment, or that such a thing seemed feasible to them. But once again, how do the escapee mutants fit into the UFO scenario?

If the footage of the flap of the preceding night is anything to go by, the UFO was not so much in trouble as gleefully advertising its presence by executing deft manoeuvres over the house tops for several hours as if defying capture, and certainly explanation. It seemed like a UFO on a mission, either to make the local inhabitants and the authorities look stupid – something it shares with many, if not most, other UFOs – or to whip up hysteria and fear in the people, as if softening them up, perhaps, for the imminent Goatsucker phenomenon. And if that was its purpose, not only are "they" (whoever or whatever "they" may be) playing with us, as Charles Fort said, but they may also be experimenting, attempting to judge the effect of weird happenings on a rational world and how far they can push us. If it is such an experiment, then the field is wide open as far as potential culprits are concerned. . . . We are back to hidden governments, secret experiments – biogenetic "black ops" of known governments – or, of course, extraterrestrials. In any case, it is not pleasant being treated as a laboratory rat.

Whatever the origins of the Goatsuckers, they appear to have more in common with the legendary predators of the mind, such as vampires and werewolves, than stalking cats or wild dogs. Although their strange red eyes may be a normal characteristic (domestic cats' eyes glow weirdly in the dark, after all), it is hard to think of any known zoological specimen that literally disappears into thin air, or has the ability to drill perfectly round holes in bone, and tear a sturdy farm animal apart while leaving no trace of blood. There are, it is true, many weird inhabitants of the earth and seas, with behaviour and abilities that even in our sophisticated age often seem almost paranormal, but the Goatsuckers, and/or the Varginha alien (now solidly associated with the phenomenon) have distinctly demonic overtones, at least in the minds of the local people.

In itself that may be one of the few genuine signs of hysteria in the same way that a seal swimming across Loch Ness in the tourist season abruptly finds itself the centre of attention. It is easy to see how the appearance of these unknown, monstrous creatures that happened to sport horns on their heads, on the outskirts of a mundane, industrial town – especially after the uncertainty and fear of the UFO sightings – might be transmuted through the instant alchemy of shared terror into a mythical nightmare.

Stranger things have happened: in the Middle Ages whole villages caught a dancing hysteria and danced themselves to exhaustion and death, especially, it seemed, if they wore red shoes; the sixteenth century saw a tulip craze in the Netherlands in which the bulbs changed hands for vast sums of money, ruining many a merchant when the market finally failed; thanks to the wild talents of metalbender extraordinary Uri Geller, spoonbending became a craze among schoolchildren everywhere in the mid-1970s, even though it is officially "impossible", as is levitating heavy people above one's head, yet school playgrounds saw this happening routinely, during the levitation crazes that do the rounds from time to time. It seems that the power of the collective *imagination* should not be underestimated, for it produces miracles besides nightmares – as we will see in a later chapter.

Yet perhaps the descriptions of demonic Goatsuckers should be taken at face value, for to dismiss them totally – or hastily seek to explain them away – is to insult the intelligence of the local people. Like Hamlet in his saner moments, it may be safely assumed that they know a hawk from a handsaw, and if anyone knows what the indigenous fauna looks like, they will certainly include farmers on the outskirts of the towns. So, suspending disbelief once more, let us consider the demonic hypothesis seriously for a moment.

Assuming by "demon" we mean some kind of lower life-form associated with evil, akin to the more amoral and mischievous fairies, we are immediately in very uncomfortable territory. Consideration of personifications of evil – real, tangible creatures – is at odds with modern ideas, not only of what is "real" and possible, but also of right and wrong. Western cultures tend to shy away from notions of absolute good and total evil, although the twentieth century produced enough evidence of the latter, if not the former. We err in the direction of *fairness*, tolerance and decency, leaving black and white notions of absolute values to the Fundamentalists. They, of course, have no problem with the concept of demons. They exist. But this strand of belief is also prevalent in what might be termed Fundamentalist Catholicism, where the traditional beliefs and practices are still honoured, with none of this mealy-mouthed liberalism. And wherever the Church is particularly strong, there will be an opposite and equal force, an anti-Church. In the case of South America, this force is a variation of the Haitian cult of Voodoo, a form of intense and powerful ritual magic, often of the darker sort, which incorporates Catholic saints into its pantheon. Were Voodoo-type practices somehow behind the manifestation of the "demonic" Goatsuckers – and the UFO phenomenon that preceded them?

SEPTEMBER 1996, FIFE, FORFARSHIRE, SCOTLAND

Towards the end of September, 1996, Mary, her ten-year-old son and a friend, Ann, drove from their home in Fife to go shopping in a nearby village. Not far into their journey, they were startled by the sight of a huge, noiseless triangular craft hovering menacingly over their car. The UFO emitted three searingly intense beams of white light from its underside that illuminated the countryside, making it look as unreal as a film set.

However, the object made no attempt to stop them or interfere with them in any way, so they drove off as fast as they could, very disturbed by the experience. Once home, Mary told her fifteen-year-old daughter about the giant triangular UFO – to be greeted, perhaps unsurprisingly, by a certain amount of scepticism. In order to prove her point, Mary drove her daughter (and the others) in the direction in which they had first seen the craft. As Brian Allan of the Scottish organization Strange Phenomena Investigations says:

> As they approached the spot, they saw fingers of brilliant white light coming from a stand of trees, spearing into the night sky like lasers. They stopped the car by the roadside and climbed out, gazing with a mixture of fear and fascination at the scene before them. As well as the "*laser*" lights, there appeared to be dozens of brilliant starlike objects hanging low in the sky. As they watched, a haze or mist developed in front of the trees, within this haze were "*hundreds*" of small grey creatures apparently lifting and carrying boxes and cylinders.[173]

Matters became even stranger. Brian goes on: "Abruptly, part of this haze containing some of these beings detached itself from the main body and drifted towards them across a ploughed field. By this time, the group had seen quite enough and quickly climbed into their car – not quickly enough, however. Just as they were on the point of moving off, the daughter screamed out that there was a creature standing next to the car '*grinning*' at them. As the vehicle moved off, an intense burst of blue light emanated from the field behind them, throwing everything into sharp relief.

"Incredibly, when they arrived at the sanctuary of their home, although alarmed by what they had seen, displaying either great courage or astonishing foolhardiness, they decided to go back for another look. This they did and again claim to have seen more small creatures moving around in the woods . . ."[174]

After investigating the case, Brian Allan noted that some of the witnesses claimed to have returned from the sighting "bearing marks and abrasions on their bodies", while one of them said she had seen a UFO hovering close to her home. Perhaps most significant of all is the claim of the young son that he has seen a "small grey creature floating outside his bedroom window and also in their bathroom" – and the being actually "followed him to school and assisted him with his schoolwork [entirely unseen, it has to be said, by any classmates or teachers]". Had the lad been unduly influenced by the movie *E.T.,* one wonders? Or perhaps he really did have an invisible friend who just happened to be a grey alien . . .

Brian Allan checked with the MoD, the Meteorological Office, airports and the police and discovered no corroborating evidence or data that would help explain this rather sensational case. As he admits, "All the evidence we have is either entirely anecdotal, or totally apocryphal. Whatever [it is], it is not the best foundation for a solid case."

Perhaps it is also relevant that Mary was a UFO enthusiast before her sighting, subscribing to periodicals such as *Flying Saucer Review* and *Alien Encounters*, although of course in itself this means little. After all, writers on ghosts can – and do – actually see them, so why shouldn't UFO buffs see UFOs? As Brian says: ". . . because the witness is familiar with UFO lore, then they would make an excellent subject, [but] because of this very familiarity they could concoct an excellent fabrication. I do not say that this is the case, but it does fuel doubt."[175]

This case embodies many of the problems involved in assessing a claim of the more dramatic type of UFO sighting, and the necessity to treat them with objectivity – although where the line is drawn between scepticism and trust is difficult to define. Brian said: "There is a feeling about the case that maybe, just maybe, it exemplifies a hoax that got out of hand and the investigators had in effect 'painted themselves into a corner'. [The witnesses] were in a position where, because of the attention they were receiving they could not deny it. If, on the other hand, it is true, then the witnesses were extremely fortunate to have experienced an encounter of this kind. For myself, I have to state categorically that I have deep reservations about the case."

It might also be added that if true, the witnesses were extremely fortunate to escape without suffering the trauma of physical injury or abduction.

8 MARCH 1997, LYMPNE, KENT, ENGLAND

At 3 a.m. on 8 March 1997, Sarah Hall, a reporter on the *Folkestone Herald*, was driving through Lympne[176] near Hythe in Kent when she heard a peculiar humming noise and stopped. There was a huge triangular object hovering low over a nearby field, which "had a large dome at one end and lots of bright lights around the sides and looked quite shiny."[177]

Sarah continued: "I had no idea what it was. I felt the hairs stand up on the back of my neck. After a few seconds it shot off and stopped 500 metres (545 yd) away. It then moved and stopped before flying off."

The interesting thing about this sighting was that two firemen reported that the UFO – described to local UFOlogists as being a triangular craft – had been seen hovering over the nearby home of the then Home Secretary, Michael Howard. But when Sarah Hall's piece about her sighting appeared in the *Herald*, the location had been mysteriously changed to the other side of the marshes, at Romney. And when she attempted to amend the mistake in a subsequent article, the story was dropped altogether. It seems that Mr Howard's staff had been busy. As the *Fortean Times* reported: "Sheila Gunn, [Prime Minister] John Major's Press Secretary, said she was aware of the UFO reports, but the Tories managed to keep them out of the national press".[178]

Perhaps this was not so much a sinister conspiracy as an awareness of what would almost certainly have happened if such a juicy story had fallen into the laps of the media.

8 AUGUST 1997, COVENTRY, ENGLAND

At 10.30 p.m. Marilyn Harris of Coventry, England, saw a brightly sparkling object in the sky that she thought must be a comet, but when she trained her binoculars on it she realised that: "It was in fact not one object but two of entirely different shapes. The one on the left was a cigar type shape which shone really brightly and was of three colours – pastel blue, red and cream – whilst the one on the right was smaller and looked like an upside down letter 'U' and was just plain cream in colour and did not shine at all."[179]

Marilyn goes on: "[The objects] seemed to be flying in some sort of formation. However, even though they appeared to get higher in the sky they were still there a couple of hours later. It got very late so I retired to bed wondering what they were, thinking that I had just witnessed my first UFOs . . ."

LATE 1997, INGLETON, ENGLAND

One cold but sunny evening in late 1997, at about 6 p.m., Jen Eve of Ingleton in the north-west of England, was out walking her dog with a friend when she suddenly spotted an oval black object, perfectly flat, heading from the direction of Ingleton towards the Lake District. She said:[180]

> I could tell it was metallic because the sun was reflecting off it. There were three planes in the area at the time, and it was about the same height as them. It was moving quite quickly and it flew steadily. I couldn't hear any noise but my friend said she could hear a humming sound. I think my dog could also hear it because he went still and looked up and cocked his ears as if he were listening.

Jen Eve thought the UFO could not be a plane because there was no vapour trail and "it didn't look like a plane". She added that it was moving too quickly to be a balloon, and in any case it was metallic.

However, despite Jen's belief that the UFO was not a plane, it could have been some kind of experimental or secret craft being test flown over the sparsely populated Lake District and neighbouring area – and such a craft may well not fit the traditional image of what a plane should look like.

24 JUNE 1998, SILVER SPRINGS, NEVADA

At 11.57 p.m. on Wednesday, 24 June 1998, three members of the UFO organization Desert Skies Research Team (DSRT) were driving northwards just south of Silver Springs, Nevada, when a massive light appeared abruptly and headed straight for them. As Harvey Caplan of DSRT wrote to the American *UFO Magazine*:[181]

> It was floating over some houses scattered on the desert floor, 200 feet [60 m] above ground level, several hundred yards from the observation position. The light was intensely bright, at least as big as a couple of houses. It had eight perfectly defined points or "spires" protruding from its center. The thing closely resembled an ornament that one would place on top of a Christmas tree.

As the team members watched, the light remained a solid shape "and floated horizontally, eastbound, behind a small mountain, where it was lost from sight. It made no sound. The DSRT vehicle was on an elevated section of highway with no way to exit to the desert floor to give chase. The entire sighting took four seconds."

The team were intrigued to note that "the same thing was seen on the following Tuesday evening by a Silver Springs resident, floating eastward from Highway 50 to the open desert."

Did the anomalous light – with its solid shape – have any connection with the nearby Area 51?

JUNE 1998, MOJAVE DESERT, CALIFORNIA

In June 1998 David Hastings of Salhouse, Norfolk, England, was co-pilot with David Patterson in their twin-engined Cessna Skymaster, on their way from New York to San Francisco when, 10,500 feet (3,330 m) above the Mojave Desert, California, they spotted a black object executing a very dangerous manoeuvre.

"We realized there was something heading straight for us," David Hastings told reporters later. 'All of a sudden the cockpit went black as the object appeared to fly overhead, blocking out the Sun. We called the radar control tower to check if they had another aircraft and the answer was 'no' ".[182]

As another object moved close to them, the witness took two photographs, one of which seems to show a disc-shaped UFO seen from the side. Hastings showed a print of the photograph to US

Navy officers who hung on to it for some time before responding with the traditional "No comment".

1998, THE RUSSIAN *MIR* SPACE STATION IN EARTH ORBIT.

In December 1998, a Moscow newspaper[183] featured claims from cosmonaut Alexandr Baladin that UFOs had flown close to the *MIR* Space Station and the Baikonur Cosmodrome, insisting: "General Vladimir Ivanov, former commander of Russia's Military Space Forces, recalls that three objects flew at a considerable altitude over the Baikonur Cosmodrome and were picked up on radar. There is no way they could have been airoplanes [sic]."

While a speaker at Brazil's First International UFOlogy Forum, Baladin also revealed that he and fellow cosmonaut Musa Manarov had had a very unsettling experience during their second space mission. During the delicate and tense docking manoeuvres between *MIR* and his space capsule, Baladin had noticed a revolving bright light a little distance away, which Manarov managed to capture on video film.

The cosmonaut also told the Brazilian UFO forum that in his opinion the Russian military could add greatly to UFOlogical knowledge, citing the case of a semi-circular UFO flying low over the Kaputsin Yar missile base in June 1989, where it was witnessed by several military personnel on the ground. Baladin said:

"Many of my old comrades, who are now working at top-secret military facilities, acknowledge having seen unidentified flying objects over manufacturing centres, gunnery ranges and military facilities", but took care to warn that "not all that can be seen should be taken for a UFO, since it is very possible that we may be facing natural phenomena which have not been properly studied".

FEBRUARY 1999, FLIGHT BETWEEN SWEDEN AND HUMBERSIDE, ENGLAND

On 3 February 1999 British Aerospace 146 Charter flight from Linkoping, Sweden, to Humberside in England, was flying just off the coast of Denmark when the plane was flooded with a strange red light coming from a "long cylindrical object . . . as big as a battleship"[184], which just missed the plane, causing a few heart-stopping moments for the businessmen on board.

However, this dramatic brush with a battleship-sized UFO appears to have a disappointingly prosaic explanation. UFOlogist David Clarke managed to secure a copy of the airline's report to the Civil Aviation Authority, which describes how, when the plane had reached 28,000 feet (7,620 m) over the North Sea, the area underneath was "illuminated for 10 seconds by incandescent light which was not considered by reporter to be an a/c [aircraft] landing light." The pilot claimed that three other planes had also seen the light, both stationary and moving fast, but Air Traffic control rejected this because there were no other planes in the area at the time, while Humberside Airport believed the light came from a "light reflection from the underside of the jet".

Meanwhile, Berthil Lindblad, an expert on meteors, stated that the short time involved and the red afterglow, in his opinion, came from a trail of ionised gas – a bolide. Swedish UFOlogist Clas Svahn consulted Swedish radar operators who had seen nothing unusual on their screens on that occasion.

David Clarke also pointed out that the phrase "big as a battleship" featured in the *Daily Mail* and the *Daily Express* on 27 April 1998 in their coverage of the report of a massive UFO being chased by Dutch F-16 fighters across the Atlantic. That story was discovered to have come from the former editor of *RAF News* – a man, apparently, with a grudge. He was busy trying to get his own back by fabricating and selling sensational tales to the tabloids.

12 MAY 1999, NEVADA TEST SITE, FRENCHMAN'S FLAT AND INDIAN SPRINGS, NEVADA

Members of the Desert Skies Research Team (DSRT) positioned themselves on the side of a mountain at Mercury, Nevada, which gave them an uninterrupted view of the skies above the Nevada Test Site, Frenchman's Flat. Harvey Caplan said in his report for *UFO Magazine* (US):[185]

> At 9:50 p.m. a large, intense yellow light appeared over Frenchman's Flat. A protrusion that could best be described as a "stem"

> was observed about 15 degrees from the top-center of this object. The light remained motionless, hanging in the sky for about five minutes, glowing. Violent winds so severe that the members' 4X would not stop shaking, made it impossible to produce useful photographs.

The UFO "blinked out", coinciding with the launching of flares "into the sky just south of the object's last position." Harvey Caplan adds wryly: "It seems apparent the flares were used for the purpose of making any observers believe that all the objects in the sky that night were flares."

AUGUST 1999, SWISSAIR FLIGHT OVER LONG ISLAND, NEW YORK

This intriguing snippet appeared in *Fortean Times*,[186] adding to the file on UFOs in near misses with civilian aircraft:

> [A] . . . potentially controversial near-miss has been reported by a Swissair pilot. He claimed that his 747 was buzzed by an unidentified object in the same area off Long Island, New York, that TWA Flight 800 came down on 17 July 1996. At 5.07 p.m. on 9 August 1997, Swissair Flight 127 was flying at 23,000 feet (7,000 m) when the pilot informed passengers that a round white object had shot past, at an 'incredibly fast' speed, 200-400 feet (60-120 m) above them. 'It was too fast to be an aeroplane,' reported the pilot. A report by the US National Transportation Board states that another aircraft in the area saw nothing unusual.

Another case of mistaken identity, or another hoax? Or, even, perhaps, another genuine UFO?

5 SEPTEMBER 1999, MOUNT CHARLESTON, NEVADA

At 7.55 p.m. on 5 September 1999 members of the Desert Skies Research Team (DSRT), who had positioned themselves so they could observe the west side of Mount Charleston, saw a dim light to the south of Wheeler Pass that could not have been a car because it moved too steadily and was not bright enough to be headlights. Neither did it seem to be any form of known aircraft.

For about two minutes it moved northwards, but then stopped over Wheeler Pass road, emitting an amber light downwards from its underside. Harvey Caplan of DSRT, wrote to *UFO Magazine* (US):[187] "Taillights from a vehicle were also observed on the road, which was partially blocked by some Joshua trees. The object continued to hover over this area for three minutes, its amber light centered on this vehicle. The light then moved off slowly behind a nearby hill, which was soon followed by a flash of bright white light that reflected off the nearby terrain".

Next day, DSRT members found recent tyre marks in that spot where the car had driven off the road. The researchers checked for radiation but found nothing, neither did they discover any footprints.

LATE 1999, GULLANE, EAST LOTHIAN, SCOTLAND

The area around the curious pyramid-shaped volcanic hill, Berwick Law, outside the town of North Berwick in East Lothian, Scotland, has long had a reputation for weird happenings, whether associated with the doings of fairy folk or, more recently, aliens in UFOs. (It is also quite close to the mysterious Rosslyn Chapel, a strange fifteenth century building, its interior covered in elaborate and symbolic carvings, which regularly attracts hordes of those interested in mysticism and magic.)[188] Therefore it came as no surprise to the locals when an American lawyer (who, for fear of ridicule, wished to remain unidentified), a guest at the Templar Lodge Hotel in the nearby village of Gullane, claimed to have found a strange fragment of cloth on Berwick Law after seeing unexplained lights on the hill in September 1999.

The hotel's head of marketing, Dr Stephen Prior (who also happens to be a parapsychologist), said:[189] "He went for a walk with his wife after a meal at the hotel. He saw lights up on Berwick Law and went to investigate. Having looked around, he discovered what appeared to be a piece of cloth at the spot where he had seen the unknown lights."

When the world's Press got to hear of the discovery, Dr Prior and his colleague Richard Taylor positioned two webcam cameras on top of the hotel, permanently trained in the direction of Berwick Law and the nearby Traprain Law, in order to monitor the area for further UFO activity. This project was specifically organized on behalf of a Japanese film crew, who, according to Richard Taylor, "love Scotland and are really into aliens – this gives them the best of both worlds."

Dr Prior, whose own views err towards scepticism – he believes the various fairy or UFO sightings on the Laws, both of which are solid granite, may be connected with natural emissions of coloured gas from the rock (see Explanations – page 451) – passed on the mysterious fragment of cloth to Joyce Cook, a textile expert from Heriot-Watt University, Scottish Borders campus, who subjected it to a battery of tests. After examining its fibres under a microscope, and testing their reaction to the process of bleaching and burning, she announced that:[190] "It's not man-made in the nylon or polyester sense. My feeling is that it is a natural fibre, and that it is leather-based. If something is nylon or polyester it melts and goes hard if you burn it.

"With natural fibres they just produce an ash. This just burned down to a pure ash. It is probably more accurate to say it has been leather.

"It's in a very high state of degradation. If you put pressure on it, it gives way very easily."

The UFO community, especially in the local area, was torn. As reporter Jack Mathieson wrote in an article on the controversy in the *Edinburgh Evening News* (20 January 2000), "In one camp are the UFO watchers who say that this small fragment of ragged material is evidence of a visit from an alien life form. In the other are the university experts who say that the truth is much simpler – the thin mustard-coloured cloth is a piece of shammy leather of the type used by window cleaners."

Although it does seem, disappointingly, that the cloth is indeed "shammy" (chamois) leather, perhaps the real significance of the controversy lies in the reactions of the UFOlogists. Does a bit of window cleaning cloth really matter all that much? Just as in the Roswell controversy, feelings run high, but surely the whole phenomenon of UFOs – indeed of paranormality in general – does not, and should not, depend on the authenticity of an object. The skies are much wider and considerably more mysterious than that.

LATE 1999-EARLY 2000, MASS SIGHTINGS OVER CHINA

Reports of sightings of UFOs over China in late 1999 and 2000, which were witnessed by thousands of people, are filtering through to western UFOlogists. Coinciding with a general decline in UFO reports in the USA and other western countries, this wave prompted aviation engineer Zhou Xiangiang, secretary-general of the Chinese UFO Research Organization, to declare: "It is the most significant number of sightings reported in China since UFOs became [a] permitted area of study in 1978."[191]

Before that watershed date, the repressive Communist regime had banned any form of "superstition", which included religion and UFOs, but now Chinese scientists are keen to investigate the phenomenon – in case they could find something they could use to give them the lead in world technology, perhaps even in the weapons race.

However, one of the most sensational of all known Chinese sightings happened back in 1981 – just after the relaxing of the rules – when the Astronomical Observatory of Zijingshan issued a Press release saying, "the population of fourteen provinces in our country sighted this celestial phenomenon".[192]

JANUARY 2000, ROTTERDAM, NETHERLANDS

At around 10.30 one night in January 2000 party-leavers noted a strange yellow object "hovering"[193] over a block of flats. At first they thought it must be the full Moon, but a "ring" that appeared to be circling the object made the revellers wonder if it could be the planet Saturn (although of course its rings are not normally visible to the naked eye). In any case, the object abruptly "dematerialized" as they watched. Perhaps this was indeed an astronomical body, made visible temporarily through some unusual atmospheric condition, although there must always be a nagging doubt about the interpretation of things seen in the sky by dedicated party-goers.

27 JANUARY 2000, HOLLYWOOD, CALIFORNIA

For about an hour on the morning of 27 January 2000, twenty or so workers at the Post Group Studios in Hollywood witnessed two UFOs high in the sky. Oval in shape, they were accompanied by eight smaller objects, which formed a ring around the two larger ones when a helicopter appeared to approach them. One witness said it looked as if the eight smaller ones acted "as if to defend it". The helicopter, after the pilot got a good look at the UFOs, departed, whereupon the smaller objects formed a "V" shape behind one of the larger UFOs, and flew off in a westerly direction. The large one they had left behind shot vertically into the sky before disappearing out of sight. But that was not the end of the aerial display.

Almost immediately another UFO appeared on the scene, being described as "oval in shape, with a yellowish chrome band through the middle. The top position was red, the lower portion green."[194]

Reporting the incident in its round-up of an unusually widespread UFO flap in the early months of 2000, *Fortean Times* noted wryly: "Coincidentally or not, at the time of the sighting, the studios were playing host to Steven Spielberg's new big budget, 14-part alien abductions mini-series. Perhaps the objects were carrying technical consultants for the production?"

6 JANUARY 2000, GROOM RANGE, NEVADA

Another case from Harvey Caplan's Desert Skies Research Team (DSRT) files[195] describes how team members ascertained that "flying/testing of UFOs has definitely not ceased at Area 51".

On 6 January 2000, together with two Japanese tourists, they took up their position by Groom Lake Road and, at 8.25 p.m., observed a white UFO "that literally appeared out of nowhere" hover over a nearby ridge in the Groom Range. As they watched, it grew a brighter white before changing to a brilliant red colour, then sped off at a phenomenal speed. As Harvey Caplan reports: "There was no gradual acceleration between full stop and gone."

Then, at 4.20 in the morning, they saw what appeared to be the same UFO, wobbling along the top of the ridge, once again changing from white to red before disappearing behind the mountains.

The UFOs may still be flying but, as Caplan reports, "What has ceased is the use, by the roving guard patrols, of the infamous white Jeep Cherokee 4Xs. During the time DSRT spent at Tikaboo Valley, members were checked once and chased once by new-model, champagne-colored, shortbed, extended-cab Ford pickup trucks."

He adds with tongue in cheek, "DSRT will miss the Jeeps. The Fords just don't seem as intimidating."

12 JANUARY 2000, NORTH-EASTERN MISSOURI

At 8.45 on the evening of 12 January 2000, a family had the astounding experience of witnessing no fewer than thirty UFOs within two hours. Having spotted a lone light, they chased it in their car, but got more than they bargained for when the UFO suddenly came close enough to hover over them, following them home as they proceeded to speed in panic. Once there, however, they saw at least 30 UFOs travelling randomly around the sky, while seven aircraft appeared on the scene, making the only noise.

MARCH 2000, MEMPHIS, TENNESSEE

At around 9 p.m. one evening in March 2000 a flight attendant on board a commercial plane flying from Memphis to St Louis, Tennessee, observed a "fuzzy ball of pinkish light" approaching at great velocity. Having reached the plane, it continued to keep pace with it. Together with a female passenger, the flight attendant noted that the metallic UFO, which seemed to emit shining sparks, suddenly disappeared from sight.

23 MARCH 2000, BELFAST, COUNTY ANTRIM, NORTHERN IRELAND

Returning home from work, a factory worker was apparently followed by an egg-shaped light. Panicking, he sped in the opposite direction to his home until he was overcome with an irresistible urge to stop the car. Just before he lost consciousness he had the feeling that his car was being lifted into the air . . . When he came to it was six hours later. He drove home, baffled and disturbed, but had no recollection of what happened during his missing time. Perhaps his encounter with the UFO had become an abduction by its occupants.

30 MARCH 2000, YUKON, CANADA

At 5 a.m. on the morning of 30 March 2000 drivers on the Klondike Highway in the Yukon, Canada, observed a 40-foot (12 m) wide, "classic disk-shaped",[196] domed UFO hovering in front of a mountain, some 300 feet (90 m) distant. It was reported as having curved windows through which "aqua-coloured" lights were shining. After about five seconds, the UFO zoomed off across the road at great speed, landing for a "split second" again, before finally shooting off into the distance. During the encounter the witnesses' headlights dimmed and the tape deck went dead.

APRIL 2000, CHILE

Four friends were dining with Jose Ignacio Prieto at his home near Tucapel near Concepcion, Chile when they experienced a sudden warmth and the lights were abruptly extinguished. A strong blue light came through the windows, briefly blinding them. The lights came back on as they regained their normal vision.

The next day they saw a triangular burn mark outside, 115 feet (35 m) at the base and 65 feet (20 m) along the sides, encompassing a high-voltage electricity tower. Plants inside the burnt triangle were found to be bent over like those in the middle of crop circles.

Outside San Pedro de Atacama, drug squad police chased unknown red lights in the sky believing them to be signals from a hidden landing strip, but found another burnt triangle, inside which were two rocks that had been so badly burnt they were still hot to the touch.[197]

Meanwhile, the area around Calama, Chile, was rife with stories of mysterious deaths of farm animals due to attacks by what appear to be Goatsuckers. Over 300 animals have been discovered, completely drained of blood, with neat incisions in their necks. After reports of unearthly cries and howls coming from outlying areas, the local police investigated, firing shots at one fleeing individual (although whether human, animal or Goatsucker is not reported).[198] Adding to the hysterical atmosphere, weird howling was heard coming from a cemetery.

Another local town, San Fernando, also suffered the mysterious loss of at least thirty animals, including pigs, ducks, dogs and hens. And not far away in Talcarehue, over ten animals were reported to have died and unknown footprints were found afterwards, coming from the area.

The Regional Governor, Francisco Segocia, commented: "The footprints were taken to the investigations lab. We cannot say what they are at this time, but at first sight, they come pretty near to dog prints, which tend to expand and acquire strange shapes."

The more rationalist of the explanations put forward for these depredations favours the wild dog theory, arguing that a pack of dogs could have turned "rogue" and begun attacking farm animals because they had acquired a taste for blood after fights within the pack. That may be the case, and the Goatsucker flap does appear to have gathered strength in a hothouse atmosphere of panic and perhaps superstition, although it seems unlikely that hard-headed farmers would be totally unfamiliar with the characteristics of wild dogs. (But given the capacity of the human mind to create the most dramatic of special effects, perhaps the unearthly howling in the graveyards was merely a normal creature of the night, such as a vixen.)

MARCH 2000, GRIMSBY, LINCOLNSHIRE, ENGLAND

At 2 p.m. on 24 March 2000 greengrocer Tony Kirby witnessed an "unearthly" looking, bright white UFO hanging in the sky over a Grimsby street,[199] then later – at 8 p.m. – a young local boy, Reshi Kurnar, saw a similar light while travelling in his father's car. The UFO hovered, then abruptly dropped, leaving a "white line" where it had been. Then it vanished from sight.

Earlier on the same day, forty-year-old Elaine King was abducted from her bed and witnessed a repulsive surgical operation on a man, before being taken safely home.

In another incident that may have been part of the Grimsby flap, sixty-eight-year old Cliff Blyth from the nearby seaside town of Cleethorpes, reported to the local paper[200] that a "red ball of fire" had hit him on the chest through a closed window, which was witnessed by his wife. The strange red light left no marks and Mr Blyth was none the worse for the experience. It has been suggested that he may have been hit by a laser pointer.[201]

Around the same time, Dorothy and John Ramsden of Humberston saw a UFO hovering in the sky, which they described as being a "lovely" silver cigar-shape, emitting a pinkish light. They watched it for about ten minutes, during which time it "hung motionless at an angle" – which is also the perfect description of a stationary airship.

The next day, back in Cleethorpes, Mrs Elaine Reed noticed a bright light moving slowly above the River Humber, "as red and as big as an aerosol can. [Whatever that may mean!] At first I thought it was a helicopter, but it moved so quietly that I knew it couldn't have been." An anonymous witness also reported the sighting.

17 APRIL 2000, LOS ANGELES, CALIFORNIA

At 10.45 on the night of 17 April 2000, an unidentified reddish light was witnessed to move swiftly across the sky in a north-easterly direction over Los Angeles. The UFO emitted a bright white light that changed to luminous red and back repeatedly. Moving at a rapid pace, the object "bobbed" along a straight trajectory, and was observed for only about three seconds before it disappeared.

8 APRIL 2000, NEW DELHI, INDIA

At 11.55 on the night of 8 April 2000, a crowd of people at the Indira Ghandi Airport in New Delhi, India, witnessed a shining orange-yellow cigar-shaped UFO come towards them from the north-east at a high altitude, with a pulsing glow that created a flashing effect along the outside. It slowed to a hover over the air cargo terminal, apparently tried to descend (or land?) but began to wobble, and disappeared.

23 APRIL 2000, NEVADA

At around 6.30 in the evening of 23 April 2000, a passenger on a plane flying from Denver to San Jose saw a "circular silvery object with a rounded top"[202] speed several thousand feet below. The passenger, who happened to be a mathematician, was able to calculate that the UFO was about 55 feet (16.5 m) in diameter and was travelling at roughly 3,000 mph (4,800 km/h).

EARLY MAY 2000, TANDIL, ARGENTINA

In early May 2000 the Argentinian newspaper *El Eco de Tandil* reported that there had been a spate of UFO sightings in the area. One of the many witnesses, sixty-year-old Hugo Macias, an employee of *La Capital* newspaper in Mar del Plata, reported observing UFOs during his newspaper delivery along the National Highway 226. He said: 'I travelled some 6,000 metres [6,540 yd] and suddenly heard a very strange noise approaching me from behind, and I felt myself 'contained' within a powerful bright light. I say 'contained' because the light covered a diameter of 50 metres [54 yd], and penetrated all sides of my vehicle. It seemed to pierce the rooftop."[203] According to the witness, his sighting lasted a few "long" seconds.

He also told reporters, "I returned to Tandil, discussed the event with a group of friends and with a gentleman who belongs to the Army Command and to whom I deliver the paper every day. When I returned to Mar del Plata, I headed toward the Gendarme Station to see if the duty guard had seen anything that night. But he wasn't there. There was another one who told me that on five or six separate occasions [strange] similar things had been seen."

The next day Macias returned to Tandil to discuss the matter with the Army man, who showed him an article in the local newspaper. Macias said: "On the same day, with a half hour difference, a similar case had been reported over Tandil. But this gentleman from the Army Command also told me that the guards on duty the very same night that I had my experience witnessed a light in the sky, with a very strange shape, which had balanced itself in the heavens and possibly descended behind Cerro Leones."

The witness went on: "On my way from Tandil I made the comment to Sheriff Lopez of El Dorado (on Route 226), who's knowledgeable about UFOs, and he advised me on why the Gendarmerie avoids getting into the subject, because it knows how NASA works: they can send an airplane with six scientists anywhere in the world and carry with them an all-terrain vehicle in which they can visit locations in person. Many locals don't like strangers visiting them."

However, Macias himself did not feel negative about his sighting, calling it "lovely, because what began as fear, at first, is now the satisfaction of having been contacted by something that researchers in the field would like to have seen and never have. I'm convinced that it wasn't from this world. The power of that beam of light is not a power produced by any luminous source that we may have . . ."

Despite his positive feelings towards the experience, Macias had suffered physically because of it. He says, "At the time I felt nothing at all, physically, but for the past two days I've felt headache, laryngitis and many other symptoms . . ." Even so, he muses happily: "nothing like this had ever happened to me before, nor did I ever think it could happen to me. I'm just happy for the chance to have lived through a contact experience like this one."

5 MAY 2000, MUNCHENGLADBACH, DUSSELDORF, GERMANY

Several people on the outskirts of Munchengladbach, near Dusseldorf, Germany, reported what they initially took to be an aircraft crashing near the A-52 autobahn.[204] According to their reports, they saw what appeared to be an aircraft moving slowly overhead, flashing bright white and red lights before disappearing from view behind some trees. A policeman who was present described the lights as being far brighter than usual landing lights – so bright that they hurt the eyes.

Believing it to be a stricken plane, 22 Red Cross personnel, 50 police officers and 12 fire fighters hurried to the site – but they found nothing, and discovered that the local airport had recorded no radar blips at that time and place. Completely mystified, the authorities experimented with car headlights to see if they could have been misidentified as aircraft, but proved beyond doubt that this was not the cause of the reports. However, to add a note of conspiratorial intrigue to the story, a local radio ham later reported that he had heard an exchange between policemen in which the phrases "green males" and "odyssey through the Universe" played a part. (Do German policemen really use such poetic terminology?)

8 MAY 2000, JUNIPER, BRITISH COLUMBIA, CANADA

A large triangular craft was observed hovering above the mountains near Juniper, B.C., which split into four small triangular objects with flashing lights. They were reported[205] to give off an "aura" of changing colours, and a "powerful strobe light" that lit up the mountains. Finally all the craft merged into one once more before shooting away out of sight.

17 MAY 2000, SYDNEY, AUSTRALIA

At 6.45 on the morning of 17 May 2000, surfing enthusiast Chris Beacham set off to find the "swell"[206] at South Avalon Beach outside Sydney, Australia. While parking his car on the clifftop he saw a peculiar orange "puff" of cloud with a "very bright light travelling alongside it". Looking round, Beacham noticed several other people scattered around the beach, including three women and a "kid about twelve years old" – who asked him what "it" was as the witness took several photographs of the anomaly.

The group of witnesses grew as a man walking a dog joined them, all looking at the UFO that could still be seen distantly. When Beacham checked, he saw that he had taken his first photograph at 7.19 and the last at 7.25, but he felt that the experience had lasted longer than a mere 6 minutes.

The next day he returned to Avalon Beach to try to encounter some of the others who had been on the beach in order to discuss the sighting with them, but it was raining and nobody was about. But on Friday, when he was making his usual round of the good surfing beaches in the locality, he saw – off Palm Beach, Sydney – three Australian Navy frigates being buzzed by a couple of UFOs, which was also witnessed by about twelve other people on the beach.

Writing of his experience on 29 May 2000, Beacham says: "It's hard to say how far out the object was or how high it was flying, given its erratic course and enormous speed bursts which left nothing but the 'orange cloud' visible, although I would guess its position was about one mile [1.6 km] out to sea at Avalon Beach."

He goes on, "It was completely silent at all times, with the 'fire' trail far more visible than the craft itself. At no time did the craft attempt to fly over land."

Beacham met the man with the dog later, mentioning that he had taken photographs of the object – which seemed to relieve him, perhaps because he had thought he was alone with a strange and awesome experience in his memory.

Notes

1. Gilbert Cornu, *Pour une politique de la porte ouverte*, in *Lumières dans la nuit* (Le Chambon, 1981-2).
2. On 6 April, 1917.
3. See p. 262 of Keel.
4. *Strange Magazine* No. 13, Spring 1994, edited by Mark Chorvinsky.
5. The names have been changed. This case is from the author's own files.
6. He sold and installed fire equipment.
7. Kenneth Arnold and Ray Palmer, *The Coming of the Saucers* (1952) p. 14.
8. In an article entitled *Flight of Fantasy* in *Fortean Times* No. 137.
9. Martin Kottmeyer 'Resolving Arnold – Part 2' in *The REALL News* (July 1997), vol 5, No. 7.
10. Via an Internet-based forum for "Pacific North-West birders".
11. See *Fortean Times*.
12. In *The Coming of the Saucers*.
13. *Ibid.*
14. Mike Havener, 'Soaring with Pelicans'. See *http//www.capistrano.com/LESC/lesca5.htm*
15. See Easton, *Fortean Times*.
16. From the *British Columbian* of New Minster, 12 July 1947.
17. By John Spencer in his *The UFO Encyclopedia*, for example.
18. *Ibid* and elsewhere.
19. A reference to the Shaver mystery.
20. See John Spencer and Hilary Evans' *Phenomenon* (London, 1988), p. 41.
21. From Ruppelt's *Report on Unidentified Flying Objects*, pp. 26-7. He changed all the names in the report. Also quoted in Curtis Peebles' *Watch the Skies!* pp. 14–17.
22. Charles Fort (see Explanations Chapter) was the great American collector of anomalous data, which he called "the damned".

23. Keel, p. 250.
24. See Curtis Peebles' *Watch the Skies!* pp. 21–5.
25. *Ibid.*
26. *Ibid.*
27. *Ibid*, pp. 25–28.
28. See Project Blue Book files, Incident 144, National Archives Case 179.
29. Air Intelligence Information Report 102-122-79, Project Blue Book Files, National Archives, Case 179.
30. To Edward J. Ruppelt of Blue Book.
31. See Ruppelt's *Report on Unidentified Flying Objects*, pp. 30, 40–1.
32. Peebles, p. 74.
33. Ruppelt, pp. 43–6. He also outlines reports of other "dogfights" between aircraft and what turned out to be balloons. Quoted on p. 76 of Peebles.
34. See p. 77 of *Ibid* and accompanying note (36).
35. By Peebles, among others.
36. Peebles, p. 79.
37. From Charles Bowen's files.
38. *Ibid.*
39. Spencer, p. 279.
40. He was assassinated in 1979.
41. On at least one occasion the guests included Brian Inglis, the former historian who was converted to the paranormal by witnessing extraordinary phenomena surrounding Uri Geller in 1973, and Matthew Manning, the most scientifically tested healer in the world.
42. A pseudonym. "Lisa" is well known to the author.
43. *Coronation Street* is the world's longest running soap opera (beginning in December 1960 and still the top-rated television programme in Britain). It is set in the backstreets ('two [rooms] up, two [rooms] down') of "Weatherfield", a northern town based on Salford, near Manchester.
44. Not least because the Nativity story is taken from other traditions, including the legends of the births of other "dying-and-rising" gods such as the Egyptian god Osiris (who was also born on 25 December, died on a Friday and rose again on a Monday), who predated the birth of Jesus by many centuries. Other elements of the story – such as the Magi returning by another way – came from the myths built up around the Emperor Nero. The Star of Bethlehem originated with these other, older traditions.
45. See Keel, *UFOs: Operation Trojan Horse* (New York, 1971).
46. *Ibid*, p. 200.
47. From the files of Charles Bowen for *The Unexplained.*
48. From Charles Bowen's file for *The Unexplained.*
49. *UFO's From Beyond the Iron Curtain* by Ion Hobana and Julien Weverbergh (London, 1974), p. 67.
50. *Ibid.*
51. *Latajace Talerze (Flying Saucers)* by Janusz Thor, (1961), quoted in *Ibid* p. 69.
52. 'Silent Saucers over Mosbro', *Derbyshire Times* (Chesterfield, Derbyshire),

June 22, 1962, p. 22, quoted in *The UFOs That Never Were* (London, 2000) by David Clarke in his chapter on the Birch hoax.

53. 'Do you think these things are flying saucers?' *Sheffield Telegraph*, August 20, 1962, and quoted in *Ibid.*
54. In a similar way, the visionary Bernadette Soubirous did not claim she had seen the Virgin Mary. It was the local priest, a fervent Marian, who told her to ask the "thing" if she was "the Immaculate Conception" – and, as is the way with 'things' in trees, it responded that it was indeed Jesus's mother. In all these cases it is the *adults* who colour the event according to their own preconceptions and obsessions.
55. From the *Sheffield Telegraph.*
56. 'Flying Saucer with an orange glow filmed by Sheffield man', *Sheffield Star,* 20 August 1962, quoted in *Ibid.*
57. *Ibid,* p. 130.
58. From Alan Watts' report in the BUFORA case file, 10 March 1962, quoted in *Ibid.*
59. From David Clarke's interview with Alex Birch, Nottinghamshire, 22 October 1998.
60. *Ibid.*
61. Internal MoD memo, 24 September 1962.
62. It is now in the National Archives in Washington. See Clarke, p. 132.
63. According to Clarke's interview with Alex Birch.
64. 'Sheffield man confesses . . . I hoaxed the world with UFO picture', Malcolm Tattersall and Heather Smith, *Sheffield Telegraph,* 6 October 1972. Quoted in Clarke.
65. From Clarke's interview with Stuart Dixon, 6 April 1999.
66. *Ibid.*
67. For example, the storm that greeted the results of the carbon-14 dating of the Shroud of Turin, which revealed it to be a medieval or early Renaissance fake rather than the miraculously imprinted burial cloth of Jesus as was widely believed, especially among Roman Catholics. It is interesting that the very people who lobbied the Church long and hard to give permission for the testing to take place – because they were so sure it was genuine – immediately began to criticize, usually violently, the whole process of carbon dating once the results did not go their own way.
68. From a letter from the witness to Nina Pendred, Editor of *Alien Encounters,* 6 November 1998.
69. *www.ufo-images.ndirect-co.uk*
70. See p. 1140 of *The Unexplained.*
71. See page 323.
72. p. 184 of Peebles.
73. *Ibid.*
74. Peebles' source was Philip J. Klass' *UFOs – Identified,* (New York, 1968), Chapter 19.
75. "Myra H." is a close friend of mine. She told me this story at the time.
76. From the Condon Report, quoted by Charles Bowen in his article 'The Rex Heflin Photographs' for *The Unexplained.*

77. *Ibid.*
78. See p. 306 of *The Unexplained.*
79. The Condon Report, quote in *Ibid.*
80. *Ibid.*
81. Presumably this is the size of an American football, not a soccer ball.
82. P. 180 of his *UFOs: Operation Trojan Horse* (New York, 1971).
83. *Ibid*, p. 179.
84. According to the London *Daily Express* of 12 April 1966.
85. One Gary Turner.
86. In a commentary in *Flying Saucer Review*, quoted on p. 1438 of *The Unexplained.*
87. *Ibid.*
88. From Charles Bowen's files.
89. From the *Western Australian*, 1 November 1967.
90. See pp. 66–68 of. Keel.
91. *Ibid.*
92. It is reproduced here in its entirety.
93. From Charles Bowen's files.
94. According to Charles Bowen (see *The Unexplained*, p. 560). It seems no one knows whether they actually followed this up or not.
95. *Ibid.*
96. *Ibid.*
97. *Ibid*, p. 64.
98. *Ibid.*
99. *Ibid*, p. 305.
100. Source: *Filer's Files* No. 6, 11 February 1999.
101. See pp. 245–6 of *Watch the Skies!* by Curtis Peebles.
102. Quoted on p. 246 of *ibid.*
103. It was identified as such by Robert Sheaffer in May 1977.
104. From Charles Bowen's files for *The Unexplained.*
105. To investigator Dr Walter Buhler.
106. 28 June 1970.
107. This is a guess on my part. Her letter to *The Unexplained* in 1980/81, on which this account is based, simply says that the experience took place "some years ago" – a subjective turn of phrase, which I am interpreting as meaning a decade or so. Of course it may have been much more recent than 1970.
108. From Charles Bowen's files.
109. *Ibid.*
110. See 'Phantom Helicopters Over Britain' by David Clarke and Nigel Watson, FUFOR, 1985.
111. See *UFO Retrievals* by Jenny Randles, (J. Blandford, 1995) p. 120.
112. See *Covert Agenda* by Nicholas Redfern (Simon and Schuster, 1997), pp.111–124.
113. Quoted in *The UFOs That Never Were* by David Clarke, Jenny Randles and Andy Roberts, (London House, 2000), pp. 150–151.
114. *Ibid.*

115. *Ibid*, pp. 160–161.
116. In a letter to *The Unexplained* in 1980.
117. Usually Venus or Sirius.
118. The case was initially investigated by the Tasmanian UFO Investigation Centre, and reported to *Flying Saucer Review* by W.K. Roberts.
119. I met "Barney Baines" after giving a talk in Wakefield some years after his experience. All attempts to contact him recently have failed.
120. In the form of a letter to *The Unexplained*, some time in 1980 or 1981.
121. Since the publication in 1982 of *The Holy Blood and the Holy Grail* by Michael Baigent, Richard Leigh and Henry Lincoln, the mystery of Rennes-le-Chateau has become something of a cult interest. It centres on the sudden and inexplicable wealth of the village priest, Berenger Sauniere, who died in 1917, and the bizarre symbolism with which he decorated his church that some have taken to be satanic. In *The Templar Revelation*, myself and co-author Clive Prince revealed the true nature of Sauniere's "heresy", which involved the Freemasons and the underground cult of Mary Magdalene. However, we discovered that not only were other local priests perhaps even more fascinating, but that the whole area has long been associated with strange practices and dark rumours. Many claim to have seen strange lights in Sauniere's graveyard and elsewhere in the neighbourhood, which some local New Agers have not hesitated to label "UFOs".
122. In this movie, in which the Underworld is reached by walking through mirrors, the poet Cégeste is plagued by nonsensical messages, delivered in a monotone, that come through his car radio. This is part of his terrorisation by a beautiful woman who is Death.
123. P. 1479 of *The Unexplained*, and in conversation with the author, June 1980.
124. In a letter to *The Unexplained*.
125. This story comes from a letter written to *The Unexplained* in 1980 or 1981, and concerns an event that took place "some years ago".
126. Cyril Picknett, MBE (1910–83), the author's late father.
127. From a letter to *The Unexplained*.
128. In a letter to the me, care of *The Unexplained*.
129. In a letter to me, in 1981.
130. To me, care of *The Unexplained*.
131. In a letter to *The Unexplained*.
132. For example, John Spencer, *The UFO Enclycopedia* (London, 1991), pp. 240–241.
133. From a letter to *The Unexplained*.
134. See p. 142 of his *Open Skies, Closed Minds* (London, 1998).
135. Called 'The Halt Package', this is now available from Quest International (see Useful Contacts).
136. Pope, p. 145.
137. See Tim Good's Introduction of Pope, p. xvii.
138. See 'Anguish of the UFO victims' by John Schuessler in issue 108 of *The Unexplained*.

139. See *Fate* magazine, May 1969 issue.
140. See Schuessler.
141. In a letter to me, care of *The Unexplained.*
142. From a letter to *The Unexplained.*
143. From a letter to me.
144. Originally from Stephen Pile's hilarious *The Return of Heroic Failures* (London), under the heading of 'The Least Successful Sunbather', this little story comprised the Afterword of John Spencer's *UFO Encyclopedia.* Spencer said: "It seems to me . . . that, if true, it is a case that nearly reconciles ground-based and air-based UFO sightings, crash retrieval stories and the UFO-related city blackouts that occur from time to time, to say nothing of bizarre entity sightings."
145. Pile.
146. The character from Greek mythology who made wings out of feathers and wax in order to reach Heaven. Unfortunately he flew too near the sun and his wings came unstuck, landing him abruptly in the sea.
147. From the author's own files.
148. See Nick Pope's *Open Skies, Closed Minds* (London, 1998), p. 131.
149. As a comparison, only 209 UFOs were reported in the UK for the whole of that year.
150. Pope, p. 133.
151. *Ibid.*
152. *Ibid,* p.134.
153. *Ibid.*
154. No. 13. Their own source was the *Fort Wayne News-Sentinel*, 27 March 1993.
155. See Pope, p. 134.
156. *Ibid,* p. 139.
157. *Ibid,* p. 140.
158. From the files of Brian Allan of Scotland's Strange Phenomena Investigations, which he kindly shared with me.
159. *Ibid.*
160. *Ibid.*
161. *Ibid.*
162. Source: the Derby *Evening Telegraph.*
163. Source: *http://www.parascope.com/articles/0997/chinaufo.htm*
164. *Ibid:* the article, by J. Antonio Humeeus, International Co-ordinator for MUFON, first appeared in *Fate* magazine, vol. 50 No. 9, September 1997.
165. As reported in *Fortean Times*, No 80.
166. From the *Manchester Evening News*, 27 January 1995 and the *Daily Mirror,* 28 January 1995, quoted in *Ibid.*
167. From Humeeus, as above.
168. From Jason Bennetto's article in *The Independent*, 23 January 1999.
169. Bennetto's article was based on research into UFO witnesses in the police force by Midlands-based researcher Irene Bott, assisted by author Nick Redfern.

170. Shown on 2 June 2000, presented by Bruce Burgess (also Executive Producer), and produced by Jackie Stapleforth.
171. See page 406 for a more detailed discussion of this theory
172. *The Island of Dr Moreau* by H.G. Wells told of an archetypal "mad scientist" who experimented with creating part human, part animal, creatures.
173. From Brian Allan's files, which he kindly allowed me to use.
174. *Ibid.*
175. *Ibid.*
176. Along Donkey Street, to be precise.
177. As reported in *Fortean Times* No. 113.
178. *Ibid.*
179. In a letter to *Uri Geller's Encounters* magazine, December 1997.
180. In a letter to the Editor of the (now defunct) *Uri Geller's Encounters* January 1998.
181. See *UFO Magazine* (US) Vol 15, No. 8, September 2000.
182. *Metro* magazine (London), 12 June 2000.
183. Source: *Inexplicata http://www.inexplicata.com*
184. Source: *Independent UFO Network News*, 18 May 1999, quoted in *Fortean Times* No. 125.
185. *UFO Magazine,* (US) Vol 15, No.8 September 2000.
186. No. 125, original source: *Toronto Sun,* 5 March 1999
187. *UFO Magazine* (US) vol 15, No 8 September 2000.
188. Rosslyn Chapel was built in the 1450s by Sir William St Clair, who had it carved with rich and mysterious symbolism because "books can be burned". It is thought to contain great – and probably heretical – secrets, and is the focus today for both the Freemasons and modern Knights Templar. It is widely thought that mystical artefacts are buried under the building. Although called a "chapel", and is today used as a church of the Church of Scotland, originally it contained no altar, and there is virtually no Christian imagery in the carvings, giving rise to the idea that it was in some way pagan or heretical. For a discussion of its significance, see *The Templar Revelation* (London, 1997), by Lynn Picknett and Clive Prince.
189. Dr Stephen Prior is a colleague and personal friend of the author.
190. See *UFO Magazine* (UK) May/June 2000, p. 38.
191. Source: *http://www.homestead.com/moreufo/ufosweepschina.html*
192. Source: J. Antonio Humeeus' article for *Fate* magazine, quoted on *http://www.parascope.com/articles/0997/chinaufo.htm*
193. *Ibid.*
194. CNI News 16 Feb 2000 and the *Fortean Times,* No. 136
195. As reported in *UFO Magazine* (US), vol 15, No. 8, September 2000.
196. See *UFO Magazine* (US) vol 15 No. 7, July 2000.
197. Filer's Files 17 *www.filersfiles.com*
198. *Cronica,* Chile, 24 April 2000 via *Filer's Files 17 www.filersfiles.com.*
199. Willingham Street.
200. *Grimsby Evening Telegraph,* 4th, 5th, 10th and 14th April.
201. By *Fortean Times,* No. 136, for example.

202. *UFO Magazine* (US) vol 15 No. 8, September 2000.
203. Source: Scott Corrales and the Institute for Hispanic UFOlogy, quoted in *UFO Magazine* (US), col 15, No. 7, July/August 2000.
204. From *Rheinische Post*, 8 and 18 May (via *UFO Roundup*, vol 5, 15 June 2000).
205. See *UFO Magazine* (US) Vol 15 No 8, September 2000.
206. From his e-mailed article for *UFO Magazine* (US), vol 15, No. 8, September 2000.

CHAPTER 3

THE TRUTH ABOUT THE ROSWELL INCIDENT

In December 1995, in a gesture of support for the fledgling peace process, President Bill Clinton made a high-profile visit to Northern Ireland where one of his most public engagements was the lighting up of a Christmas tree before a huge and enthusiastic Belfast crowd. The President's speech fit the festive occasion: no heavy politics, but wishes for a peaceful future interspersed with light-hearted asides. One of the biggest laughs came when Clinton referred to a letter he had recently received from a thirteen-year-old Belfast boy, adding: "Ryan, in case you're out there, here is your answer: No! As far as I know, no extraterrestrial spaceship crashed at Roswell, New Mexico, in 1947." After a pause while the laughter subsided he went on, "If the Air Force really recovered any extraterrestrial bodies, they did not tell me – and I want to know!"

The story illustrates the status that the "Roswell Incident" has achieved in recent years. (And maybe the President really *did* want to know: one of the less attention-grabbing revelations of the Monica Lewinsky affair was that, at the time of that embarrassing episode, Clinton had been reading one of the many books on Roswell.) Thanks to a host of books and television documentaries – but mainly to popular TV series such as *The X-Files*, *Dark Skies* and, most recently, its own teen sci-fi drama *Roswell High*, and to hugely successful movies such as *Independence Day* – Roswell is now a household name. Whether or not they actually believe it, to the public the name of this small New Mexico town is today inextricably linked with tales of crashed alien spaceships, autopsies of alien bodies, and a massive official cover-up that has lasted for half a century. Many believe that Roswell was the place where the "first contact" between extraterrestrials and the ruling powers of planet Earth took place in July 1947.

It wasn't always that way. Strange as it may seem today, for decades the Roswell case was written off by even the most ardent UFO enthusiasts. In 1967, when the United States Government appointed a committee from the University of Colorado to evaluate the UFO enigma, the leading American UFO research groups were asked what they considered the best cases. None of them put Roswell forward. In the same year, in his *Report on the UFO Wave of 1947*, which listed every report that appeared in the media in that first summer of the UFO age, UFOlogist Ted Bloecher omitted Roswell entirely. Margaret Sachs' comprehensive *UFO Encyclopedia*, published in 1980, has no entry for "Roswell", nor is the case so much as mentioned under "UFO Crashes". It was one of the few cases that passionate believers and dyed-in-the-wool sceptics agreed had been explained away, the culprit being a weather balloon and an over-enthusiastic US Army Air Force Press officer.

That was soon to change. In the same year as Sachs' encyclopaedia, the first book chronicling the event, *The Roswell Incident* by Charles Berlitz and William L. Moore, was published. Since then several major books have appeared, as well as a number of ratings-grabbing television documentaries (not to mention the fictional versions). Jerome Clark has described it as "the most important case in UFO history". Stanton Friedman – one of the leading advocates of Roswell – dubs it the "Cosmic Watergate": the case that will not only finally confirm the reality of extraterrestrial visitation, but also expose the US government's conspiracy to hide this from the American people.

The tale itself has grown in the telling. What began as the simple claim that a "flying disc" had been recovered from a remote New Mexico ranch has now become an exciting story of multiple crashes, government cover-ups, alien autopsies and even, in some versions, the capture of living extraterrestrials. More energy has been expended by UFOlogists on investigating and debating (often turbulently) this incident than any other single case. And no case has attracted more scorn from sceptics, who, of course, dismiss the whole thing as a fabrication based on a fifty-year-old mistake, the main beneficiaries being the authors of serial books on the subject and the economy of the city of Roswell.

So what really happened at Roswell in 1947? Is it possible now to sift through the claims and counter-claims, the contradictory stories, and the downright fabrications (of which there are many). Can we get to the truth of what happened?

Astoundingly – given that the incident is supposed to have inspired the greatest military cover-up in history – the reason anybody heard about it in the first place is that the air base involved actually put out a press release about it. The historic announcement, issued in the early afternoon of Tuesday, 8 July 1947, began:

> The many rumors regarding the flying disc became a reality yesterday when the intelligence office of the 509th Bomb Group of the Eighth Air Force, Roswell Army Air Field, was fortunate enough to gain possession of a disc through the co-operation of one of the local ranchers and the Sheriff's office of Chavez county.
>
> The flying object landed on a ranch near Roswell sometime last week. Not having phone facilities, the rancher stored the disk until such time as he was able to contact the Sheriff's office, who in turn notified Major Jesse A. Marcel, of the 509th Bomb Group Intelligence office.
>
> Action was immediately taken and the disc was picked up at the rancher's home. It was inspected at the Roswell Army Air Field and subsequently loaned by Major Marcel to higher headquarters.

(Significantly, there is nothing in this release that says the object crashed or that it was in pieces. It says the disk *landed*, and that it had been stored and then picked up – by implication, intact.)

Not surprisingly, the press release caused a sensation, not only locally but also worldwide. The news went out over the Associated Press (AP) wire service, and within hours enquiries had been received not only from across the United States but also the world. As the town's morning newspaper, the *Roswell Dispatch*, noted the morning after, the longest call had come from the *Daily Mail* in London.

But the excitement was over within a matter of hours. That same evening a hastily-convened press conference at the Eighth Army

Air Force's headquarters in Fort Worth, Texas, announced that the "flying disc" had been identified as nothing more than the wreckage of a weather balloon. The story deflated equally quickly . . .

SETTING THE SCENE

In 1947, the future was being forged in the wilderness of New Mexico, a state called by its inhabitants – appropriately, as rumour would have it – the "Land of Enchantment". It was here that the violent birth of the nuclear age had taken place, with the development of the atom bomb at Los Alamos, home of the Manhattan Project. The Trinity Site, where the world's first atomic fireball and mushroom cloud had ushered in the stark uncertainties of the Cold War, lies some 120 miles (192 km) from Roswell, within the vast tract of desert given over to the White Sands Missile Range. And in 1947 White Sands also played host to the last and most fearsome weapon of random terror from the Second World War. Captured Nazi V-2 rockets were being test fired, as another piece of the modern world took shape, those launches representing the first steps into the Space Age – but also towards the Damocles Sword of the Intercontinental Ballistic Missile. (The city of Roswell itself has its own proud connections with the history of spaceflight. It was there that rocket pioneer Robert H. Goddard, known as the "Father of the Space Age", carried out his first test launches in the 1930s – there is a museum to him in the town – and it was where Apollo moon astronaut Edgar Mitchell attended primary school.)

The many sensitive and top-security nuclear and military facilities in the state included the one that was to become the focus of the Roswell Incident, Roswell Army Air Field (AAF). (The US Air Force did not become an independent army until later in 1947; in July the nation's air force was a branch of the US Army.) Now long closed, and today the site of a business park, in 1947 Roswell AAF was one of the most sensitive and highly protected places on the planet. It was the home of the 509th Bomb Group, which was then the world's only nuclear air force. The crews that bombed Hiroshima and Nagasaki a bare two years earlier had been drawn from its ranks.

Naturally, because of the presence of so many classified and top-secret facilities, security was tight throughout the state. With a world war freshly fought and won, the United States was shaping up for what it knew would be the next phase in world power-politics: the struggle between the "free world" and the Communist bloc. The first tensions in what would soon become the Cold War were already apparent, and the fear of Communist spies, gathering information for their masters in Moscow, was rife. There was much in New Mexico that had to be protected from prying eyes. America was then the world's only nuclear power, but knew that it was only a matter of time before it would be joined by the new enemy. Secret projects had been set up to watch for the tell-tale signs of a Russian atomic bomb, using instruments carried in airplanes or suspended from high-altitude balloons, and many of these, too, were being testing in the skies over New Mexico. That was another reason for the heightened state of security.

The proliferation of military bases and nuclear facilities in that barren and underpopulated state cuts several ways as far as the Roswell Incident is concerned. If there was one place on Earth that watching aliens would be keeping a very careful eye on, it would be New Mexico, where mankind was both taking its first tentative steps into space and developing the means to wipe itself out.

On the other hand, with so much hardware – much of it secret – being flown over New Mexico, there were many things in the sky that would have been unfamiliar to the general populace. As many were experimental, mishaps were not unusual: missiles could and did go astray, landing outside the closed perimeters of White Sands. And, for those seeking a natural explanation for any mysterious lights seen in the sky, New Mexico is not only known for wild and sudden thunderstorms, but its clear skies are perfect for watching meteors and other celestial events. Indeed, scientific teams regularly used the state for studying meteors.

But it was not just the place that was right, but the time, too. The Roswell story broke just fourteen days after Kenneth Arnold's seminal "flying saucer" sighting in Washington State's Cascade Mountains. In those two weeks America had gone saucer crazy. Reports of sightings poured in from every corner of the country – peaking, perhaps ironically, on Independence Day, Friday, 4 July (presumably because more people were out and about during the national holiday). Not only the skies, but the media, buzzed with flying saucers.

The novelty of the enigma places another important context around the Roswell Incident. Flying saucers had not yet acquired many of the trappings that are familiar to us today, when "UFO" is largely regarded as a synonym for "extraterrestrial spacecraft". Just days after the term had been coined, the idea that flying saucers might not be of this world had not yet taken shape. The prevailing view was that – if they existed at all – the saucers were some secret Russian device, or (the view initially taken by Kenneth Arnold himself) experimental American flying machines.

One of the few people in America who hadn't heard of flying saucers was William W. "Mac" Brazel. Described by his son as "an old time western cowboy", forty-eight-year-old Brazel was a foreman working on the sheep ranch owned by J.B. Foster, where he lived in a lonely shack which had no water or electricity – let alone a telephone, television or radio. The Foster ranch was about 60 miles (96 km) from the town of Roswell, nearer to the town of Corona, although it was here that the "Roswell Incident" – another piece of the future – was born.

According to the one surviving public statement made by Brazel, on 14 June 1947,[1] while out with his eight-year-old son Vernon moving sheep some 7 or 8 miles (11 or 12 km) from his shack, he came across an area strewn with strange and unfamiliar material that was clearly the wreckage of something that had fallen from the skies. But it was not until 4 July that Brazel returned with Vernon, his wife and fourteen-year-old daughter Bessie, and together they gathered up as much as they could of the debris.

Over the Independence Day holiday weekend, Brazel told some of his neighbours (the nearest of whom lived some 8 miles (12 km) away) about the find. They told him about the flying saucer mania that was filling the newspaper columns – and that some papers had offered a reward of the staggering sum of $3,000 for the first person to find one. Mac Brazel decided that perhaps he ought to tell somebody in authority about what he had found.

He made the journey into Roswell on Monday, 7 July, and informed the town sheriff, George Wilcox, who telephoned the nearby US Army Air Force base. Major Jesse A. Marcel of 509th

Bomb Group's Intelligence Office was called from his lunch to see Brazel. He went to the Sheriff's office, spoke to the rancher and examined some pieces of the debris that he had brought with him, and decided it was worth looking into. Major Marcel arranged to meet Brazel back at the Sheriff's office in an hour or so, and returned to the base to enlist the help of Captain Sheridan Cavitt of the Army Counter-Intelligence Corps (CIC). (For security reasons Cavitt's name was kept out of the newspaper reports at the time.)

They arrived at the ranch late, and had to spend the night in the rancher's spartan shack, with a meal of cold pork, beans and crackers. Early the next morning Brazel took them out to examine the debris field. They collected some more of the wreckage and brought it back to the house where they tried to fit it back together to see what it had originally been, but were unable to make any sense of the jumble. Marcel and Cavitt loaded the wreckage into their cars (a Buick convertible and jeep respectively) and took the pieces away, and Brazel heard no more until the furore broke the next day.

At the air base the discovery was reported to the commanding officer, Colonel William H. "Butch" Blanchard. As a result, the find was reported to the headquarters of the Eighth Army Air Force (of which 509th Bomb Group was part), which ordered that the wreckage be brought to them immediately. Marcel had it loaded aboard a B-29 bomber that took off for Fort Worth.

One of Roswell AAF's Public Information Officers, First Lieutenant Walter G. Haut, wrote the now-famous press release. In the early afternoon he drove into town to deliver it in person to its two newspapers and two radio stations. The sensational news was in time to make that day's edition of the town's evening paper, the *Roswell Daily Record* – where, not surprisingly, it was given a banner headline: 'RAAF [Roswell Army Air Field] captures Flying Saucer on Ranch in Roswell Region'. The paper also put the story out over the AP wire service.

While he was at the *Record*'s office Lieutenant Haut supplied a few more details at the request of the reporters.[2] He told them that he had been authorized to make the release by the 509th's intelligence office, specifically by Major Marcel, and that he not been given any details of the disc's design or construction. The paper linked the news with a sighting of a flying saucer by two Roswell inhabitants, the Wilmots, six days earlier (2 July).

By the end of the afternoon, after Marcel arrived at Fort Worth, the story was already beginning to cool. A reporter and photographer from the *Fort Worth Morning Star-Telegraph*, James Bond Johnson, was among those summoned to a press conference in the office of Brigadier General Roger Ramey, commander of the Eighth Army Air Force. Here it was announced by Ramey that the object was simply a weather balloon that was fitted with a Rawin (ray wind) target – a reflector that enabled it to be tracked by radar in order to measure wind speed and direction. The reflector was a simple construction, made of thick paper covered with tin foil.

The wreckage was laid out on the floor of Ramey's office, and Johnson took several photographs of the General and Major Marcel with it. The wreckage being displayed clearly was from a weather balloon, and not a spacecraft. But here, proponents of the UFO crash theory assert, the cover-up had already begun. What was being shown was not what Mac Brazel had found on his farm.

Johnson's story also went out on the AP wire service, just a few hours after the "flying disc" announcement. The Roswell papers of 9 July reflected the disappointment that the story had such a mundane explanation. The *Roswell Dispatch*'s headline read "Army Debunks Roswell Flying Disk" and the *Roswell Daily Record*'s "Gen. Ramey Empties Roswell Saucer".

The *Roswell Daily Record* also carried the only surviving interview with Mac Brazel, which he had given at the newspaper offices late the evening before. (After this, Brazel went to the KGFL radio station where he gave an on-air interview, but this was not recorded. However, by all accounts he said much the same as he did to the *Record*.) His description of the debris is disappointing. It consisted of "rubber strips, tinfoil, a rather tough paper and sticks",[3] all of which Brazel estimated weighed no more than 5lb (2.25 kg). Brazel also described a considerable amount of scotch tape with flowers printed on it, and eyelets in the paper. On the other hand, he did say that he had found weather balloons on the ranch on two previous occasions, and that this wreckage looked nothing like them. Brazel's description tallied with that given the same evening in Fort Worth by Major Marcel: "tinfoil and broken wooden beams . . . and the torn synthetic remains of the balloon".[4]

All in all, given Johnson's photographs and Brazel's description, it is not surprising that few people – if any – took the story seriously. Apart from the press release, there was nothing to suggest anything remotely unusual, let alone otherworldly, about the episode.

On the other hand, there must have been *something* odd about the affair. After Marcel and Cavitt had arrived back at the base with the debris, and the base commander had seen it, a press release announcing the "capture" of a flying disc had been issued, so clearly there was some uncertainty about what they had found. And it had obviously been thought unusual enough to report to the Eighth Army Air Force's headquarters, where it had aroused enough curiosity for them to request that it be brought to them, post haste. Fort Worth had no difficulty in identifying it – why, then, if it *was* just a weather balloon, had Roswell AAF not recognized it?

THE REVIVAL

The revival of the long-forgotten story – which, if it appeared in the UFO literature at all, was included as a cautionary tale in how easy it was to let enthusiasm get the better of judgement – began in 1978 with Jesse Marcel, the intelligence officer at the centre of the events of 1947.

A year after the flying disc publicity, Marcel was transferred to USAF Headquarters Command in Washington, D.C., where he worked on the highly classified project, AFOAT, to watch for signs of Soviet atomic tests. He left the USAF in 1950 and returned to his home town in Louisiana, where he opened a TV and radio repair shop. By 1978 he was retired.

In February of that year he was tracked down by UFOlogist Stanton T. Friedman, a former nuclear physicist, who since leaving that field in 1971 had become one of that rare and courageous breed, a professional UFOlogist. In 1978, Friedman was particularly interested in cases involving the retrieval of crashed UFOs and alien bodies, and it was this research that led him to Marcel.

Friedman, with a fellow UFOlogist, former school teacher William L. Moore, made several trips to Louisiana to interview Marcel in 1978 and 1979. The first interview was disappointing: Marcel was dismissive of the whole episode, saying that he could not even remember the year in which it had occurred. Nevertheless, Friedman arranged for the former Major to appear in a Chicago radio show in April, and to be interviewed for a TV documentary, *UFOs Are Real*, (co-written by himself) early the following year. In these, and further interviews with Friedman and Moore for the book they were planning, Marcel began to speak of the wreckage – and the whole incident – in terms that suggested that something more than a weather balloon was responsible.

Of the wreckage he said: "There was all kinds of stuff – small beams about three-eighths or a half-inch [12 mm] square with some sort of hieroglyphics on them that nobody could decipher. They looked something like balsa wood, and were of about the same weight, except that they were not wood at all. They were very hard, although flexible, and would not burn. There was a great deal of an unusual parchment-like substance which was brown in colour and extremely strong, and a great number of small pieces of a metal like tinfoil, except that it wasn't tinfoil."[5]

He also mentions Captain Cavitt picking up a black, metallic box several inches square – the only thing that resembled any kind of instrumentation or electronics.

The unusual features of the material were that the "balsa wood" could not be burned and the "tinfoil" could not be torn. He also reported that 'one of the boys' at Roswell AAF told him that they had tried out a sledgehammer on the foil, and couldn't even dent it.[6] (In later interviews Marcel used phrases such as, "We even tried . . .", implying that either he wielded the hammer or that he saw it being done, but his first accounts make it clear that he was only *told* about it.) He said that the material was impossible to bend – although it has been pointed out that his qualification of what he meant by this is seldom quoted: "Now by bend, I mean crease. It was possible to flex this stuff back and forth, even wrinkle it, but you could not put a crease in it that would stay."[7]

Marcel said that the reporters at Fort Worth had not been allowed to see some of the more significant parts of the wreckage, especially those marked with "hieroglyphics". He also intimated that he had been warned by General Ramey not to say certain things at the press conference. And in a later interview he told William Moore that the material photographed at the press conference had been substituted for the real wreckage – although he maintained that one of the photos does show pieces of "the actual stuff we found".

By February 1980 Marcel was telling the *National Enquirer* that the material was "nothing that came from this earth".[8]

An obvious question is, if Marcel and his superior really thought they had got hold of the remains of a flying saucer, why did they authorize the issue of a press release saying so? Ironically, the existence of the press release can be seen as evidence *against* it really being a UFO – especially in the scenario that has taken shape since 1980, in which it is argued that the whole episode was the subject of an immediate and extensive cover-up. Indeed, several leading Roswell researchers now suggest that the alien nature of the event was known several days *before* 8 July, which makes the press release even harder to understand.

How did Marcel explain this? Despite Walter G. Haut's statement to the Roswell press that he had issued it under Marcel's instruction, in 1979 Marcel claimed that Haut had acted on his own initiative and had been severely reprimanded for doing so. He describes Haut as an "eager-beaver" who took it upon himself to issue the news.[9] (A few weeks after the incident, Haut was transferred to the commander of the base security guard – although whether this amounts to a demotion is open to question. He resigned from the USAF in August 1948, the same month that Marcel was transferred to Washington.) Marcel even claimed that Haut issued the release before he and Cavitt had arrived back at the base, which clashes with all the evidence that the wreckage, and Marcel, were already en route to Fort Worth when the announcement was made.

Marcel's testimony – he was, after all, one of the primary witnesses to the events of 1947 – resurrected the Roswell Incident as a fit subject for UFOlogists. Very soon it would become the most widely-publicized case in the history of UFOlogy, and the cornerstone of belief in the government conspiracy.

Although Stanton Friedman had found Marcel, and brought William Moore into the investigation, Moore teamed up with Charles Berlitz (then best known as the author of *The Bermuda*

Triangle) to publish the results in *The Roswell Incident* (1980). This caused a rift between Friedman and Moore (one of many among Roswell researchers). Friedman, with Don Berliner, later wrote his own reconstruction of the event, *Crash at Corona* (1992).

In the late 1980s two UFOlogists, Kevin D. Randle and Donald R. Schmitt, persuaded the Center for UFO Studies (CUFOS), of which Schmitt was Director of Special Investigations, to fund their research to produce the "definitive" account of the Roswell Incident. Randle is a former Vietnam helicopter pilot and officer with the USAF's counter-intelligence branch, the Air Force Office of Special Investigations (AFOSI). He had written two books about his UFO research before becoming involved in the Roswell enigma, and was largely sceptical about the subject, arguing in his 1988 book *The October Scenario* that *all* UFO reports made before 1973 were the result of misidentifications or hoaxes. The results of Randle and Schmitt's investigation appeared in *UFO Crash at Roswell* (1991) and *The Truth about the UFO Crash at Roswell* (1994).

THE ORIGINAL WITNESSES

The Roswell story with which most people are familiar today – with multiple crash sites, alien autopsies, and a high-level conspiracy to suppress the event – is a far cry from the original 1947 tale, and comes mainly from new witnesses that have come forward since 1980. But before looking at their account, what have the new generation of researchers been able to establish about the original players – those *known* to have been involved from the start? Was there anybody who could corroborate Marcel's story?

The problem was that many of those key players had died long before any UFO researcher could question them. Most significantly, Mac Brazel died in 1963. Colonel Blanchard passed away three years later. Cavitt, the CIC man, remained an elusive figure for several years and was only to make one, dismissive, statement on the affair in 1994. Apart from Marcel, of the 1947 cast of characters only Walter Haut, the writer of the press release, was still alive in the late 1970s.

Is there any independent testimony to support the crash, or fall, of an object near Roswell, that might, perhaps, give us some idea what it was? Were there any living witnesses who could corroborate Marcel's description of the wreckage, or shed further light on Mac Brazel's discovery?

A number of separate testimonies are offered as a record of the object's fall. There is the flying saucer report by the Wilmots on 2 July, which was linked with Brazel's find in the press at the time. Most specifically, two nuns at the Franciscan hospital in Roswell, who routinely made observations of the night sky, recorded in their logbook for 4 July 1947, between 11.00 and 11.30 p.m., a brilliant light plunging earthwards somewhere to the north. (There was, presumably, something about this that marked it out from the many other falling lights on America's equivalent of the quirky British festival of Bonfire Night.) Among others who recall seeing something strange in the New Mexico skies around the time of the incident, a Roswell resident, William Woody, remembered seeing a meteor-like object fall, as did a former corporal at one of Roswell AAF's outlying facilities. These have all been put forward as witnesses of the crash.

The problem is that, with the exception of the Wilmots' report and the nuns' log entry (which can hardly describe the same object as they were separated by two days), these reports were culled from the memories of the witnesses some forty years after the event. They can't be specific about the date, or even year, so there is no guarantee that they are talking about the same event. And there were plenty of things in the sky over New Mexico that could account for them. As we have seen, apart from the various items of military hardware that frequently generated UFO reports, the state's clear skies are renowned for meteor-spotting. And 1947, it must be remembered, was a major "flap" year. It should also be remembered that Mac Brazel himself said that he found the debris some three weeks before, but only gathered it up on 4 July.

Turning to the actual recovery of the wreckage from the Foster ranch, there are discrepancies in the chronology of events which, while they may appear trivial in themselves, do have important implications for some of the later reconstructions.

Mac Brazel's 1947 interview makes it clear that he came into Roswell to report his find on the Monday, 7 July. This makes sense, as Brazel came into town for other business as well, which means that it was a working day. This was also what Major Marcel told reporters at the Fort Worth Press Conference, and William Moore in 1979. However, most recent books shift it back a day, to 6 July, as only by doing so can certain problematic elements of post-1980 accounts be reconciled.

As already noted, Brazel, Marcel and Cavitt arrived at the Foster ranch – a drive of some two or three hours – late and did not visit the crash site until the following morning. All accounts stress that the three men were up and awake before sunrise. However, Marcel says that he and the CIC man spent the whole of that day surveying the site and loading as much of the wreckage as they could into their vehicles, and that they left to return to their base as night fell, stopping off at Marcel's house on the way so that he could show the peculiar material to his wife and eleven-year-old son, Jesse Jr. Since the press release went out shortly after noon on 8 July, this sequence can only be correct if Brazel came into Roswell on the Sunday, 6 July. This is the date favoured by most Roswell researchers. They do not claim that this was a mistake by Brazel and Marcel, but part of a deliberately contrived cover story (although what difference it makes is unclear).

Knowing the correct sequence of events has implications for the size of the debris field. If it had been relatively small, and only a small amount of material had to be collected, Marcel and Cavitt wouldn't have spent very long on the task, and so would have been back on base by late morning on 8 July, and had therefore left the day before. If, as Marcel maintains in his 1979 interviews, there was a huge field, and it took them most of the day to collect samples, then they must have left the base on 6 July. Which is correct?

Brazel, in his 1947 interview, said that the debris covered an area of about 200 yards (180 m) in diameter. In 1979, Marcel told UFOlogist Leonard Stringfield that it formed a long, thin strip some three quarters of a mile long (1,188 m) and 2-300 feet (60–90 m) wide, and that at one end there was a gouge in the ground some 500 feet (150 m) long and 10 feet (3 m) wide.[10] If true, there must have been a considerable amount of debris. Cavitt's one and only statement, in 1994, says that the wreckage covered only about 20 square feet (1.8 m), and that it was easily loaded into Marcel's car (and none into his own jeep).[11] (Cavitt also stated that the material in the photographs of the Fort Worth Press Conference was the same that he and Marcel had collected.)

Support for Marcel's version comes from his son, Dr Jesse Marcel, Jr., a physician living in Montana. First interviewed in 1979, he recalls being shown the unusual material, although he has given conflicting times in some interviews saying that it was in the early evening and in others that he was awoken by his father in the early hours of the morning. He described to William Moore how his father had spread the wreckage on the kitchen floor, and that: "The material was foil-like stuff, very thin, metallic-looking but not metal, and very tough. There was also some structural-like material too – beams and so on. Also a quantity of black plastic material which looked organic in nature."[12]

A few weeks later Marcel Jr., after speaking with his father, wrote to Moore that he had remembered something else about the wreckage: that "hieroglyphic-type" characters were imprinted on some of the beams.

In subsequent interviews, Marcel Jr. described an "I-beam", about half-an-inch (125 mm) thick and 12 to 18 inches (30 to 45 cm) long. (All other witnesses, including his father, say that all the struts were rectangular in section.) In a live TV interview in 1988, he said: "I held this material in my hand. It was not from this earth or this universe. That's what I think. We're not alone."

Ironically, the major proponents of the Roswell UFO crash have to discount the testimony of the primary witness: the interview that Mac Brazel gave to the *Roswell Daily Record*. In that, as we have seen, Brazel described very terrestrial debris, even though he said that it wasn't like either of the two weather balloons he had found on his ranch before. Maybe not, but it was made up rubber, wood and sticky tape, which hardly sounds like an alien spaceship.

As new witnesses came forward with evidence of a more sensational kind, including that of the military cover-up that had held since 1947, researchers began to reconsider Brazel's testimony. Some have suggested that Brazel was intimidated by the military into changing his account to support the weather balloon cover story, or (to the anger of his family) that he was bought off, citing statements by his neighbours that the rancher suddenly seemed to have a lot more money, as he bought a house and a new truck.

And yet – as the same researchers point out – Brazel ended the interview by saying that it was "no balloon" and that, in future, ". . . if I find anything else besides a bomb they are going to have a hard time getting me to say anything about it". (This statement is often taken as an indication that Brazel had been pressurized by the authorities. However, it is clear from the interview that it was because of the publicity at which the quiet and unassuming rancher had found himself at the centre.)

Although no recording was made of the radio interview that Brazel gave to KGFL radio immediately after speaking to the *Record*'s reporters, it seems that he told the same story. However, the KGFL presenter who interviewed him, Frank Joyce, told researchers in the 1980s that he had spoken to the rancher the day before, and that the story he had given then was *different* from that recounted in the studio.[13] Joyce said that he had phoned the Sheriff's office to see if there was any news for the station, that Brazel had been there reporting his discovery, and that Sheriff Wilcox had put him on the line. Unfortunately, Joyce would not elaborate on Brazel's original story, saying that it would be unfair as Brazel was no longer alive. (In which case, why say *anything* to suggest that Brazel lied in his second interview?)

Clearly, if these researchers are right, Brazel was barefacedly lying in the two interviews. But why? Friends of his say that, in the days following the breaking of the story, he was accompanied by USAAF officers wherever he went. In particular, they went with him to the *Record*'s offices and the radio station. There are also reports that the rancher was held by the military for several days. The clear implication is that Brazel was forced to change his story to support the "cover" that had been devised at Fort Worth. So what did he really find?

Although living some 8 miles (12 km) from Brazel's shack, Floyd and Loretta Proctor were his nearest neighbours, and were among the people with whom Brazel discussed his discovery over the holiday weekend. They were obviously prime witnesses for researchers, and were first interviewed by William Moore in 1979. Floyd died soon afterwards, but Loretta continued to give interviews and to appear on radio and TV programmes about the incident.

In their first interviews neither Loretta nor Floyd Proctor said that they had seen the debris, but only that Brazel had described it to them. They declined Brazel's suggestion that they go out and see the debris field, as they were not particularly interested and "gas and tires were expensive then".[14] However, Loretta has since – from 1989 onwards – said that Brazel *did* show them a piece, which she described as a four-inch long rod or strut, tan-coloured and light like balsa wood, although it appeared to be plastic. She said that, "We cut on it with a knife and would hold a match on it, and it wouldn't burn."[15]

Loretta Proctor also claimed that her son, Nathan (nicknamed "Dee") was with Brazel when he found the wreckage on 4 July, although Dee himself says that he has no memory of the event. This, of course, clashes with Brazel's 1947 testimony, in which he said that he found it three weeks earlier, and only gathered up some of the pieces on 4 July. (Some researchers, puzzled by the fact that Brazel didn't mention Dee's presence, have suggested that the kindly rancher wanted to spare the young boy from the attentions of the military. This may be so – but why, then, did he say that his *own* son Vernon was with him at the time?)

Another important witness was Brazel's oldest son, Bill Brazel, who was in his twenties and living in Albuquerque at the time of the incident. Bill had naturally spoken to his father about it, and was able to add some details, although he said that his father would not speak much about it: "My dad found this thing and told me a little bit about it. Not much, because the air force asked him to take an oath that he wouldn't tell anybody in detail about it. And my dad is such a guy that he went to his grave and he *never* told anybody."[16]

According to Bill, the night before he found the wreckage, during a thunderstorm, his father heard an exceptionally loud explosion. The next morning he found the wreckage, but did not collect any for a day or two.

More significantly, Bill Brazel claimed to have found small pieces of the material himself over the next two years.[17] His description of the unusual material tallies with that given by others – like tin or lead foil, but could be neither torn nor creased, and which when folded, or wrinkled would return to its original, smooth, shape. Unfortunately, his small collection was confiscated by USAF officers in 1948 after he had unwisely talked about it in a bar in Corona. (Although this date seems unlikely. Bill Brazel claimed that, of the two officers who came to take the material, one was white and the other, a sergeant, black. In fact, racial segregation was the norm in all the US armed forces at the time – it was not until May 1949 that the USAF permitted mixed units.[18] Records show that there were no black sergeants stationed at Roswell AAF in 1948.)

Contradicting her brother's account, Mac Brazel's daughter Bessie – who was fourteen at the time of the incident, and who helped her father gather pieces of the debris – says that the family found *no more* fragments after the military had cleared the site. She had, however, seen the material, and told Moore in 1979:

> There was what appeared to be pieces of heavily waxed paper and a sort of aluminum-like foil. Some of these pieces had something like numbers and lettering on them, but there were no words that we were able to make out. Some of the metal-foil pieces had a sort of tape stuck to them, and when these were held to the light they showed what looked like pastel flowers or designs.[19]

Bessie Brazel confirmed that the military had sworn the family to secrecy.

Another important player in the events of 1947 – although he never claimed to have seen the wreckage himself – was the author of the press release that started the ball rolling, Walter Haut. He continued to live in Roswell after he left the USAF in 1948, running a series of small businesses. When first contacted by researchers, Haut maintained that he had written and issued the release because he had been instructed to do so by Colonel Blanchard and that, as far as he was aware, the weather balloon story was the correct one. In 1980 a television documentary reunited him with Marcel after forty years, and he said that this occasion was the first time he heard the former intelligence officer reject the official explanation.[20] However, Haut too changed his tune, writing in an affidavit in 1993, "I am convinced that the material recovered was some type of craft from outer space."[21]

In contrast to Haut's press release, and the testimonies of Brazel, Marcel and Cavitt, which either state or imply that all the debris was collected by the two intelligence officers on their one visit to the ranch, there is evidence of a military presence at the crash site in the days after Marcel's visit. A reporter from Albuquerque, Jason Kellahin, says that he visited the site, which was still littered with debris – he described as looking like it came from a balloon or kite – and that Mac Brazel was there with five or six military personnel.[22] The officers did not prevent him from talking to Brazel, but would not give any information themselves. (Although Kellahin said that this happened on the day of the press release, this seems to be a mistake in his forty-year-old memories. He said that he drove from Albuquerque after the news came through on the wire and took a detour to the crash site on the way. However, as the news did not break until early afternoon, and did not say where the "disc" had been found, it seems more likely that it was the following day.)

But the most interesting confirmation comes from one of Roswell's former Counter-Intelligence Corps officers, Master Sergeant Lewis Rickett. We have seen that, when Marcel went to the site, he was accompanied by a CIC man, Captain Sheridan Cavitt, although his name was kept out of the papers at the time. The reason for this was that the CIC, being counter-intelligence, were a more secret outfit than Marcel's USAAF intelligence unit. Marcel's job was to gather information, whereas CIC's role was to prevent enemy agents learning US military secrets, and stop classified information from reaching the public. CIC officers went in plain clothes, did not use their ranks, and did not report through USAAF's normal chain of command but direct to headquarters in Washington, in case they were required to investigate any senior officers at the base who fell under suspicion. Clearly, Cavitt's role in visiting the site was to determine whether something of a classified nature lay behind it.

Lewis Rickett told researchers that he had returned to the Foster ranch with Cavitt the following day, where a team were collecting the rest of the debris. He said of the wreckage: "There was a slightly curved piece of metal, real light. It was about six inches by twelve or fourteen inches . . . I crouched down and tried to snap it. My boss [presumably Cavitt] laughs and said, 'Smart guy. He''s trying to do what we couldn't do.' I asked, 'What in hell is this stuff made out of?' It didn't feel like plastic and I never saw a piece of metal this thin that you couldn't break."[23]

In his one and only statement on the affair, to USAF investigators in 1994, Cavitt was dismissive of the whole episode, saying that it was obvious to him from the start that the debris was from a weather balloon. However, Cavitt introduced one inconsistency by saying that he, Marcel *and Rickett* accompanied Brazel to the crash site. This is clearly a slip of the memory, as neither Marcel nor Brazel mention a third officer, and seems to be a result of Cavitt confusing his two visits to the site. The significance of this is that it confirms Rickett's account of visiting the site on 8 July, at which time not all the wreckage had been collected.

If a UFO had crashed, the Fort Worth Press Conference that announced that the debris was a weather balloon had, of course, to be part of a cover story: the wreckage that was displayed and photographed was that of a real Rawin balloon and target – but was not what Marcel had brought from Roswell.

One of Fort Worth's meteorologists, Warrant Officer Irving Newton, was summoned by his commanding officer to make the identification. Newton told researchers that he had immediately recognized the material as a Rawin balloon, adding, "While I was examining the debris, Major Marcel was picking up pieces of the [radar] target sticks and trying to convince me that some notations on the stick were alien writings. There were figures on the sticks lavender or pink in colour, [which] appeared to be weather faded markings with no rhyme or reason. He did not convince me these were alien writings."[24]

But there is evidence that the real wreckage was sent on to either (or both) USAAF headquarters in Washington or Wright Field in Dayton, Ohio. Wright Field was the headquarters of the Air Technical Intelligence Center (ATIC), which would be the logical place for unknown material to be sent for analysis. Now known as Wright-Patterson Air Force Base (AFB), this facility had already achieved a prominent status in UFO lore by the time of the Roswell "revival", because of claims that the preserved bodies of aliens were stored there (see below).

It is clear from 1947 press reports and wire messages that the original plan was indeed to send the wreckage from Fort Worth to Wright Field. However, when the material was identified, the onward flight was supposedly cancelled. But there is evidence that it was not . . .

An hour or so after the Fort Worth Press Conference, the FBI office in Dallas sent a teletype message[25] to J. Edgar Hoover in Washington and the nearest FBI office to Roswell, which was in Cincinatti. Timed at 6.17 p.m. on 8 July, the teletype was not made public until the 1970s, when it was unearthed by UFOlogists using the Freedom of Information Act. Like many pieces of the Roswell puzzle, the teletype contains both good and bad news for the supporters of the UFO crash.

The FBI had been called by Fort Worth AAF to advise them of the discovery, the reason being, the message said, to forewarn the Bureau of the anticipated press interest in case it received any enquiries. The FBI itself would not be conducting any further investigation.

The FBI reported: "The disc is hexagonal in shape and was suspended from a balloon by cable, which ballon [sic] was approximately twenty feet [6m] in diameter." It went on to say that Fort Worth AAF had advised them that the object "resembles a high altitude weather balloon with a radar reflector" but that when they had described it to Wright Field the latter "had not born [sic] out this belief". In other words, it appears that the personnel at Fort Worth *had* believed it to have been a weather balloon, but that Wright Field had disagreed.

Most significantly, the FBI teletype (which was sent out *after* the Press Conference) states: "Disc and balloon being transported to Wright Field by special plane for examination."

Support for this comes from Brigadier General Thomas DuBose who, as a colonel in 1947, was General Ramey's Chief of Staff. DuBose, then in his eighties, told Roswell researchers that the weather balloon story *was* a cover, and that the special flight had *not* been cancelled. DuBose says that the wreckage was despatched to USAAF Headquarters in Washington, but it may then have been sent on to Wright Field.

DuBose said that he received a call from Major General Clements McMullen in Washington at 2 or 3 o'clock in the afternoon saying that unusual wreckage had been recovered from the Roswell area, and that it was being flown to Fort Worth in sealed containers.

DuBose was ordered to arrange its transport on to McMullen. However, some of what DuBose said clashes with other evidence. He said that the call from Washington came "two or three days" before the press conference – i.e. on 5 or 6 July.[26] Although some researchers maintain that DuBose was referring to the original fragments that Mac Brazel had brought into Roswell, which were sent ahead of the wreckage collected by Marcel, this seems to be a slip in the octanagerian's memory, as he makes it clear that this was the same material that was sent on to Washington after the press conference.

To add further – apparent – confusion, when interviewed by Jaime Shandera in 1990, DuBose was adamant that the material displayed in Ramey's office was the same that had come off the plane from Roswell[27] – in fact, he was the one who collected it from the plane and took it to the office. Yet he told Randle and Schmitt that the weather balloon story was a cover.[28] And he also told Shandera that the special flight to Wright Field had not been cancelled, and that on orders from the General McMullen, he had placed the material in a sealed pouch and given it to a specially selected and trusted courier, who he had escorted out to a B-25 that was bound for Washington.[29] Surely all three of DuBose's statements cannot be correct? And yet, as we will see, there is one scenario that does allow for this.

In 1989 the widow of Oliver "Pappy" Henderson, a legendary Second World War pilot who was based at Roswell in 1947, claimed that, shortly before her husband's death in 1986, and after he had seen some of the Roswell coverage, he had confided to her that he had flown a large amount of the wreckage direct to Wright Field in a C-47 transport.[30] As this was not Marcel's flight to Fort Worth, it has been surmised that this was the bulk of the wreckage, which was cleared from the crash site in the ensuing days. Other witnesses claim that parts of the wreckage were flown to the nuclear weapons facility at Los Alamos.

From all this testimony, it does seem that something odd was going on. There *was* a military presence at the crash site in the days following Marcel's visit – a presence that seems more to do with the Counter-Intelligence Corps – when more wreckage was collected. There *is* evidence that Mac Brazel was, as one of his friends put it, being kept on a "tight leash" by the military in the ensuing days. And it seems that, despite the debunking press conference, the wreckage collected by Marcel *was* sent on either to Washington or to ATIC at Wright Field – as was the rest of the material that was collected later by CIC.

THE SECOND CRASH SITE AND THE BODIES

Nobody involved during the original events in 1947 had mentioned bodies. The only such suggestion comes from KGFL newsman Frank Joyce, who told Randle and Schmidt in 1989 that, after his interview, Brazel had made an off-air remark as he left the studio: "Frank, you know how they talk of little green men? They weren't green."[31]

However, this is unlikely. The term "little green men" didn't become popular until the 1950s – and by his own admission Brazel hadn't even heard of flying saucers until two days before coming to Roswell. None of the original first- or second-hand witnesses to the recovery of material from the Foster ranch said anything about bodies – most significantly Jesse Marcel, who, as the main "whistle blower" of the Roswell Incident, surely would have if he had seen them. However, gradually stories of alien corpses associated with the Roswell crash began to emerge.

The claim that the US Government possessed corpses of one or more extraterrestrials, preserved or cryogenically frozen, had been circulating in the UFO community for most of the 1970s. The leading proponent was the veteran UFOlogist Leonard H. Stringfield, who had been an investigator since the early 1950s. Throughout that time Stringfield had worked closely with the USAF and other official investigations into UFOs – for example, acting as a co-ordinator for the Condon Committee. In this time he built up many contacts in military and governmental circles. In 1978, Stringfield published a paper entitled *Retrievals of the Third Kind: A Case Study of Alleged UFOs*, in which he presented the evidence – largely based on the testimony of his military informants – that the US Government was secretly guarding the remains of dead aliens.

In 1974 there was a flurry of reports in the US media, begun by UFOlogist and TV researcher Robert Carr, that a UFO and the frozen bodies of its crew were being held in Wright-Patterson AFB, in a building known as Hangar 18. It was even claimed that President Gerald Ford was about to make a public announcement about it, Carr himself declaring in October, "Five weeks ago I heard from the highest authority in Washington that before Christmas the whole UFO cover-up will be ended". Although this revelation never happened, Carr's claims were apparently confirmed by Stringfield's informants.

Carr claimed that he had received this inside information from people who had worked at Wright-Patterson. Unfortunately, he linked the bodies with another alleged flying saucer crash in New Mexico, which had supposedly happened in 1948 near the town of Aztec (around 300 miles (480 km) from Roswell) and was the subject of one of the first bestsellers about UFOs, Frank Scully's *Behind the Flying Saucers* (1950). The story was exposed as a hoax two years later, when it was revealed that it had been fed to Scully by two known, and convicted, confidence tricksters.

However, when the Roswell case was revived and began to be taken seriously by the UFO community the two stories seemed to dovetail neatly. As we have seen, there is evidence that the wreckage from the Foster ranch was taken to Wright-Patterson AFB. If the corpses of the alien crew had also been recovered, this would explain the preserved extraterrestrials in that base. But how good is the evidence for aliens at Roswell?

In October 1978, after a lecture in Minnesota, Stanton Friedman was approached by Mr and Mrs Vernon Malthais. They recalled being told of a saucer crash in New Mexico by a now-deceased friend, Grady L. "Barney" Barnett, in 1950. Barnett was a civil engineer based in Socorro and employed on soil analysis work. According to the Malthaises, while working somewhere in the vicinity of the towns of Magdalena and Datil, in an area called the Plains of San Agustin some 150 miles (240 km) from the Foster ranch – he had come across the wreck of a crashed disc-shaped craft and the bodies of several tiny creatures with "small and oddly spaced" eyes. He had been joined by a team of archaeologists from the University of Pennsylvania. As they took in the eerie scene, a military unit arrived in jeeps and ordered them all to leave, instructing them to keep quiet about what they had seen.[32]

William Moore was able to track down Barnett's supervisor at the time, who confirmed that Barnett had told him the same story on the day it happened (but that, when he had dismissed it as "bullshit", Barnett, understandably, had lapsed into silence).[33]

Barnett's story links only approximately with the Roswell Incident in time and place: it happened during the late 1940s in New Mexico. There is nothing in the recollections of those to whom Barnett told his story to pinpoint the year or day, and the diary kept by Barnett's wife does not place him in the Plains of San Agustin in the days around the Roswell crash. Although most Roswell researchers link Barnett's experience with the wreckage on the Foster ranch, could this be an entirely separate incident, unrelated to the "flying disc" debris found by Mac Brazel?

As the investigation developed, some researchers began to argue that, while Barnett's story was true, those who had transmitted it had become confused about where it had happened, and that he was actually nearer to the Foster ranch. However, this is simply an assumption made in order to make the story fit into the Roswell scenario.

The riddle of the archaeologists seen by Barnett has long intrigued researchers, who realize that it might fill one of the most important gaps in building a credible case for the recovery of alien corpses. As we will see, there seems to be no shortage of people who claim to have seen the wrecked UFO and bodies, but none of these witnesses give any corroboration to the others. Not only do they give widely differing descriptions of the craft, the bodies and the sequence of events, none of them appears to have noticed the other witnesses.

Despite their best efforts, Roswell researchers have been unable to track down Barnett's archaeologists, or to find any record of an archaeological dig in the San Agustin area in July 1947 by the University of Pennsylvania or any other institution.

In 1990, after seeing a TV documentary that featured the Barnett story, a Missouri man, Gerald F. Anderson, called the station to say that he and four members of his family (all, unfortunately, dead by then) had seen the same thing. But – sensationally – Anderson claimed that one of the aliens *was still alive.*[34] The TV company put him in touch with Friedman and Randle, who interviewed him.

Anderson was only five years old at the time, but remembered that it happened on the Plains of San Agustin. He also remembered the archaeologists and – remarkably for such a long time before when he was so young – could recall the name of the team leader, Dr Buskirk. Almost incredibly, he even produced an identikit image of the man. UFOlogist Tom Carey was able to track down a Dr Winfried Buskirk, who lived in Albuquerque and published a book in 1986 whose jacket bears a photo strikingly similar to Anderson's identikit picture. However, Buskirk flatly denied being in the San Agustin area in 1947, or any knowledge of a crashed flying saucer. And Carey also found that Buskirk had been a teacher at Albuquerque High School when Gerald Anderson was there in the 1950s.

A few months later, Stanton Friedman received a letter that was supposedly from a cousin of Anderson's, a nun called Vallejean Anderson, enclosing pages of what purported to be the diary kept by her father Ted – Gerald Anderson's uncle – for July 1947. The pages gave details of the UFO crash that backed up Anderson's story. However, an analysis arranged by Las Vegas millionaire and UFO enthusiast Robert Bigelow found that the ink used was of a type not manufactured until the early 1970s. Ted Anderson died in 1965.[35]

During a later public dispute with Kevin Randle over a telephone interview, Anderson was proven to have produced a forged telephone bill to support his version of events.[36] Anderson's credibility collapsed, and it became clear that his tale was not the confirmation of the second-hand Barnett story that it had once seemed to be.

In the early 1990s, paleontologist C. Bertram Schultz added two new pieces of information.[37] Schultz remembered, many years before on a fossil-hunting trip to the Roswell area, seeing a military cordon as he drove north out of Roswell on Highway 285. And he also remembered later talking to members of an archaeological expedition who told him of finding a crashed flying saucer. Most exciting of all, he could remember the name of the team's leader: W. Curry Holden.

Although Schultz could not be sure of the date or year that he had seen the cordon, he was sure about the place – north of Roswell. This was too far from the Foster ranch for the cordon to have been related to Mac Brazel's discovery. However, as we have seen, as there were many experimental aircraft, rocket and balloon projects in New Mexico in the 1940s and 50s, there were just as many reasons for the military cordoning off an area, and what Schultz saw was not necessarily connected either to the Roswell Incident or to the story he later heard from the archaeologists. But the naming of W. Curry Holden opened a new avenue for researchers.

In 1992 Kevin Randle managed to find Holden, then ninety-six years old and living in a nursing home, and interviewed him. Frustratingly, when Randle questioned him about the incident, although Holden repeatedly said, "I was there. I saw the whole thing",[38] he was unable to give any details. Holden's wife and daughter said that he was easily confused, and his memory unreliable – and that he had never said anything to them about finding a crashed UFO. Nor do Holden's personal or professional papers place him in the Roswell area in July 1947. Holden died a few months after Randle's visit, adding another frustrating dead end to the Roswell investigation.

The San Agustin event made only a tentative link between alien bodies and Mac Brazel's discovery. However, some ten years after the revival of Roswell, the first eye-witness accounts explicitly connecting aliens with that event began to surface. The decade-long delay is significant. As we have seen, nobody connected with the Foster ranch crash – most importantly Jesse Marcel – had mentioned bodies. (Marcel died in 1986, before the first of the new claims appeared.) But the story of Barney Barnett and the San Agustin crash had featured in Berlitz and Moore's 1980 book and in several TV documentaries, introducing the concept of alien bodies into the story. Sceptics have pointed out that it is curious that it took a decade for anyone to come forward with information about bodies at Roswell itself.

Something else that had happened in the intervening years was the appearance of the controversial "Majestic 12" papers (see Chapter 7). First surfacing in the mid-1980s, these documents described the formation, on President Truman's orders, of a secret group to oversee studies of the UFO and alien corpses recovered from Roswell. This was the first suggestion that bodies *had* been found there.

The catalyst for the first of the eye-witness accounts appears to have been Robert Shirkey, the officer who in 1947 had organized the B-29 bomber that transported the wreckage to Fort Worth. In the late 1980s Shirkey, who still lived in Roswell, began to tell people that he had heard a tale involving mysterious, non-human bodies from his good friend, Roswell's retired mortician, W. Glenn Dennis.[39] (Walter Haut had been a friend of Dennis's for 40 years, and expressed surprise when he first heard this claim in 1988 or 89.)[40] Shirkey put Stanton Friedman in touch with the former mortician, whose tale added considerably to the developing story.

In 1947, Dennis ran a funeral home in Roswell that provided ambulance and funeral services to the air base. He says that one afternoon he received a call from the base asking some very unusual questions: how many sealed caskets could he supply, and what were the smallest he had that could be hermetically sealed, along with questions about embalming methods. Later that afternoon he had to drive an injured airman to the base hospital, and found a fleet of military ambulances parked outside. There was a heavy presence of Military Police in and around the hospital, and he realized that something out of the ordinary was going on. After delivering his patient, he went in search of a nurse that he knew. As he wandered around the hospital, his nurse friend appeared and, surprised and worried at seeing him there, asked him how he had got in, as outsider visitors were being kept away. She told him, "My God, you are going to get killed." At that moment he was seized by two MPs who escorted him off the base, with bloodcurdling threats that, if he spoke about what he had seen, he would be "picking his bones out of the desert" and that he would turned into "dog food". A few hours later he received a phone call threatening him with jail if he didn't keep his mouth shut.

The next day the nurse invited him to lunch in the base's officers' club. Clearly shaken, she told Dennis that she had been called on to assist at the autopsy of three strange creatures, one of which was badly mutilated. She drew him sketches: the beings had large heads with big black eyes, and four-fingered, thumbless hands with suction pads at the tips. The bodies, she said, were then flown to Wright Field. She swore Dennis to a "sacred oath" that he would never reveal her identity – although he did later name her as Naomi Maria Selff. Dennis was unable to produce her sketches, having lost them some forty years before, but he drew them from memory for Stanton Friedman.

Dennis was the first person to directly link Roswell AAF – and by extension the wreckage from the Foster ranch – with alien remains. Could they have come from the other crash at San Agustin? Perhaps, although it is more likely that they would have been taken to the much nearer air base at Albuquerque.

Several details of Dennis's account simply fail to gel. For example, he said that, of the two MPs who threatened him, one was white and the other black. As already noted, US Army Air Force units, in common with all the US armed forces, were racially segregated in 1947. He also claimed that his visit to the hospital took place on 7 July – the day before the press release – as he specifically remembers seeing the evening Roswell paper's headline story *after* his lunch with Naomi Selff. As Marcel and Cavitt were then either (depending on which version of the story is correct) on their way to the Foster ranch or actually examining the crash site, how could the bodies already be at the base?

And what of the nurse, Naomi Selff? Clearly, as Dennis had not seen the bodies himself, she would be a vital witness, if she was still alive and willing to talk. But here the story began to take some very strange twists and turns.

In 1991, Glenn Dennis told Philip Klass that shortly after the event the nurse had been transferred to a base in Britain, and that a letter he wrote to her was returned marked "Deceased". He later learned that she had been killed in an air crash.[41] (Investigations were unable to find any military air accidents in 1947 or 48 in which a nurse was a casualty.) But later the same year Dennis told another researcher that she had died just three years before (i.e. in 1988), and that she had been in an order of nuns since leaving the Army.

Adding to the mystery, the five nurses stationed at Roswell AAF in 1947 appeared in the base yearbook – and none were named Naomi Selff. Then Donald Schmitt claimed that, not only had his extensive searches of military archives failed to find any trace of a Naomi Selff, but that all records relating to the five known nurses had disappeared. Schmitt said that he had been in touch with several organizations, including the Department of Defense and the

National Archives in Washington, but there was no record of any of the nurses ever having existed. This introduced a very sinister note into the investigation: had they all been silenced, and their very existence erased, because of what they had seen?

One person who was impressed by this claim, and its implications, was *Omni* journalist Paul McCarthy. In 1995 he suggested to the magazine's editors that he write an article about the "nurse"s story'.[42] Instead, the editors suggested, McCarthy should try to replicate Schmitt's attempts to locate the Roswell nurses. McCarthy was pessimistic: Schmitt had apparently been trying for five years without success, and, as he lived in Hawaii, McCarthy was restricted to using the phone and mail. Yet, to his surprise, he was able to locate the records for all five nurses in the National Personnel Records Center in St Louis *in just three days*. Not only that, one of the nurses, Rosemary McManus, was still alive, and McCarthy was able to track her down and interview her. She knew nothing of anything out of the ordinary happening in Roswell AAF's hospital during her tour of duty there, nor did she remember a Naomi Selff. She did, however, remember having recently been asked the same questions by two UFO researchers . . .

When McCarthy faced Schmitt with this, he admitted that he *had* found the nurses' records in the St Louis archives, and had himself located and interviewed Rosemary McManus. His explanation for omitting this from his and Randle's book was that he suspected that McManus might actually be the nurse described by Dennis and hidden behind the "Naomi Selff" pseudonym, and did not want to cause trouble for her. McCarthy, among many others, found Schmitt's explanation deeply unconvincing. (This episode, and other damaging revelations about, in Kal K. Korff's words, Schmitt's "flawed research and dishonesty",[43] led his co-researcher Kevin Randle to end their association, although Randle continues to support Glenn Dennis's testimony.)

In 1993 another witness came forward. This was Jim Ragsdale, who claimed that, while camping out in the desert one night in early July 1947, he and his girlfriend Trudy Truelove (since deceased) had seen a UFO crash. He was unsure of the exact date, but it was during the holiday weekend of 4-6 July. A large bright object – Ragsdale described it as like that from an arc welder – flashed overhead and plummeted to the ground a mile or so away. At first light they went in search of the object, and were greeted by a startling scene.

Some kind of craft, like an aircraft with narrow wings, was sticking out of a small cliff at an angle. Pieces of debris were scattered around. According to Ragsdale, "You could take that stuff and wad it up and it would straighten itself out". This matches Jesse Marcel's description of the debris from the Foster ranch. However, other fragments found by Ragsdale had exactly the opposite quality – "You could bend it in any form, and it would stay. It wouldn't straighten out."[44]

But eclipsing even that were the bodies. In a notarized affidavit sworn at the request of Donald Schmitt, Ragsdale wrote: ". . . I and my companion came upon a ravine near a bluff that was covered with pieces of unusual wreckage, remains of a damaged craft and a number of smaller bodied beings *outside* the craft [his emphasis]. While observing the scene, I and my companion watched as a military convoy arrived and secured the scene. As a result of the convoy's appearance we quickly fled the area."[45]

In an interview with Schmitt, Ragsdale elaborated: ". . . it was either dummies or bodies or something laying there. They looked like bodies. They weren't very long . . . over four or five feet [1.2 or 1.5 m] long at the most. We didn't see their faces or nothing like that but we had just got to the site and heard the army, the sirens, all coming and we got into a damned jeep to take off."[46]

He described the military convoy as consisting of two or three trucks, weapons carriers, a pick-up, a "wrecker" and a car full of Military Policemen. Ragsdale and his girlfriend watched under cover from a distance as the team cleared the site of all traces of the crash, even raking the ground. He says that it didn't take them very long. However, he said that the area was still cordoned off by MPs a month later.

Ragsdale claimed to have kept some of the pieces of debris and shown them to friends – all of whom, unfortunately, had since died. Some of it was lost when his car was stolen five years later, and the remainder in a burglary in 1985.

There is some uncertainty about the date of Ragsdale's experience. Although he placed it on the Independence Day holiday weekend, therefore on the night of the 4, 5 or 6 July, he also said that it was "two or three weeks"[47] before the press furore over the Roswell flying disk. And the site identified by Ragsdale was neither that found by Mac Brazel or the far-away Plains of San Agustin, but some 35 miles (56 km) north of Roswell.

Some UFOlogists were quick to accept Ragsdale's story, as it helped explain the anomaly of the bodiless crash at the Foster ranch and the corpses allegedly seen at Roswell by Glenn Dennis's nurse. In this scenario, an explosion had occurred aboard the UFO over the Foster ranch, scattering wreckage over the ground below, but the craft itself had crashed some 35 miles (56 km) further on.

However, two years later, Ragsdale swore another affidavit that placed the crash site some 50 miles (80 km) west of Roswell, not 35 miles (56 km) north. Ragsdale also changed several other aspects of his story, which became considerably more dramatic: he and Trudy had gone immediately to the scene of the crash, in the darkness; they *had* gone close to the craft, in fact had ventured inside, which is where they discovered the bodies; they *had* got a good look at the aliens – Ragsdale had even tried to remove a helmet from one of them; their most prominent features were huge, black eyes; he had removed two sackfuls of debris, and so on.[48]

The reason for the change of location seems all too clear: the Roswell International UFO Museum and Research Center, (one of three rival museums in the town dedicated to the crash, which was co-founded by Walter Haut and Glenn Dennis) had tried to buy the first site in order to charge admission, from which Ragsdale would receive 25 per cent of the profits. But when the rancher refused to sell, the location was shifted to another site which was on public land.[49] The museum was also to sell the book and video of *The Jim Ragsdale Story*, from which he would take a share. (Ragsdale's motives may not be as self-serving as they first appear; at that time he was diagnosed with terminal cancer, and the money was to go into a trust fund for his grandchildren.)

Researchers such as Kevin Randle have accused Ragsdale of changing his story for money. However, Randle still maintains that his first version is reliable. Assuming this is the case, how does it add to the unscrambling of the Roswell Incident?

Is this a third crash site, or was what Ragsdale saw the same thing that Barney Barnett reported to his friends? Was the locating of

Barnett's experience on the Plains of San Agustin a mistake by those who had passed the story on – had he really been much nearer to Roswell? However, both Ragsdale and Barnett claimed to have been at the site when the military vehicles arrived – yet neither mentioned the other. Nor did Ragsdale report seeing the archaeological team that Barnett saw. As Ragsdale claims to have witnessed the whole clear-up of the site, from the arrival of the convoy to the raking of the ground, it is odd that he makes no mention of any other civilians being present. But despite these discrepancies, some Roswell researchers have inextricably linked the Barnett and Ragsdale stories.

After the claims of Glenn Dennis and Jim Ragsdale, Roswell researchers were able to find several more witnesses who confirmed the recovery of alien bodies. Sergeant Melvin Brown claimed that he drove the bodies from the crash site to Roswell AAF.[50] Although ordered not to look under the tarpaulin on the back of his truck, he did sneak a peek and saw small bodies with leathery, orange-yellow skin. (Some books omit Brown's description of the skin colour, since, of course, all aliens are grey . . .)

Like so many "witnesses", Brown was dead by the time his story came to light, and his story actually comes from a member of his family, in this case his daughter Beverly Bean. She has given three different versions of how she first heard of it. In the first, told to British researcher Timothy Good in 1988,[51] she said that her father had been reading an article in the London *Daily Mirror* about Roswell in 1970, when he suddenly exclaimed, "I was there!", and proceeded to tell her the story. In the second, to Stanton Friedman in 1989, she said that Brown had frequently told the story to the family when they were children.[52] In the third, to Randle and Schmitt a year later, the revelation had come as he was watching the Apollo moon landing in 1969.[53] Significantly, Brown's widow and other daughter have refused to confirm the story. Roswell researchers may put this down to their fear that the authorities can still do some posthumous harm to Brown – but why should they maintain the pretence when he has already been "outed" (without official repercussions) by Beverly?

Another question mark about Brown's involvement was raised by the discovery that, although he was certainly based at Roswell AAF in 1947, he was actually one of the base's *cooks.*[54] Would a member of the catering staff really have been chosen for such a sensitive mission?

Another claim that has been heavily promoted by Roswell researchers dates from 1993. Frankie Rowe related that her father, one of Roswell's city firemen, had been called out to the crash site and had seen two dead and one living alien, which were taken away to the air base. A few days later, one of the state troopers who had been at the scene came to the fire station while she was there, bringing pieces of the wreckage, which "spread out like liquid or quicksilver" when he dropped it on the table. Frankie Rowe also said that Air Force officers had come to her house and threatened her family with imprisonment, even telling the little girl that she would be taken out into the desert and killed if she ever spoke about it.[55]

The problem here is that none of the other alleged witnesses to the event say that fire trucks were among the convoy that arrived, or that the wreck was burning or even smouldering. And Roswell AAF – as would be expected for a major air base – had its own fire crews. Why call out the civilian Fire Department?

Dr John Kromschroeder, a friend of the late "Pappy" Henderson – the Second World War hero who claimed to have flown some of the wreckage to Wright Field – said that in 1978, during a fishing trip, Henderson had confided in him about seeing alien bodies. (Henderson's widow, whom he told about flying the wreckage out of Roswell, knew nothing about this.) Henderson did not give Kromschroeder any details about when or where he had seen them, or any description other than that they were "little guys".[56] But Kromshroeder also claims that Henderson showed him a piece of the wreckage. However, Henderson's friends attest that the pilot had a great reputation as a practical joker, and that he had a fragment from a V-2 rocket, a souvenir from his wartime exploits, that in later years he would present to people as a piece of a crashed flying saucer.[57]

One of the names that was top of the list for Roswell researchers to interview was Major Edwin Easley, Roswell AAF's Provost Marshall (head of Military Police) in July 1947. The Provost Marshall would be bound to have been involved in any operation to secure the crash site or sites, and his presence at the Foster ranch on 8 July 1947 was mentioned by CIC officer Lewis Rickett.

Randle and Schmitt tracked Easley down, but he refused to talk about the incident. Intriguingly, he would not deny that it had happened, but deflected all questions by saying that the episode was still classified.[58]

Easley died in 1990. Since then Kevin Randle has asserted that, on his deathbed, he did admit his part in the operation, that there had been bodies of "creatures", and that he had seen them. Unfortunately, Randle himself has refused to say how he knows this, or given any source.[59]

One of the most sensational – some would say literally incredible – sources of information on the Roswell Incident is a character named Frank J. Kaufman, who first made himself known to Randle and Schmitt in 1990. Kaufman appears in their writings under his own name and two different pseudonyms, Steve MacKenzie and Joseph Osborn, creating the impression of three witnesses backing up each others' claims, whereas it actually all comes from one and the same person.

When first interviewed by Schmitt and Randle in January 1990, Kaufman described a role that was, in their words, "on the outside for most of it".[60] He said that a friend of his, an officer named Robert Thomas (who has never been identified) had flown into Roswell from Washington in order to look for the downed craft. A crate was later brought to one of the hangars and kept under heavy guard before being flown out.

By 1992, when he was interviewed again by Randle and Schmitt, Kaufman's role had expanded enormously. He was now the only witness who claimed to have seen the whole thing through from beginning to end, playing a major role in every sequence from first detection of the UFO to the retrieval of the alien bodies – then going on to join the super-secret team, which he named the "Unholy 13" (a variation of the Majestic 12 theme) set up to oversee the study of the UFO and aliens, and to safeguard the secret of their existence.

According to Kaufman,[61] the military knew that something strange was happening in New Mexico before the crash, as a UFO had been repeatedly tracked on radar over the White Sands Missile Range from 1 July. Kaufman was stationed at Roswell AAF, but on 2 July received a personal telephone call from Brigadier General Martin F. Scanlon of the Air Defense Command in Washington, ordering him to White Sands to monitor the UFO's movements on radar and to personally keep Scanlon informed of any developments. Why White Sands' own radar operators were judged not to be up to the task is left unexplained. Questions are also raised by the realization that, while General Scanlon *was* one of the top brass at the Air Defense Command in 1947, he was actually the USAAF's head of public relations.

Kaufman says that he was on duty in the radar room for 24 hours straight. According to Randle and Schmitt, he even rigged up a system of mirrors so that he could keep the screen in view when he went to the bathroom. (Later, Randle would claim, bizarrely, that he had made a mistake, and confused Kaufman's testimony with a science fiction movie he had seen, which is where the "mirror" story came from.)[62] He was relieved from this duty on 3 July, and ordered to return to Roswell AAF to meet a special team that were flying in from Washington. Clearly, the authorities were expecting something to happen.

Kaufman relates that, at around 11.20 p.m. on 4 July, the object's signal on the radar screen began to pulsate, then grew into a "sunburst" before disappearing. This, claimed Kaufman, showed that the UFO had been struck by lightning – which reveals that, for someone apparently judged to be more reliable than White Sands' radar personnel, he has a very slender grasp of the workings and capabilities of radar. The object had disappeared somewhere north of Roswell, and a search was immediately started for the wreck.

Kaufman joined the convoy carrying MPs and the special Washington team that went out in search of the UFO. When it was located, the area was sealed off by a ring of MPs, and only a team of nine with the very highest security clearances – which included Kaufman – were allowed to approach the crashed vessel. He describes it as triangular, 25-30 feet (7.5–9 m) long and about half that in width. There were five bodies, three inside and two outside the craft, all dead. They were around five feet (1.5 m) tall, slender, with disproportionately large heads. Their eyes, too, were larger than a human's, and – unlike the classic "grey" – had pupils.

The corpses were zipped in body bags, and the craft and wreckage cleared from the site. All the alien material was soon on planes out of Roswell, destination Wright Field with a stop-over at Andrews AFB near Washington so that the bodies could be shown to Dwight Eisenhower, then Army Chief of Staff and later President. In subsequent interviews Kaufman indulged in more name-dropping, saying that Dr Werner von Braun, Charles Lindbergh and President Harry S. Truman also saw the bodies. Finally, Kaufman claims that he was selected as a member of the "Unholy 13", and continues to be part of that super-secret project to this day.

Kaufman is something of an enigma. He has admitted that he actually left military service in 1945 – but then, with someone so highly trusted by his government as to be one of the select few assigned to the Unholy 13, perhaps that was just a cover to allow his involvement in more clandestine work.

Leaving aside Kaufman's more exotic claims, his story does fit with the main elements of the others, particularly Jim Ragsdale's, and his description of the bodies does correspond to the sketches made by Glenn Dennis – but then Kaufman only revealed these details after they had appeared in Randle and Schmitt's first book. And while Kaufman is precise as to time and date (fitting the chronology in Randle and Schmitt's book) he is vague as to place, except that it was somewhere north of Roswell.

But the big question is, if Kaufman really is part of the Unholy 13, whose purpose was to conceal the Roswell crash, why did he give all this information to Randle and Schmitt?

In much the same vein are the claims by retired Army intelligence officer Colonel Philip J. Corso in his 1997 book *The Day after Roswell* (co-written with William Birnes). Corso's involvement did not begin until 1961, when, while working for Army Research and Development in the Pentagon, he was placed in charge of the secret programmes to exploit the technology recovered from Roswell. According to Corso, such developments as lasers, transistors, microchips, Velcro and microwave ovens all owe their origins to "back-engineering" from equipment found aboard the Roswell UFO. Eventually Corso's teams were able to create weapons that finally allowed Earth to stand up to the aliens, negotiating an interplanetary truce that also brought about the end of the Cold

War. Corso literally (but with all due modesty) proclaimed himself to be the hero who had saved the world.

It has been shown that, leaving aside the extraordinary nature of Corso's claims, his book is littered with factual errors and anachronisms, such as the episode in 1961 in which, after being harrassed by the CIA, he walked into the agency's headquarters in Langley, Virginia, to brow-beat its Director of Covert Operations Frank Wisner into backing off.[63] Unfortunately, this confrontation was with a man who had left that post three years earlier and took place in the CIA headquarters a year before it was built.

But when it comes to claims that the US Government established a secret project to analyze the Roswell wreckage, one of the most impressive, yet perplexing, sources is Brigadier General Arthur E. Exon. Not only is he the highest-ranking figure to have confirmed the reality of the UFO crash – his illustrious career includes a period as commanding officer of Wright-Patterson AFB (the former Wright Field) in the 1960s – but he had also, as a Lieutenant Colonel, served at Wright Field in July 1947.

In 1990 Exon told Randle and Schmitt an extraordinary story from his earlier posting.[64] He confirmed that wreckage and bodies from the Roswell crash had been brought to the base. Exon also confirmed that there were two impact sites: the Foster ranch, where only debris had been found, and the main crash site some miles to the south west where the bodies were recovered. Indeed, he claimed that four months later he had flown over the area and seen both sites, and that the signs of the impacts were still clearly visible.

Exon said that the specialists at Wright Field had performed a number of chemical and mechanical tests on the wreckage, but had been unable to identify it. He said that parts of it "were very thin but awfully strong and couldn't be dented with heavy hammers." He had little information about the bodies, other than that they were found outside the craft and in a relatively intact condition. But Exon was emphatic and unequivocal about the event: "Roswell was the recovery of a craft from space."[65]

Exon went on to say that the weather balloon story *had* been concocted as a cover up, and that there were no classified balloon projects at that time. (This is incorrect – see below.) And as a result, Exon confirmed, a top secret committee – as in Kaufman's "Unholy

13" and the better-known "Majestic 12" – had been established to oversee the scientific study of the craft and aliens and to control access to the data.

Although he never claimed to have seen the wreckage or bodies himself, Brigadier General Exon's credentials make him the star witness in the Roswell case. He was careful to point out that everything he said was based on what he had heard on the rumour mill at the time and over subsequent years – even writing to Randle and Schmitt to complain that they had not made this clear in their book – but even confirmation that such rumours were circulating within the USAF in 1947 is a major step forward, since sceptics have maintained that there was not even a whisper about alien bodies until the end of the 1980s. And Exon's status surely rules him out as a hoaxer or practical joker – especially as he was effectively accusing the service and government that he had served so faithfully of conspiracy. He is one of the major problems for sceptics.

Critics have charged Randle and Schmitt with deliberately misrepresenting Exon's testimony as first-hand instead of – as Exon himself made clear – as based on the rumours he had heard. But this is missing the point. That someone of Exon's standing should give credence to such apparently off-the-wall stories is significant in itself. Added to which his career is such that any rumours that came his way would have come from very reliable sources, especially when they include such specific details of the testing of the material, the recovery of alien bodies, and the setting up of the secret oversight group. Would anyone pass made-up tales like this on to someone of his rank?

But this highlights the major paradox about Exon's testimony: why should he pass on rumours at all, since he was in a position to *know* what really happened at Roswell? He later served as the commanding officer of Wright-Patterson AFB, where, according to those "rumours", the UFO and bodies were taken. Surely he would have access to the records? Surely only someone with clearance for such sensitive material would have been given the job?

Added to this is the fact that, as we have seen, Wright-Patterson AFB has long been the focus of claims that it housed preserved alien bodies – bodies that many researchers believe came from Roswell. But if that is the case, they must have been there when Exon was in charge of the base. Why, then, did he apparently not know about them? There is a Catch-22 situation for Roswell researchers here: if Exon's testimony is accepted, then it effectively disproves the claims of alien bodies at Wright-Patterson. But if his testimony is rejected, so is his support for the Roswell crash.

Exon's background makes him the most reliable and credible informant on the Roswell affair – and yet the same background should make him one of the *last* people to speak out about it. He should be part of the cover-up, not collaborating in exposing it; he should be denying, not confirming . . .

IMPLICATIONS

One result of all the new testimony that emerged in the late 1980s and early 1990s was that the discovery of the crash by the military had to be shifted back several days before the "official" breaking of the story on 8 July 1947. Glenn Dennis said that the bodies were being autopsied the day before. Frank Kaufman said that the main crash site was discovered on the night of 4–5 July. Jim Ragsdale also placed the recovery of the bodies during the holiday weekend of 4–6 July. The only way that these accounts can be reconciled with Mac Brazel's discovery is by assuming that the main crash site – with the body of the craft and corpses – was found a day or two before Brazel turned up in Roswell with his pieces of wreckage, and that personnel from Roswell AAF had already located and secured that site, warning off several civilians in the process. Therefore the base must already have been aware of the non-terrestrial nature of the event, and the cover-up was already under way.

However, this conflicts with the original evidence of a crash – Walter Haut's press release of 8 July, which can only otherwise be explained by initial confusion and mystification about the true nature of the crashed object. If the main wreck had been discovered days before, we have to believe that when Brazel turned up in town, the USAAF went to pieces and accidentally told the world the very thing it was trying to hide.

If the base was on high alert because of the find, surely the first people to have been briefed – if not on what had actually happened, but on what *not* to tell the public – would have been the Public

Information Officers. And clearly Jesse Marcel knew nothing of the discovery of a UFO and bodies when he was despatched to the Foster ranch, otherwise he would have said so when he blew the whistle on the story thirty years later. Up to his death, Marcel never said anything about bodies or a second crash site.

The original case for the Roswell cover-up – which attracted the interest of UFOlogists in the first place – was based on Mac Brazel's discovery on the Foster ranch and Walter Haut's press release. However, in order to reconcile the conflicting claims of alleged eyewitnesses that have come forward since 1980, the original event has had to be rewritten to the point that those original pieces of evidence have not only been marginalized, but actually rejected. For example, Brazel told the press that he had found the debris on 14 June and held on to it for three weeks until his next trip into town. In order to square this with the witnesses to the main crash site on or around 4 July, most pro-Roswell researchers have rejected Brazel's own testimony on the grounds that he was either bullied or paid to make this statement by the military as part of the cover-up. And no explanation has been offered for the press release.

To further add to the confusion, the Majestic 12 documents – accepted as genuine by some, though by no means all, Roswell researchers – state that the crashed UFO was recovered on 7 July, as in the original version of the story.

THE MOGUL BALLOON

The Roswell Incident, as it stands today, is effectively a combination of two stories. The first is of the discovery of strange material – subsequently explained as a weather balloon – from the Foster ranch. The second is of the recovery of a more intact craft and bodies from a second crash site. The two are only linked by what are, to say the least, highly questionable sources, such as Glenn Dennis's mystery nurse and the testimony of Frank Kaufman.

Looking at the first story – and assuming for the moment that it is independent of the second – much of the aura of strangeness could be explained on the grounds that what Mac Brazel found, and what Jesse Marcel collected, was not a weather balloon, and that this was indeed a cover story – but one designed to conceal not a UFO crash, but some highly classified but nevertheless terrestrial device.

For example, if a V-2 rocket had veered off course and crashed outside the secure zone of the White Sands Missile Range, just such an operation to secure the site and prevent the news reaching the public would have been set in motion. However, from the description of the debris, and records of V-2 test firings, that particular explanation can be ruled out.

In the 1960s, John Keel suggested that what Mac Brazel found was the remains of a Japanese Balloon Bomb from the Second World War. These peculiar devices were used by the Japanese in a somewhat unsuccessful attempt to spread random terror in the US in the closing months of the Second World War. They were rice-paper balloons carrying high explosives, some 9,000 of which were floated towards America. Only one of these claimed any victims – six members of a group of picnickers in Oregon were the only casualties on the US mainland during the entire war. Keel theorized that one of these balloons had been kept aloft by freak air currents for two years, finally descending on the Foster ranch where it had detonated. However, there is no reason for the authorities to have wanted, or needed, to conceal this from the public in 1947. In fact, they are more likely to have issued warnings for civilians to watch out, and not to touch, any more that were found.

There is, however, one candidate that offers a very neat explanation both for Brazel's discovery and for the ensuing cover-up. This is the "Project MOGUL" theory, which was the conclusion of an official USAF investigation into the Roswell Incident in 1994. It was, however, first arrived at by UFOlogist Robert Todd, who pieced the story together by painstaking detective work and judicious use of the Freedom of Information Act.

MOGUL was a sub-project of one of the most highly classified military projects of the time, known as AFOAT, whose mission was to monitor the Soviet Union for signs that it had developed its own atomic bomb. (AFOAT was the project to which Jesse Marcel was assigned in Washington a year after the Roswell Incident.) The importance of AFOAT is demonstrated by the fact that it was given a security rating equal to that of the wartime Manhattan Project, Top Secret A-1. It was not declassified until 1973.

The purpose of MOGUL was to develop ways of detecting Soviet nuclear tests using instruments carried by very high-altitude balloons that could be flown over Soviet airspace. It was assigned to a team from New York University who were given the task of devising a method of keeping the instruments at a constant high altitude. Although based in New Jersey, it also carried out test flights from Alamagordo in New Mexico – 90 miles (144 km) south of the Foster ranch.

In July 1947, the team at Alamagordo were experimenting with a 600-foot (180 m) long chain made up of more than twenty linked weather balloons to carry the detection instruments, along with other equipment to keep the balloons at a constant altitude and to transmit radio signals back to base. The balloons were fitted with radar targets, and were similar to Rawin balloons, although various new designs of both balloon and targets were tried out. In assessing MOGUL as a candidate for the Roswell object, it is important to understand the protocols and procedures under which the MOGUL team were operating. Although MOGUL was classified at the highest level, the fact that it used adapted weather balloons was not the sensitive part – it was the *purpose* of the project and the equipment that the balloons carried that fell within the top-security rating.

Realising that this could offer an explanation for the Roswell Incident, Robert Todd managed to tracked down the MOGUL project engineer, Charles B. Moore, and sent him a copy of Brazel's July 1947 description of the debris. Moore confirmed that this matched a MOGUL balloon – in particular, the scotch tape used to construct the radar targets was, bizarrely, supplied by a children's toy manufacturer, and bore flowers and other symbols,[66] accounting for the "little flowers" seen by Brazel and the "hieroglyphs" described by Marcel. Brazel's daughter Bessie also said that there were "pastel flowers" on the tape. While supporters of the UFO theory argue that the whole of Brazel's description was fed to him by the military, this is surely an odd and unnecessary detail – especially as the use of flower-bedecked tape on the MOGUL radar targets was not known until the 1990s.

Moreover, Charles Moore also found from his records that a MOGUL balloon chain had been launched on 4 June 1947 and tracked to within 17 miles (27 km) of the Foster ranch – after which it disappeared. It was never recovered. To compound the irony, Moore and his team left Alamagordo to return to New Jersey on the morning of 8 July – just hours before news of the "flying disc" discovery was released.

The MOGUL explanation fits many of the anomalies in the 1947 accounts:

- While describing the material in mundane terms – as rubber and scotch tape – Mac Brazel also stated that it was unlike the weather balloons he had previously found on his ranch.
- The debris was apparently not readily identifiable to Marcel, which led to both the press release and the passing of the news up the chain of command. Clearly, by informing Fort Worth, someone at Roswell AAF – either Marcel or his commanding officer – must have been aware that what they had was *not* a conventional weather balloon.
- Brazel's son said that his father had been made to swear an oath to the USAAF not to speak about the incident, and other witnesses have said that the rancher was accompanied by a military escort in the ensuing days. If, by the end of the day, it had been realised that what he had found was part of a top-secret project whose very existence was classified, this makes sense. It was not the discovery of the balloon that was the cause for concern, but anything else that Brazel might have found or said that could potentially give away the purpose of the device or the existence of the MOGUL project. The treatment of Brazel does seem a little heavy-handed all the same, although it must be remembered that this was an era of extreme tension and suspicion, as America and Russia positioned themselves for what was clearly going to be a lengthy power struggle.
- On the MOGUL hypothesis, the "flying disc" press release was designed not only to offer an explanation for the military's sudden interest in the Foster ranch, but also to discourage members of the press and public from going out to see it for themselves. The wreckage from a train of MOGUL balloons would be extensive, and more than Marcel and Cavitt could have carried. We have seen that military personnel were

engaged in clearing the site at least a day later. But the press release did not speak of a crash and wreckage – it said that a flying disc had *landed* and been recovered, by implication intact, so that there was nothing left to see at the site. It did not give any details of the location or the name of the rancher – in fact it employed a little misdirection by paying tribute to the co-operation of the Sheriff's office of Chavez County. Although the Roswell office, to which Brazel reported his find, was in Chavez County, the Foster ranch itself was in the neighbouring Lincoln County.

Roswell AAF had taken advantage of what was then the two-week-old flying saucer craze in order to throw local people off the scent – not appreciating the interest that it would attract from around the world. As it became obvious within hours that the cover story had backfired, a hasty press conference was arranged to try to kill it off.

From the testimony of Sheridan Cavitt and Lewis Rickett, the recovery operation seems to have been under the control of the Counter-Intelligence Corps, rather than Marcel's USAAF Intelligence unit. This is what we would expect with the MOGUL equipment, as part of CIC's job was to prevent information about secret projects from leaking out.

- There is an obvious contradiction between the Fort Worth Press Conference, where the debris was shown and dismissed as an uninteresting weather balloon, and the evidence that it was flown on to Andrews AFB in Washington and/or Wright Field. This has led many to conclude that the material displayed in General Ramey's office was a substitute for the original debris. On the MOGUL hypothesis, both can be correct. The harmless balloon and radar targets could be displayed, but any of the instruments relating to MOGUL would be secretly sent on to Andrews or Wright Field. This scenario would also explain the additional wreckage that was flown direct to Wright Field in the ensuing days.

As we have seen, Brigadier General Thomas DuBose – in 1947 General Ramey's Chief of Staff – gave what appeared to be conflicting testimony to Roswell researchers. He told Jaime Shandera that the debris from Roswell was the same that was displayed in Ramey's office, scornfully dismissing any suggesting of a substitution. Yet he told Randle and Schmitt that the weather balloon story was a cover, designed to put the press and public off the trail. On the MOGUL hypothesis, both statements – at first seemingly contradictory – can be reconciled.

DuBose stated that his orders to arrange the transfer of the material came from General Clements McMullen at USAAF headquarters in Washington. Sceptics have cast doubt on DuBose, asking how Washington should have been aware that the wreckage was worthy of interest even before it had arrived in Fort Worth. However, the CIC unit at Roswell reported direct to Washington, bypassing the base commanders. If Cavitt had recognized the debris as being more than just a weather balloon, he would have reported this to Washington and *not* informed either Marcel or Colonel Blanchard.

In the MOGUL scenario, the "weather balloon" story *was* a cover-up, but not of an alien spaceship. And there is more evidence to support this hypothesis.

On 10 July 1947 – the day after the photographs of the debris in General Ramey's office had been published – the *Alamagordo News* carried a front page photograph of a complete MOGUL balloon train.[67] The accompanying story said that this was new kind of system being used for weather monitoring. There is only one possible reason for this sudden splashing of a key component of a Top Secret project on the front page. There was a possibility that Russian agents monitoring the US media might have been intrigued by the Roswell story and suspected that something more lay behind it. It was also possible that Communist agents may have been able to find out locally that there was more material than had been admitted, and perhaps even to piece together a description of a complete MOGUL balloon train and begin to work out what it was really for. There was also a risk that the USAAF team had missed some of the Foster ranch debris. To cover such eventualities, the *Alamagordo News* story provided a satisfactory – but entirely false – explanation for the debris, passing it off as new way of weather monitoring. Now that the MOGUL team had left Alamagordo it was perfectly safe to do so. And not only had that team been operating under the cover story that they were part of a meteorological project, one of its members confirmed in 1994 that this *had* been the purpose of the 10 July press report.

The question comes back to Major Marcel's role. Did he know – or quickly learn about – the MOGUL explanation, or did he genuinely believe that what he had found was "nothing of this world"? And what of his description of the unusual properties of the material, which are hard to reconcile even with the MOGUL explanation?

Unfortunately for the Roswell mystery, research by other UFOlogists has cast some doubt on Marcel's overall credibility.

Marcel told Friedman and Moore that he had qualified as a pilot in 1928, and that he had been a fighter pilot in the Pacific during the Second World War, during which time he was credited with five "kills". His service record, however, shows that he never even qualified as a pilot, still less took part in aerial combats. He also claimed that he had a BA in physics from George Washington University. The University found no record that Marcel had even been a student there.[68]

A year after the Roswell Incident, Marcel was transferred to the Strategic Air Command headquarters at Andrews AFB near Washington D.C., where he was assigned to the AFOAT project. Marcel claimed that, when the first Soviet bomb was exploded in August 1949 – a shocking event that meant that the United States had lost its lead in the atomic arms race – he had personally written the report that a sombre President Truman read to the American people on the radio. In fact, Truman did not broadcast to the nation; the news was broken by a White House press release. While the team for which Marcel worked would certainly have provided the information for this, it does appear that Marcel was prone to inflate his personal role in such events.

However, in 1947, Marcel does seem to have been genuinely mystified and excited by what he had found. But could he really have thought that a bundle of tinfoil, rubber and sticks, no matter how unfamiliar, came from a "flying disc"? From today's standpoint the suggestion seems absurd. Would anyone think that a spacecraft would be constructed from such flimsy material? But it must be remembered that, in July 1947, "flying saucer" and "flying disc" were not yet synonymous with "alien spaceship". They were part of a very new phenomenon, barely two weeks old, which had not yet been associated with outer space. The prevailing view was that they were some secret terrestrial device – whether Russian or American – and therefore could have been made of anything.

At Fort Worth, Marcel – who genuinely seems to have thought he had made an amazing discovery, and appears to have inspired the "flying disc" press release – ended up with egg on his face, in front of the head of the Eighth Army Air Force. It may well be that Marcel was being used as a "fall guy" by that stage, to protect the MOGUL project, which would then have been classified far above his level. The world was told that he, an experienced intelligence officer in one of the most important military bases in existence, had failed to identify a simple weather balloon. Irving Newton, the AAF meteorologist called in to identify the wreckage, states that, even in Ramey's office, Marcel was desperately trying to convince others that there really was something odd about what he had found . . . One can understand why, and sympathize. Others who were stationed with him later in Washington attest to the "ribbing" that he continued to get about the flying saucer story.

So, in 1978, did Marcel see the opportunity to redeem himself, getting one over on his old ridiculers by justifying his actions in 1947?

It has to be admitted that the MOGUL theory fits the original Roswell story perfectly. However, critics complained that it totally ignores all the other testimony relating to a second crash site and the alien bodies. This is true – but, as we have seen, those claims are only tenuously linked with the "Foster ranch" story, often by "witnesses" whose accounts are highly suspect.

The claims of bodies are based on a nucleus of first- or second-hand witnesses – such as Glenn Dennis's mystery nurse or the enigmatic Frank Kaufman – who gave specific (but conflicting) details of time, place and the sequence of events. Surrounding these is a number of vaguer and less evidential reports, such as fossil-hunter Bertram Schultz's memory of having been told about a flying saucer crash by Curry Holden's archaeological team. However, major questions hang over all of the first group of witnesses, and the others not only give very little in the way of detail, but also are only associated with Roswell by implication.

THE OFFICIAL INVESTIGATIONS

Although it was a "pro-UFO" researcher, Robert Todd, who first put the MOGUL explanation forward in 1993, it was not until the USAF published *The Roswell Report: Fact vs. Fiction in the New Mexico Desert* the following year that the theory received widespread attention. Those who believed in the UFO crash, of course, were convinced that this investigation was yet another part of the long-running cover-up.

The USAF's report was not the only official investigation into the Roswell Incident that year. The Air Force's sudden interest had, in fact, been inspired by the news that Congress' General Accounting Office (GAO) – the "watch dog" that looks into allegations of irregular and unconstitutional activity by government agencies – was to carry out its own investigation into the cover-up claims.

The GAO investigation came about as the result of lobbying by UFOlogist Karl Pflock (an ex-CIA agent who was later employed by Congress and the Pentagon). He persuaded Congressman Steven Schiff, the representative for Albuquerque, to press for such an enquiry, in order to find out whether the US Government was guilty of dishonesty, and even intimidation of innocent civilians.

The GAO's task was not to pronounce on what really lay behind the incident, or to offer any explanations for the crash. However, the USAF decided that a parallel investigation to try to offer such an explanation was also necessary, eventually arriving at the MOGUL conclusion. The investigation was headed by Colonel Richard Weaver – who was part of the oversight committee for USAF "black budget" projects, and consequently held the highest security clearances. Inevitably, any scientific studies of the Roswell craft would have ended up with such a black budget project.

The GAO team examined the written records to try to establish what had happened around the time of the incident – what orders had been issued, what messages had passed between Roswell AAF and its headquarters in Fort Worth and Washington, and so on – in order to determine whether there were any signs of unusual activity related to the crash. Obviously, if such a historic event as the capture of an alien craft together with the bodies of its occupants had happened – especially in a part of the country packed with top-secret laboratories, test sites and military installations – it would have generated a frantic exchange of signals (as well as a great deal of paperwork). Alarm bells would have started to ring throughout the United States, if not beyond. Urgent messages would have flashed between Roswell AAF, the Eighth Army Air Force headquarters in Fort Worth, and Washington, as Roswell's superiors demanded more information and ordered an immediate clampdown. Bases throughout the country would have been put on red alert and told to watch for similar craft. After all, it might have been the prelude to an alien invasion . . .

Once again, although this reasoning is credited to an official investigation, it had already been arrived at by UFO researchers. For example, Jacques Vallée had reached the same conclusions, and had examined the records for *other* military bases across the USA for signs of a sudden red alert in July 1947. He found none.

The GAO examined the records for Roswell AAF, Fort Worth AAF and the Pentagon around the time of the incident. It also looked at the minutes of meetings of the National Security Council, which would obviously have been at the hub of assessing this potential new threat and deciding what to do about it. The GAO report, delivered in July 1995, concluded that it had found no evidence to support either an unusual flurry of activity around Roswell AAF in July 1947, or of orders for a large-scale cover-up. Perhaps that was not surprising, if it is assumed that the GAO, as an official government agency, was acting as *part* of the ongoing cover-up, or that records that would have shown the existence of such a cover-up had been withheld by the agencies that *were* responsible. However, one aspect of the report did raise suspicions.

The GAO team found that all Roswell AAF's administrative records for the years 1945 to 1949 had been destroyed. In particular, outgoing teletype messages – which presumably would have contained instructions concerning the transfer of the wreckage, and of the steps being taken to cover up the incident – had been lost. It was this aspect of the report that attracted most press coverage, since it was consistent with the idea of a cover-up.

Like so much of the evidence in the Roswell affair, the GAO report cuts both ways. Although the gap in the records is suspicious, the GAO had failed to find evidence of a cover-up in the Fort Worth and Washington records – which included *incoming* messages from Roswell – that were, apparently, intact.

Critics of the 1994 USAF report had complained that it dealt only with the recovery of material from the Foster ranch, and had ignored the testimony relating to a second crash site, in particular the testimony about the bodies. However, in 1997 – timed to coincide with the fiftieth anniversary of the Incident and optimistically entitled *The Roswell Report: Case Closed* – the USAF produced a second report intended to answer those criticisms.

The 1997 report concluded that what the witnesses had seen were "crash-test" dummies from high-altitude parachute experiments. Such dummies – designed to articulate as much like real humans as possible and clothed in one-piece suits – were used for research into parachute design and the development of ejector seats. The dummies were carried on specially-designed racks suspended from high-altitude balloons, from which they were dropped from as high as 98,000 feet (29,400 m). Such tests were carried out over New Mexico, and there were occasional mishaps that would send the entire rack plummeting to the ground with its "crew" of dummies still on board. Convoys of military vehicles would chase the balloons to recover the dummies. It was just such accidents, the 1997 report concluded, that were responsible for the reports of crashed UFOs, alien bodies and military cordons.

The report is extremely unsatisfactory, the main reason being one of the criticisms often levelled at UFO researchers: setting out from a preconceived position and then selecting facts to fit the preconceptions. The investigation started from the premise that there had been no UFO crash, and set out to find an explanation that would fit the "profile" of witness reports: unfamiliar wreckage surrounded by what appear to be bodies of an non-human type, and the rapid appearance of military vehicles that sealed off the area and recovered the wreckage and bodies. The USAF team found that only one type of activity carried out in the area fitted this profile: the "dummy drop" experiments. Therefore, they declared, this must be the solution.

The major objection to the "dummy drop" theory is that the records show that these tests did not begin in New Mexico until

1954. The USAF investigators therefore argued that the witnesses's memories of such an event so long ago were inaccurate, or that Roswell researchers had simply assumed that the event they described had happened in July 1947.

Having arrived at this explanation, the investigators then forced the theory to fit other aspects of the witness reports. For example, witnesses describe the bodies as being smaller than the average human, whereas the dummies were a standard 6-foot (1.8 m) height, and as having four fingers, whereas the dummies were designed to be as like a real human as possible. To explain such discrepancies, the investigators argued that the dummies had been damaged during the fall – parts of their legs might have been broken off, making them appear smaller, and they may have lost fingers. Where the "dummy drop" theory did not work, the USAF team sought other explanations. For example, the autopsies witnessed by Glenn Dennis's nurse are put down to the post mortems of the victims of an air crash in June 1956.

However, such criticisms aside, the question is whether the USAF investigation set out to explain things that didn't need to be explained in the first place. We have seen that the claims of bodies are based on a small nucleus of first-hand witnesses supplemented by a larger body of less detailed testimony which is often based on something that the interviewee had been told about. The USAF report could explain some of the second group – for example, Bertram Schultz's recollections of having been told about a UFO crash by a group of archaeologists – but its explanation for the first is highly contrived. However, the first-hand testimony used by the USAF included that of Glenn Dennis's nurse, Gerald Anderson and Jim Ragsdale. (Ragsdale did originally describe the bodies he saw as "like dummies", although he later changed his description to make them more obviously alien.) As we have seen, there are serious question marks over all this testimony – so did the USAF really need to explain it at all?

PHYSICAL EVIDENCE

A weakness in the case for the Roswell crash is that it is based only on the uncorroborated testimony of a few individuals; there is no physical evidence to support their claims – at least, none that has stood up to scrutiny.

Although several people claim to have kept pieces of the remarkable material, none has been able to produce it. In every case, they say that it has either been confiscated by the military or disappeared in mysterious thefts. In Jim Ragsdale's case, the last of his samples were, he claimed, stolen from his house in 1985. This is particularly frustrating for researchers, as this was several years after the Roswell affair had achieved widespread publicity. Why hadn't Ragsdale produced it earlier? And if as is implied, a government agency was behind the theft, why did they leave it so long to recover such compromising evidence?

On the other hand, several pieces of material that are *claimed* to be from the Roswell crash have surfaced over the years. These have always been handed on anonymously – so there is no proof that they really came from Roswell – and have all turned out to be of terrestrial manufacture. One such piece of metal was sent to American radio host Art Bell, whose show was devoted to UFOs and similar subjects, in 1996.[69] It came with an anonymous letter, in which the writer claimed that his grandfather had been part of the "recovery team" and had secretly kept the metal fragment. When tested it was found to be almost pure aluminium, with nothing to suggest that it had been manufactured anywhere but on Earth.

Like the fragment sent to Art Bell, all other pieces of alleged Roswell wreckage have been made of hard metal. None of the tin foil-like material, with the near-indestructible qualities, resistance to flame and remarkable properties of shape retention, has ever surfaced. Although a CUFOS team carried a three-day survey and excavation of the Foster ranch site in September 1989 in the hope of finding any pieces that might have remained beneath the soil, they were unsuccessful.

However, the hope that some hard evidence would surface appeared to have been answered with the discovery of the famous "Roswell autopsy" film. News of the film's existence first broke in Britain in January 1995, when former rock star turned UFO and crop circle researcher Reg Presley announced on a morning TV show that he had been shown film footage of the autopsy of one of the aliens killed in the Roswell crash. A few days later a British film and video producer named Ray Santilli announced that he had the film, although it was not shown publicly until – after a careful campaign to build up media interest – May 1995, when it was screened to members of the international media and a group of UFO researchers in London. After this Santilli set about auctioning the broadcast rights. The first showing, to an audience of many millions, was in a Fox TV special, *Alien Autopsy: Fact or Fiction?*, in August 1995, which was followed by broadcasts around the world. Santilli also produced a retail video which has sold widely.

Ray Santilli claimed that he had acquired the film while on a research trip to the US in 1993, trying to find rare footage of Elvis Presley. While negotiating with a retired newsreel cameraman for such footage, the man announced that he had the film of the Roswell alien autopsy. According to Santilli, in 1947 the cameraman (who asked to remain anonymous) had been employed by the USAAF, and on 2 June 1947 (a full month before the crash is supposed to have happened) he was ordered to the White Sands Missile Range where the bodies of the crash victims were being examined.

On first viewing the grisly film is very disturbing. Two masked surgeons dissect the corpse of a light-skinned and large-headed corpse that, although humanoid, is clearly not human. It seems uninjured apart from what appears to be a large burn on one leg. The surgeons open up the chest and stomach cavities and remove various organs, placing them in metal dishes. When the eyes are examined they are found to be covered by a dark membrane which is also removed. Finally, the scalp is peeled off to expose the skull, which is sawn open so that the brain can be extracted.

However, from the outset serious questions were raised about the film's authenticity. It was pointed out by sceptical UFO researcher Kal K. Korff that the film uses "jump cuts" between individual sequences, meaning that a new scene begins on the very next frame to the previous one.[70] The model of camera that Santilli said was used was a hand-wound, not electrically driven, model. Because of the manual operation such jump-cutting is impossible, as the first few frames in each new scene are always over-exposed. This shows that either the film had been taken using a more modern, electrically driven, camera, or that it had been edited and was not the original footage that Santilli claimed. Other observers noted that there was a suspicious tendency for the picture to go out of focus at particularly significant moments, as for example when one of the "doctors" begins to saw open the skull.

Other objections were raised about the way the autopsy was conducted. In 1947 it was more usual for a film record to be taken using a stills camera, even if a movie was also made, and the film shows no stills cameraman in the autopsy room. And not only do the "doctors" hold their instruments in a manner unlike that of trained surgeons, but the way they perform the autopsy is decidedly irregular: for example, placing the organs in metal dishes rather then the customary glass jars. And the whole process is carried out with too much haste. In a normal, human, autopsy the organs are carefully and individually examined and weighed, and a written record made, and given these unique circumstances, when an entirely new, alien species was being examined, even more care would be taken. The wall clock in the operating theatre shows that the whole process took some two and a half hours – too short a time for such a unique and historic process.

Claims that the film has been verified as genuine 1947 stock are overstated. Santilli gave a sample of the "leader" – the blank strip at the beginning of the roll which is wound around the spool – for Kodak to analyze, and it did prove to be 1940s stock. When it was pointed out that there was no proof that the piece of leader came from the same reel as the autopsy footage, Santilli produced two of the film frames which were also found to be of the correct age. However, these frames did not show the autopsy, but what appear to be fuzzy shots of a flight of stairs which do not relate to anything in the "alien" footage, so once again researchers cannot be sure they come from the same film.

Santilli's cameraman had originally asked for anonymity, but the producer later named him as Jack Barnett. Unfortunately, Barnett – a genuine newsreel cameraman who had indeed filmed the early Elvis footage also acquired by Santilli – died in 1967. Santilli countered that he had given Barnett's name in order to protect the real cameraman's identity, as he was afraid of the repercussions from the government. As Kal Korff, among others, has pointed out, this is nonsensical. As the government employed the cameraman they must know who he is. It would be safer from him to raise his profile as high as he possibly could, since if anything happened to him, it would be obvious who was behind it.

Strangely enough, one of the most damning pieces of evidence against the film is one of the most minor details. On the wall there is a "Danger" sign. The design of official US signs conform to standards laid down by the Occupational Safety and Health Administration (OHSA). According to a former OHSA member, John English, in a letter to the *Skeptical Inquirer,* the font and size of this sign were not approved until 1967 – twenty years after the event supposedly caught on camera.

In December 1998 the Roswell autopsy film was featured in a Fox TV documentary, *World's Greatest Hoaxes,* whose Executive Producer, Robert Kiviat, had also produced *Alien Autopsy: Fact or Fiction?* three years earlier.

It seems very clear that the Roswell autopsy film is a fake, although whether this was a hoax perpetrated by or on Ray Santilli is unknown. Despite this, many UFOlogists – most particularly Michael Hesemann and Philip Mantle in their 1997 book *Beyond Roswell* – continue to promote it as genuine, and as part of the evidence for the Roswell crash.

SUMMING UP

The evidence for a UFO crash near Roswell in 1947 is by no means as compelling and clear-cut as many people believe. Despite the intensive research devoted to this incident – which is probably the most comprehensively studied in UFO history – much remains frustratingly elusive and insubstantial. Something unusual certainly did crash on the Foster ranch, but most of the elements of mystery and conspiracy surrounding this event can be explained by the MOGUL hypothesis. On the other hand, descriptions of the material consistently testify to its unusual properties, in particular its resistance to creasing and near-indestructibility, which does not easily fit the MOGUL explanation. However, without a sample of the actual material itself the accuracy of the descriptions remains an open question.

The evidence for a second crash site from which alien bodies were recovered is more tangled. We have seen that the main evidence for this comes from a few individuals, some of whom have demonstrably made up their stories. Of all the testimony, that of Brigadier General Arthur Exon, although second-hand, is the most compelling but also the most puzzling. Even the major proponents of the Roswell crash disagree on which witnesses are to be taken seriously.

Although he has ended his association with Donald Schmitt, Kevin Randle continues to promote the idea of two crash sites, resulting from the explosion and crash of a single UFO. However, in 1994 he admitted to Karl Pflock that, "The evidence is not strong and it's not compelling."[71] Stanton Friedman believes that there were two crashes, at the Foster ranch and the Plains of San Agustin, the result of two UFOs colliding, although he has rejected the tall tales of Gerald Anderson that he originally believed. On the other hand, Friedman's associate and co-author Don Berliner no longer believes in the San Agustin crash at all. Karl Pflock, originally a believer in the "two crash" scenario, initially accepted the MOGUL explanation for the Foster ranch debris but argued that a UFO was downed at the second site (perhaps after colliding with the MOGUL balloon train) but has since moved to the position that *no* UFO was involved at all.

The warring factions cannot agree on the date of the crash: Randle and Schmitt favour 4 July, whereas Friedman argues for 2 July. (It is the latter that has been declared the official anniversary by Roswell's mayor.) Both sides ignore Mac Brazel's assertion that he found the wreckage some three weeks earlier.

Other UFOlogists who were originally persuaded about the Roswell Incident have changed their minds – for example, former airline pilot Kent Jeffrey, who in 1994 organized the "International Roswell Initiative" to persuade the US Government to come clean about the affair. Over 20,000 people signed Jeffrey's petition that he planned to hand to the White House on the fiftieth anniversary of the incident. However, by the time that the hand-over took place Jeffrey himself, based on his further investigations, had become convinced that there had been no UFO crash. The chief reason for his change of mind was his contact with veterans of the 509th Bomb Group, none of whom knew anything about the incident, and who had not been aware of the claims of until the "revival" in the late 1970s.[72]

Cynics point out that the main beneficiary of the incident has been the city of Roswell itself. Undeniably – and understandably, given that it is in a poor region, with incomes some 25 per cent below the national average on the nationwide scale of prosperity – the city has capitalized on the event, which brings in an estimated $5 million a year. Souvenir shops abound, and local farmers charge admission fees to the various crash sites. There are three competing museums in Roswell devoted to the "UFO crash", all promoting the idea that a real alien spacecraft crashed near the town. In April 2000, the largest of them, the Roswell International UFO Museum and Research Center – co-founded by Walter Haut and Glenn Dennis – celebrated its 750,000th visitor since opening less than nine years earlier, giving an impressive average of well over 80,000 visitors a year. As Max Littell – the third founder of the museum – told Philip Klass shortly before its opening in 1991:

> The jury is going to be the public and they're gonna decide what they want to believe. We hope 10,000 of 'em come here to find out because while you're here – over this weekend – you're gonna drop 200 or 300 bucks here in town. Alright, times 10,000, is good for our economy.

Whatever the truth about the Roswell Incident, it is surely ironic that what is described by many as the greatest official cover-up in history is paraded on every street corner, and has achieved such renown throughout the world.

The Roswell Incident incites such strong feelings that to some, to deny there was a UFO crash there is almost akin to committing sacrilege. Yet one has to wonder why. What difference does it make to the huge, fluid and ever-exciting subject of UFOlogy if some flimsy pieces of wood came from a balloon and not a spaceship? Just like a belief in God should not depend on whether the Shroud of Turin is genuine, so a recognition that one famous case dwindles to nothing when investigated objectively should not ruin the excitement of chasing the truth about UFOs and the possibility of alien contact with the Earth.

Notes

1. The 9 July 1947 article is reproduced in full in Klass, pp. 20–22.
2. *Ibid*, p. 20.
3. *Ibid*.
4. *Ibid*, p. 18.
5. Korff, p. 59.
6. From the 1979 documentary *UFOs Are Real*, quoted in Klass, p. 31.
7. Klass, pp. 65–66.
8. *Ibid*, p. 67.
9. Bertliz and Moore, p. 67.
10. Randle and Schmitt, *UFO Crash at Roswell*, p. 50.
11. Cavitt's statement is reproduced in full in Korff, pp. 135–136.
12. Berlitz and Moore, p. 70.
13. Randle and Schmitt, *UFO Crash at Roswell*, pp. 135–136.
14. Korff, p. 46.
15. Friedman and Berliner, p. 72.
16. Randle and Schmitt, p. 127.
17. *Ibid*, pp. 126–131.
18. McAndrew, p. 86.
19. Berlitz and Moore, p. 96.
20. Klass, p. 67.
21. Randle, *Roswell UFO Crash Update*, p. 178.
22. Randle and Schmitt, *The Truth about the UFO Crash at Roswell*, pp. 162–163.
23. Korff, p. 135.
24. Klass, p. 124.
25. *Ibid*, pp. 16–17.
26. Randle and Schmitt, *UFO Crash at Roswell*, pp. 74–75.
27. Korff, p. 129.
28. Randle and Schmitt, p. 75.
29. Korff, p. 152.
30. *Ibid*, pp. 82–83.
31. Randle and Schmitt, p. 79.
32. Klass, pp. 41–45
33. *Ibid*, p. 42.
34. Friedman and Berliner, pp. 185–191.
35. Klass, p. 75.
36. Korff, pp. 80–81.
37. Randle and Schmitt, pp. 108–109.
38. Randle, *Roswell UFO Crash Update*, pp. 31–35.
39. Friedman and Berliner, pp. 115–119.
40. Klass, p. 68.
41. *Ibid*, p. 72.
42. See McCarthy, 'The Missing Nurses of Roswell'.
43. Korff, p. 229.
44. Randle and Schmitt, *The Truth about the UFO Crash at Roswell*, p. 7.

45. Randle, *Roswell UFO Crash Update*, p. 180.
46. McAndrew, p. 216.
47. *Ibid*, p. 218.
48. Klass, p. 145.
49. *Ibid*, p. 146.
50. Randle and Schmitt, *The Truth About the UFO Crash at Roswell*, p. 162.
51. Good, *Alien Contact*, p. 99.
52. Friedman and Berliner, p. 128.
53. Randle and Schmitt, p. 175.
54. Korff, p. 84.
55. Klass, p. 103.
56. Randle and Schmitt, p. 139.
57. Korff, p. 95.
58. Randle and Schmitt, p. 75 and p. 142.
59. Korff pp. 91–92.
60. Randle and Schmitt, *UFO Crash at Roswell*, p. 166.
61. Korff, pp. 95–97.
62. *Ibid*, p99.
63. Sparks, 'Colonel Philip Corso and William Birnes' Bestselling Book Exposed as a Hoax!'
64. Randle and Schmitt, p. 112.
65. *Ibid*, p. 112.
66. Klass, pp. 118–119.
67. Korff, pp. 249–250.
68. *Ibid*, pp. 60–68.
69. *Ibid*, pp.111–118.
70. *Ibid*, pp. 208–209.
71. *Ibid*, p. 228.
72. Klass, pp. 212–214.

CHAPTER 4

UFOS: ASSOCIATED PHENOMENA

CATTLE MUTILATIONS

Strange attacks on livestock have been reported, mainly from America, in association with unknown lights in the sky and UFO sightings, and often with mystery black helicopters. Although known generically as "cattle mutilations" – because most of the depredations are inflicted on cows – occasionally other animals are selected for the grisly attentions of the unknown intruders.

It seems that the infamous Goatsuckers or *Chupacabras*, whose activities qualify as animal mutilations of Brazil and neighbouring countries may not operate exclusively in South America. A report from the Bulgarian city of Plovdiv claimed that in April 1993 no fewer than sixteen people were killed by a unknown vampire-like creature that drained the blood of its victims.[1] A local detective, Igor Tolkov, said, "'It looks like a cat, but has huge sharp claws and glowing red eyes". Its first victim was Scottiz Karpulsky, who was attacked at daybreak – a nice touch in a vampire story – on 25 April 1993 as he shut the door of his nightclub. His widow said: "Perhaps a big cat jumped on my husband from a window sill on the first floor and bit through his neck. I took an umbrella in my hand and ran towards the beast which was sucking his blood. Then I ran away filled with panic."

Perhaps, as Madam Karpulsky suggested, the mystery blood-sucker really was a large cat – possibly a wild cat turned "rogue", having somehow acquired a taste for human blood and developed homicidal tendencies. And although there are no known connections between the Bulgarian vampire cat and UFOs, unlike the South American Goatsuckers, the reports of strange glowing red eyes of the eastern European creatures suggest some kinship, if only in the burgeoning folklore.

Although there have been few actual witnesses to strange cattle deaths or mutilations, one rancher of Chacon Canyon near Las Vegas – and close to Area 51 – saw something very odd happening. At around 5.30 p.m. on 14 September 1994,[2] Larry Gardea checked on the cattle belonging to Estevan Sanchez when he heard a loud, anomalous humming sound, which he first thought must emanate from "someone working nearby or something". But clearly this was not a workaday event but a most unusual experience: all the cows started galloping in the opposite direction to the noise – except for three of them, which were apparently being drawn, struggling violently and "bawling", through the trees by a beam of light towards the origin of the noise.

Gardea said: "I shot at the beam with a 30.06 rifle, and the hum stopped and the cows stopped bawling." But whatever the nature of the beam, its effect was deadly: one cow was dead and strangely mutilated, while another was hurt – and the third nowhere to be found. Terrified, Gardea called the Sheriff's department and Deputy Sheriff Greg M. Laumbach came out to investigate.

The two men went over the ground where the incident took place together, examining the dead cow, whereupon Laumbach discovered that its jaw had been cleanly skinned on one side and the tongue, anus and reproductive organs had been removed.

Laumbach stated that: "The wounds looked like they were done with a really sharp instrument or something. There was no blood or anything. They [the wounds] weren't jagged." He found no gunshot wound, and could offer no theory as to what the strange beam of light could have been. Undersheriff John Sanchez added: "I haven't heard of anything like this since the late 70s or early 80s. I have heard of the same kind of things being reported but not recently."

Sanchez said that although there had been a number of reports of cow deaths, there had been none concerning cattle mutilations, adding – perhaps significantly – that, "Lighting and things like that have been common recently."

From a distance, it may have looked as if the cattle were being drawn towards the noise by a beam of light, but that could have been an illusion, aided by memories of lurid cattle mutilations stories the ranch-hand may have read and half forgotten. Instead, the "beam" could have been a lightning strike, resulting in the unusual injuries of one cow while driving another away. But if that somewhat sceptical explanation rings a little hollow, what other hypothesis could possibly explain such a scenario? Were the cattle interfered with by aliens or, as in *The X-Files*, by a very terrestrial scientific/military "black ops" team as part of some hideous long-term experiment – not so much into the workings of the bovine body, for surely they must know them literally inside out by now, but into the effect of such weirdness on the owners of the cattle? Certainly, the ubiquity of the mystery black helicopters suggests some kind of government involvement and monitoring. Perhaps, as some have suggested, government officials are clandestinely taking samples from livestock to check the levels of pollution or radiation in their blood and tissue, something that they would be reluctant to do openly for fear of causing panic, should the true condition of American meat be known. But in that case, why the cloak and dagger approach? Why not simply instruct vets to conduct regular tests, claiming they are for something else, such as BSE status?

The most famous cattle "mutes" researcher is the television documentary maker Linda Moulton Howe, whose interest in the subject was kindled back in 1979 when she heard about an orange disk "the size of a football field" that had been seen in the sky following the disturbing spate of cattle mutilations in Wyoming. Her initial reaction was that the deaths were caused by some form of chemical contamination, but soon she became convinced that an alien intelligence was responsible. Six months later her Emmy-award winning *Alien Harvest* was broadcast on CBS, earning the praise of UFOlogists and laymen alike.

In a later interview for a British magazine,[3] this is how Linda Moulton Howe described the defining characteristics of a typical animal mutilation: "Ear missing, eye missing or circle of flesh around the eye removed, jaw flesh usually removed. You're left with about half of the animal. The tongue is removed in a cut that usually goes all the way to the larynx; in many cases, not only is the larynx taken, but 8 inches [20 cm] of trachea. The genitals are removed, male and female. Law enforcement officers in the 1960s started calling these cuts 'cookie cutters" because that's what they look like. The rectal tissue is usually cored out, sometimes going up to 14 inches [36 cm] into the body.'

Sceptics suggest that the injuries could be caused by natural predators such as coyotes and large birds, but Moulton Howe rejects this utterly, saying, ". . . all the tissue is dry and there are no tracks around the animal. How could a 2,000lb [900 kg] bull in Colorado end up on its back with its horns 6 inches [15 cm] into the ground with its four legs up, both eyes removed, the tongue removed, the genitals removed, and the rectum cored out?"

The film-maker points out that the earth around the animal corpse reveals that it has been exposed to a very high temperature, adding that the phenomenon may be "related to crop circles. When we look at the grass around some of these animals – and sometimes even inside the animals – we find changes in the cell metabolism. It's the same with crop circles: altered cell structures."

Linda Moulton Howe believes that, "The only way you could effect such changes would be in the presence of microwaves and an intense vortex. Intensely focused energy suggests some sort of technology, so, if you tie that into the eyewitnesses who said they had seen craft in the pastures where some of these animals had been, that's the link."

So far, however, the vast majority of reported cases come from the United States. While a few are reported from elsewhere (see below), this is pretty much an American phenomenon, just as crop circles seem to be peculiar to England – which in itself suggests that aliens may *not* be responsible. If they were, surely they would have taken samples from across the world, and we would know of examples of cattle mutilation on a much larger scale from many different countries. With such a relatively parochial spread of cases, the finger seems to be pointing remorselessly at the American government or their close allies. (For a discussion of Linda Moulton Howe's involvement with the intelligence agencies, see Conspiracies section page 419.)

However, a few cattle and other animals *are* being strangely killed and mutilated elsewhere. Christopher Hughes of "Cobbie" Farm at Mutdapilly in Queensland, Australia, had a shock when, just two days after moving into the property in July 1994, he discovered two $600 steers dead in the paddock, one of which had its rear end cut out, and the tongues, muzzles and one ear removed from both the dead beasts. Although there were no signs of gunshot wounds, during the police investigation it emerged that other cows had been found in a similar condition – although they had been shot – several years previously.

Mr Hughes did not connect the occurrence with anything other-worldly or with UFOs, although he acknowledged that others would be quick to make the connection. Like the police, he simply condemned it as a criminal act.

Then there was the horrific spate of attacks on horses in the dead of night in England in the mid- to late 1990s. The animals, usually thought safe in their stables, were found to have been "ripped" by a sharp instrument like a Stanley knife, their genitals butchered or even severed, eyes mutilated or gouged out, and occasionally left stricken but alive – although most of the poor beasts had to be put down. As far as can be ascertained, despite the best efforts of the police and horse-loving community, no one has been caught. Although there is a suggestion of occult ritual involved because the attacks took place on the night of the full Moon (which could simply be because the attacker needed to see without the aid of a torch), there have been no allegations of alien involvement.

There are other possibilities. Some writers have linked the cattle mutilation phenomenon with the old fairy realm – indeed, traditionally, the fairy folk took cattle, and occasionally drained them of blood, either as a due offering or as vengeance for some act against them by humans. And of course the link with vampires is all too obvious.

There may be yet another class of entities – invisible vandals and hooligans – whose greatest delight is swooping on cattle and ripping them apart, just for kicks. Are they interdimensional (or ultradimensional) tourists, using the Earth as their recreation ground, seeking to play sick pranks on us? Perhaps it is relevant that other forms of paranormal entity appear to be becoming increasingly violent and vicious. In Glasgow in the 1980s Professor Archie Roy, an eminent astrophysicist and psychical researcher, actually witnessed a young woman screaming with pain and terror as an invisible attacker slashed her torso, cutting her badly, with what appeared to be a razor. He discovered that this was only one of several such attacks by poltergeist-like entities: perhaps they also amuse themselves by cutting cows to pieces?

Notes

1. Source: *Dracula* (Romania) No. 31, April 1994 and reported in *Fortean Times* No. 90.
2. Source: *Nexus* magazine, December 1994-January 1995.
3. *The X Factor*, No. 31.

CROP CIRCLES

Since the mid-1980s, mysterious and increasingly complex patterns have been found in fields of crops, and although widely dismissed as the work of clever and artistic hoaxers, the subject of crop circles is still hotly debated.

The phenomenon began with plain, swirled circles in the middle of grain crops, mainly wheat, in the west and south of England, particularly in remote areas or in the neighbourhood of ancient "places of power", such as Silbury Hill, the huge neolithic earthern mound near Avebury. (Wiltshire and Hampshire remain the favourite locations for the yearly crop of over 100 circles.) Because there were several reports of "UFOs" being spotted at roughly the same time that the circles were formed, in those early days they were known as "UFO nests".

Mysterious globes of light were seen – and on a few occasions, filmed – gliding over the crops. Some of these were later dismissed as birds, bits of litter blowing in the wind, or kites, but some still mystify. Certain scientific researchers call these spheres "plasma balls", but doing so brings us no nearer to the heart of the mystery.

Suddenly crop circles seemed to appear everywhere, but as their numbers mushroomed they departed from the simple format and expanded into the now-familiar huge, complex designs, often extending over hundreds of feet (much to the dismay of the farmers, who retaliated by charging visitors a fee). Very rapidly the phenomenon became so embedded in the collective consciousness that crop circles featured in television commercials and on record albums. But are crop circles anything more than a natural phenomenon? Are they, as many still believe, created by aliens who wish to communicate some message to us?

Various hypotheses have been suggested to explain the circles, from Dr Terence Meaden's almost equally mysterious "plasma vortexes" to the sceptics' somewhat wild notion of circularly running bunnies. Meaden's idea of localized mini tornadoes, however, remains a persuasive explanation for the simple circles. (Plain circles have also been found in other substances, such as ice and sand, see illustrations.)

But what of the increasingly complicated, massive designs that continue to brighten the English countryside every summer? It seems unlikely that a tornado could etch with such precision patterns that have included a Celtic cross, a fractal star of David that appeared on 23 July 1997 behind Silbury Hill, and a 400-foot (120 m) snowflake found at Winterbourne Basset, Wiltshire, on 4 June that same year. And with each summer season, new and exciting patterns appear, each more elaborate and stunningly beautiful than the last: in July 2000 a 200-foot (60 m) square appeared in a field, once again in the shadow of Silbury Hill, Wiltshire. The *Daily Mail*[1] described the formation as "shredded wheat" – a reference to the breakfast cereal of roughly the same shape – and as "a baffling maze of flattened pathways and swirls", noting that "once you take to the air to get a clearer view, its geometrical precision becomes clear."

Since the early days of "UFO nests", mystics, New Agers and psychics have flooded to meditate in the circles in what has become something of an annual pilgrimage, claiming to "tune in" to spiritual or alien forces that seem to concentrate in the patterns. Unfortunately, on several occasions the "aliens" were revealed to have very terrestrial origins. In 1991, two pensioners from Southampton, Doug Bower and David Chorley – the infamous "Doug and Dave" – confessed to making patterns in the fields since the late 1970s with "stompers" or wooden planks that they use to flatten the crops. But there were other culprits: an increasingly sophisticated wave of young hoaxers took up the challenge with enthusiasm, such as Rob Irving and Rod Dickinson, who see their hoaxing as a serious art form.

Although many previous "True Believers" – such as Nick Pope – are now convinced that only the original plain circles are genuinely unexplained, acknowledging the complex formations as the work of hoaxers, some still maintain that there is something disturbingly weird about the phenomenon that is not so easily explained away.

One such is Lucy Pringle, who has researched the circles for many years, and believes that they are the product of extraterrestrials, that, "It must be some sort of means of communication. Whatever intelligence is behind it is probably in advance of our own." She cites[2] the peculiarly disorienting effect that the circles have on many who enter them: people keel over and lie prostrate, suffer violent nausea and headaches and other symptoms similar to those experienced by UFO witnesses. She writes: ". . . whatever force causes circles to form hits the ground vertically creating a huge electrical discharge. This results in microwave activity which softens the base of the corn stalks, allowing them to fall over but not break."

She also notes: "The majority of crop circles in this country occur on the chalk lands of the Wessex triangle. Under the chalk lie aquifers – underground water sources – which almost certainly account for the high incidence of formations in this part of the country, as water attracts electricity." It may also be the reason why crop circles have occasionally been spotted in paddy fields, which are of course permanently water-logged.

Some have suggested that those circles that are not the work of the likes of Rob Irving or Doug and Dave have been created by government agencies using high-frequency sound equipment that leaves the crops strangely swollen and swirled. The agencies involved, it is claimed, are producing the formations in order to monitor the reaction of the masses to yet another unexplained phenomenon, which might explain the odd characteristics of the crops involved. Many of them have been demonstrated to have unnaturally swollen stems, which can hardly be the result of a few hoaxers with wooden planks. And, assuming that the peculiar abreactions of visitors is not merely subjective, what can possibly explain the spate of headaches and instant nausea that is so often reported? Even more bizarrely is the evidence that some people are healed, not hurt, by the phenomenon (although once again there may be an element of "faith" or self-healing through expectation, involved).

Judging by all the hyperbole and controversy, one might be forgiven for believing that crop circles are a very modern phenomenon, but this is not so. In fact, they have a much longer history than might be supposed. A sixteenth century woodcut shows the Devil mowing a circle that closely resembles the plain swirled circles of the 1980s, and "fairy rings" have been written of since time immemorial. It was said that if you stepped inside a fairy ring terrible things would befall you: certainly you would feel disoriented and lose all sense of time, not to mention being carried off to the fairy kingdom or killed by an "elf bolt" (which mostly struck down cattle, as in the phenomenon of cattle mutilations, see above). But even the age-old fairy rings have not escaped the attentions of science.

Dr Tom Gaynard[3] wrote to the *Fortean Times*:[4] "Such rings of vegetation are common and are due to the activity of a fungus growing just under the soil surface. The mycelium ('body') of the fungus comprises many fine hyphae (thread-like structures) which grow out in a circle from a central point of origin, much like a colony of mould on the surface of jam in a jar. The older hyphae towards the centre of the colony die, leaving the younger ones around the circumference."

He goes on: "These feed by decomposing dead vegetation underground, releasing nutrients in the process. The uptake of these nutrients contributes to luxuriant growth of vegetation, producing the ring of dark green grass . . . Superstitious folk have believed these rings to be the site of 'fairy dances', hence their name. However, their true nature is revealed sooner or later when a circle of fruiting bodies (mushrooms, but not necessarily edible) appears." But how does this explain the phenomena associated with the circles? There are genuine mysteries still to be uncovered, as many researchers admit. Indeed, American billionaire Laurance Rockefeller has donated what is believed to be a sum of many thousands of pounds towards the furtherance of this research. (He has also funded Dr John Mack's Starlight Coalition, which comprises ex-US military personnel and intelligence officers – which some may think interesting in itself . . .) The project is being co-ordinated by veteran "cereologist" (as crop circle enthusiasts are known) Colin

Andrews, and will involve the police, the National Farmers' Union and much dramatic rushing about in helicopters. However, even Andrews admits that 80 per cent of the formations each year are caused by hoaxers, although he believes the other 20 per cent to have an unknown origin.

Others are sceptical about the new project: circle-maker Rod Dickinson told *Fortean Times*:[5] "Along with the other participants in the crop circle phenomenon, Mr Rockefeller will find only what he expects to find. The phenomenon is regulated by the desire and belief of each individual recipient. This nebulous work of art continues to penetrate and extend its hold, like a form of mind virus that feeds on the visions, dreams and perceptions of others."

Avid "croppie", Andy Thomas, countered:[6] "The evidence, circumstances and the complicated geometry of the circles shows that they are not simply constructed in a few hours by people . . . There is a much more intelligent consciousness behind it. I have a feeling that they are precursors to some events or new period of time. It is no coincidence that there has been an increase in numbers over the last 10 years."

Whatever the truth about crop circles, the dangers remain the same for the over-credulous who maintain not only that virtually every formation is of extraterrestrial origin, but that they have inside information about the superior beings behind them. There was an astonishing scene at BUFORA in the mid-1990s when the speaker, London-based film-maker – and follower of the Council of Nine, (see The Chosen Ones) – David Percy, announced that a certain formation contained special information from the aliens.[7] At question time Rob Irving announced that he had created that particular pattern – to which, after a very short pause, Percy replied that he may have *physically* hoaxed it, but the aliens must have provided the inspiration!

Notes

1. 14 July 2000.
2. See Pringle, *Crop Circles*, (London), 2000.
3. TJGaynard@aol.com
4. No. 122.
5. No. 125.

6. *Ibid.*
7. He also claims, in his odd tome, *Two Thirds*, co-authored with David Myers, that an alien penchant for fried eggs surfaced in a particular formation (which he insists on calling "crop glyphs").

THE MEN IN BLACK

Despite the flippant, fun and hugely popular Hollywood movie *Men In Black*, starring Will Smith, the MIB are a real phenomenon, often reported as an integral part of UFO sightings, presenting researchers with yet another layer of possible conspiracy theory – and all too often also certain elements of flagrant absurdity. Typically one, two or three men in black (although some women are reported) with slant-eyes and Oriental-looking complexions, visit the home of an individual who has been witness to a UFO sighting, where they proceed to behave very strangely – using stilted "B" movie language – warning the witness never to tell anyone or something dreadful (usually unspecified) will befall them. Then they leave, never to be heard of again. And the great paradox is that the only reason we know about the MIB is that their threats proved to be hollow: the witnesses talked and got away with it.

With the outward appearance of FBI agents (or Mormon missionaries, perhaps), it has long been suggested that the MIB are government agents endeavouring to suppress stories of UFO sightings and contact. A more specific claim was that they were agents of the Atomic Energy Commission trying to suppress news of the alleged leakage of radioactive material in the Maury Island case (see page 49), although this is unlikely, as many believe that whole story to have been a hoax.

More intriguing is the theory that the MIB are really aliens, intent on suppressing information about themselves in order to have a free hand to carry out their sinister secret agendas. There are tales of "high strangeness" surrounding them: sometimes the men turn up wearing make-up, as if not understanding the difference between the sexes in earthly society. (It is certainly one way of being remembered.) Typically the MIB drive old models of cars that are nevertheless brand new (as are their clothes), which suggests that they had been specially made for the occasion. On the whole, their purpose appears to be to warn off those who come too close to the UFO mystery: as Jenny Randles says, "It seems that the MIB have a job to do, and we are in their way".[1] But their failure to follow up their threats is almost universal, indicating that they are *not* government agents – despite their occasional assertion that they are from the CIA or the Ministry of Defence in Britain. Much as this is an unwelcome thought, surely it is not beyond the combined brains of the military and intelligence agencies of the most developed nations on Earth to silence powerless individuals so comprehensively that outsiders *never* get to hear their stories? But are the MIB just oddball characters enjoying a prank at the expense of solid citizens? The MoD's former "UFO man", Nick Pope, has suggested that the majority of MIB cases are simply "Walter Mitty types" indulging their bizarre sense of humour, but that fails to explain their frequent ability to defy natural laws – walking through walls, appearing and disappearing in an instant, possessing uncannily detailed knowledge of UFO witnesses and so on. Certainly *some* alleged MIBs may have been simply hoaxers – or even occasionally rival UFOlogists taking their opposite numbers for a ride – but most cases are far too surreal to be explained quite so easily.

The archetypal MIB case is that of Albert Bender in the early 1950s, at the beginning of the Contactee Era. In 1952 he was Director of the International Flying Saucer Bureau, which operated from his home at Bridgeport, Connecticut, and which boasted an impressive membership. But all that was to change rapidly – because, or so it was claimed, of intervention from the MIB.

The story began when Albert Bender claimed that he knew the "secret" behind the UFO phenomenon, announcing that he was about to publish it in the Bureau's *Space Review*, but made the mistake of sending his report to a colleague first. That was when the MIB called – but not politely ringing his doorbell, as in most other cases. These three "shadowy figures" materialised in his bedroom as he was resting: their appearance was accompanied by an overwhelming sense of nausea. Bender described the scene:

> The figures became clearer. All of them dressed in black clothes, they looked like clergymen, but wore hats similar to Homburg style. The faces were not discernible, for the hats partly hid and

> shaded them. Feelings of fear left me . . . The eyes of all three figures suddenly lit up like flashlight bulbs, and all these were focused upon me. They seemed to burn into my very soul as the pains above my eyes became almost unbearable. It was then I sensed that they were conveying a message to me by telepathy.[2]

The black-garbed bedroom visitors' message confirmed that Bender was on the right track with his discovery of the UFO secret, but characteristically they refused to allow him to make it publicly known. That, apparently, is the function of the MIB. It seemed that Bender had really ruffled their feathers, for they also ordered him to close down the Bureau and cease publication of *Space Review* (with the clear implication of "or else . . ."). But they did unbend enough to confide yet more of the UFO secret in him – which terrified him so much he was only too willing to forgo any further research or publicity connected with the subject. Bender promised the MIB to keep quiet about the matter "on his honour as an American citizen".[3] Which of course prompts the central question of how we know about it, but like many other early contactees, Bender felt compelled to rush out a popular book detailing this and other UFO-related experiences.[4] (So much for his honour as an American citizen.)

Strangely, this immediate defiance of the MIB is extremely common, but unfortunately seems to have little to do with the witnesses's courage and more with either their conscious or their unconscious recognition that despite the threats they will not be harmed. Of course if they had fabricated the whole MIB saga, then they would know from the outset that they had nothing to fear, but the evidence suggests that many witnesses had quite genuine encounters with weird strangers. But even so, they still sensed something about them that was insubstantial and basically ineffectual., something that sometimes seemed almost to *encourage* defiance. Could it be that although the MIB warn against making the subject of UFOs known, in fact they operate on the principle of reverse psychology, actually intending to inspire the witnesses to speak out? Is the MIB's real agenda the *spreading* of rumours of conspiracies about the UFO mystery?

The threats that the MIB utter often have the melodramatic ring of a "B" movie script, just as their clothes and cars seem to come from Central Casting: they are brand new, but at the same time resolutely out of date. This quaint aspect of the MIB phenomenon became very evident to Robert Richardson of Toledo, Ohio when in July 1967 he found an anomalous lump of metal on the road after hitting an object (which immediately vanished). Suddenly he found himself the central character in a MIB drama.

Finding the metal puzzling, he sent it to APRO, whereupon he was visited late at night by two twenty-something men who interrogated him on the doorstep for just 10 minutes. Afterwards, Richardson realized an odd thing: the men did not offer to show any I.D. – but then neither did he ask to see any. It was all very strange. The two men issued no threats and their behaviour was neither overtly hostile nor obviously sinister. Their car, although a black Cadillac (what else?) in mint condition, was a 1953 model, and the registration number (of which the resourceful Mr Richardson made a note) proved to belong to no known vehicle.

Seven days afterwards, two more black-suited and dark-complexioned[5] MIB arrived, this time in a contemporary Dodge car. They both seemed somehow foreign – indeed one had an unplaceable accent, although the other spoke perfect English – and their mission appeared to be to persuade the witness that he had seen nothing odd on the night he hit the disappearing object. After putting pressure on him for a while, they suddenly changed tack and demanded that he give them the piece of metal. He explained that he had sent it to APRO, to which they responded with extraordinary examples of scriptwriting that would do justice to the worst dialogue of Ed Wood,[6] such as "If you want your wife to stay as pretty as she is, then you'd better get the metal back!"[7]

But how did they know about the metal? Only Richardson, his wife and two APRO officers knew – at least officially. Had the MIB tapped telephones, like any self-respecting sinister CIA agents? Or had they no need of such clumsy terrestrial technology to know what was going on? And, once again, there is no suggestion that Richardson got the metal back, or that – having defied the MIB – his wife lost her good looks through their evil ministrations.

An even weirder case is the story of fifty-eight-year-old Dr Herbert Hopkins of Maine, who in September 1976 was acting as consultant (he was also a hypnotist) on a UFO case, when the telephone rang and a man claiming to be the Vice President of the New Jersey UFO Research Organization asked if he could call round to have a conversation about the case. Without a second thought, Hopkins agreed, and the next thing he knew the man had arrived on the doorstep – impossibly – dare one say, *paranormally* – fast. The only callbox was some distance from the Hopkins' home. The caller also appeared to have come on foot.

Yet on that evening, it all seemed perfectly normal. Dr Hopkins showed the stranger into the house – his wife and daughter were out, so they had the place to themselves – and he had a good opportunity to look at him. The witness's first impression was that he looked like an undertaker. He was smartly dressed in a white shirt, black suit, shoes and tie. He also wore grey suede gloves and a hat. There was not a wrinkle or bit of fluff on his clothes, and his shoes shone brightly. Then he took off his hat . . . He had "dead white" skin, a completely bald head and no eyebrows or eyelashes (as if suffering from *alopecia totalis*), and his lips were unnaturally red – the reason for this was soon revealed when he wiped his lips with the back of his hand and a smear of lipstick came off on his suede gloves. Not surprisingly, Dr Hopkins was most astonished, yet it was only afterwards, when he was trying to digest the details of the visit, that the full surreality of it finally dawned on him. At the time, he merely discussed the UFO case with his guest in a fairly normal fashion.

Having listened to the doctor's summary of the case, the visitor instructed him to erase all the tapes of the hypnosis sessions he had conducted for the UFO investigation – before suddenly announcing that he (Dr Hopkins) had two coins in his pocket, which proved to be true. He suggested that he put one of them on the palm of his hand, asking him not to take his eyes off it. As he looked, the outline of the coin seemed to become indistinct, then it began to dematerialize. The MIB told him that, "Neither you nor anyone else on this planet will ever see that coin again".[8] Neither man seemed to think this abrupt switch of mood into impromptu prestidigitation at all odd – at the time.

As they continued to discuss the case, Dr Hopkins began to notice that the MIB's words were noticeably grinding to a halt. Indeed, very soon the visitor rose to his feet with great difficulty, saying (extremely slowly), "My energy is running low – must go now –

goodbye", before jerkily exiting towards a bluish-white light in the driveway, which Dr Hopkins put down to unusually bright car headlights. He did not actually see a car, however.

Once he had time to think about his encounter, Dr Hopkins was understandably greatly disturbed by it, and was only too happy to erase the hypnosis tapes as instructed by the weird visitor. Yet it is difficult to think of any other circumstances in which he would so unquestioningly have obeyed the totally unreasonable and unauthorized demands of a complete stranger. Moreover, he discovered that the "New Jersey UFO Research Organization" did not exist, and therefore had no vice-president. But the story did not end there.

Within days Dr Hopkins's son John and daughter-in-law Maureen were telephoned by a stranger who asked to bring a colleague to meet them, which they agreed to (although, once again, it was only afterwards that it dawned on them just how peculiar the whole set-up was). When they rendezvoused at a local restaurant, the newcomers – a man and a woman – seemed to be normal enough. Closer inspection, however, revealed some marked oddities: although both the strangers were in their thirties, they wore quaintly old-fashioned outfits. The female had peculiarly low-slung breasts and there was something unusual about the movement of her hips – indeed, both she and her companion took very small steps while leaning forward, as if scared of falling over. And although they accepted the offer of drinks, they ignored them completely, making no move to taste them, after which they sat together gauchely on a sofa, closely questioning the Hopkins family about their lives, sometimes interjecting extremely personal questions. However, it was rather difficult to concentrate because all the time the male MIB was fondling the female, occasionally pausing to ask if he was doing it right! Clearly the weird creatures were experimenting with erotica, with which they seemed obsessed: when John briefly left the room, the male stranger asked Maureen if she had any naked pictures of herself. When it was time to go, something went badly wrong with the mobility of the "man", who stood stiffly without moving. The woman had difficulty with manoeuvring round him, finally appealing to the Hopkins family for help, saying: "Please move him. I can't move him myself."[9] Then, as if uncorked, they both moved off – in straight lines – without uttering another word.

This astoundingly surreal scenario was witnessed by persons of good standing in their community, with no interest in promoting themselves or seeking publicity, so assuming that there was no outright fabrication involved, what was going on? It seems that the MIB behaved in a distinctly robotic fashion – Dr Hopkins witnessed his visitor's speech "running down" as if the MIB needed winding up again or perhaps plugging in to some energy source, and the younger Hopkins' dining companions could only walk in straight lines, and even then, at the end of the evening, with difficulty. All three of the Hopkins's MIB seemed like rejects from the clone factory – almost, but not quite, passable as human beings: their clothes, speech and behaviour was all very much off-centre, as if they were trying, but failing, to ape the everyday behaviour of modern western people. But if they were robots, who or what had created them – and why? And if they were *not* robots, what on earth were they?

However, although most known MIB cases follow that pattern to a greater or lesser extent, there are entities that sound suspiciously similar that turn up in non-threatening roles in UFO encounter stories: for example, Luciano Galli's humanoid guide on the tour of the universe to the "mothership" was olive-skinned with Oriental features (see page 65).

Another apparently benign MIB – although one that was strikingly similar to Dr Hopkins's visitor – turned up at the Scarborough, North Yorkshire, home of "Adele" in May 1968. Then a sixteen-year-old schoolgirl, today a successful businesswoman and mother of two, she answered a knock at the door and found herself looking into the face of an extraordinary individual. "He was tall and lanky, with a black suit and tie, white shirt and a very small 'pork pie' hat," recalled Adele in the early 1990s.[10] "He had an extremely florid complexion and an almost insanely beaming smile. Although he was decidedly odd, I wasn't frightened of him at all – more plainly taken aback!

"After grinning madly at me for what seemed like ages – but probably only a few seconds – the man's whole body jerked, then he said: 'Have you got insurance? Is it now?' His voice was most odd. Like a robot's – jerky and without feeling. Looking back, I'd say it was more like a computerized voice. You know, the sort that says, 'Printing completed' ".

Adele thought there was something very peculiar about this ("Is *what* now?" she thought, mystified), but politely said that her parents would know about insurance but they were out, suggesting that he came back later to talk to them. At that he seemed, quite suddenly, to "sweat from every pore", removing his hat to wipe his forehead with the back of his hand – revealing a completely bald, and totally white, head. The florid "complexion" was now revealed to be a thick layer of badly applied stage make up, some of which came off on his hand. Still smiling fixedly, he looked her in the eyes and said: "Can I see a glass? Of water?"

Completely bemused, but worried in case the stranger might collapse on the doorstep, she asked him in. He followed her with strange jerky steps, his head thrown back, just like a marionette walking. Adele noticed that his "brand new" shoes were on the wrong feet and his trousers and sleeves were far too short, revealing completely hairless, "dead white" skin. She showed him to a chair in the lounge, and got a glass of water. When she came back from the kitchen she found him standing in front of the fireplace, staring at the carriage clock on the mantlepiece.

"He made me so nervous I started to blather," she says. "I told him that the clock was my father's retirement present, which seemed to be some kind of huge revelation to him. He stared at me – still smiling – and said: 'It is your father's time? Is it here and now?' Then he took the glass of water and just looked at it. I realized that he'd asked if he could *see* a glass, and that's what he was doing. I was flabbergasted. After scrutinizing it in a sort of polite way he handed it back, having not taken even the smallest sip. I began to think I was dreaming, or had gone mad – or that he had. But I never felt threatened in any way, just bewildered. Completely bewildered."

The visitor then stood staring at the clock, tapping it repeatedly and saying over and over again: "Your father. Your father. His time, his time". Turning round to face Adele with some difficulty – he had to use his hands to swing one leg round – he said, 'Watch the lights!' before patting the clock once more and walking in the direction of the door. Adele got there just in time to open it for him, otherwise he may have simply walked into it. She followed him to the front door, babbled something about it being "nice to meet you", but her visitor walked out without a word or a backward glance. Consumed with curiosity, as soon as she had closed the front door, Adele rushed into the front room to look through the window, but there was no sign of the weird creature anywhere, although she had a clear view down the street in both directions. This was very strange because, as it was a very long avenue, there was no way he could have disappeared from view in such a short time. But he had vanished.

Adele told her parents that a strange "insurance man" had called, but added no other detail because she felt unsure of her ground. They might have thought she was exaggerating – "who wouldn't!" she laughs – or disbelieved her. However, there were several unexplained incidents that happened in the wake of the visitor calling: the clock on the mantlepiece in the lounge mysteriously stopped, although it had worked perfectly for two years. Adele's father put it away, only bringing it out again as an experiment when they, like millions of others, sat enthralled watching the debut of young Israeli psychic Uri Geller on the *David Dimbleby Show* in October 1973. The viewers were invited to put metal objects or broken clocks and watches on or near the television set: a great deal of cutlery was bent, and many timepieces rejuvenated – Adele's father's clock among them . . .

Another weird thing happened almost immediately after the man had left the house. Within two hours several small but very bright white lights appeared and began to dance around the living room before exiting through the window – "as if it wasn't there" – into the garden, where they disappeared. Adele was still on her own, and thought it best not to mention them to her parents. But presumably they were what the man was referring to when he urged her to, "Watch the lights!"

What does one make of all that? What was the *point* of it? Nothing of a paranormal, UFOlogical or spiritual nature happened to Adele afterwards as a result. The lights came and went within seconds, leaving no apocalyptic visions or message for mankind. She visited no other planet and experienced no missing time. All that happened was that a very strange MIB came to call . . .

He seemed completely artificial: his voice sounded like robotic or computerized, and his whole body jerked before he uttered a word, as if he were being wound up or "rebooted". He was covered in make-up, and his smile had been fixed in place. His clothes were new, but the trousers and sleeves were too short, revealing artificially hairless and "dead white" skin, and his shoes were on the wrong feet. He was completely bald. Walking was a problem – he had to move his legs manually – and, but for Adele's quick thinking, would have walked into a closed door. All of which is very reminiscent of many other MIBS, even the more overtly menacing kind.

One young man – whom I will call "Peter"[11] – had a similarly bizarre experience on a train pulling into Victoria Station in London in the late 1970s. Settling down to read one of the most high-profile UFO books of that era (possibly even John A. Keel's *UFOs: Operation Trojan Horse,* which would have been particularly apposite*)* in one of the old-fashioned carriages, he was extremely disconcerted to find a strange character sitting on the opposite seat, staring quite blatantly at him without any particular expression, but somehow still with an air of menace. Although it was a warm day, he was wearing a thick black suit and black polo-neck sweater, very dark sunglasses and a black hat. On reaching the rail terminal, Peter stood up to gather his belongings for a few brief seconds, but when he looked up, the MIB had already gone. This was strange, but he was glad to see the back of him. However, once on the platform, the man suddenly reappeared, tapping him on the shoulder and saying, in a mechanical-sounding voice: "*Can you spare your life*?" before disappearing into the crowd. Understandably, Peter was shaken by his experience and pondered for some time about what the "message" could possibly mean.

Perhaps, though, there was no message. Maybe the MIB mean nothing, just as their threats prove to be empty bluff. Perhaps they have no logic where they come from – wherever that might be.

Yet, nonsensical though they appear to be, like not very convincing robots with faulty logic circuits, all the reported MIB appear to come from much the same mould. The new black clothes, the problems with mobility, the poor grasp of idiom (or a propensity for spouting nonsense), and the association with UFOs and paranormal events are found time and time again in the literature – although rarely in recent years. They seem to have been a central feature of the Contactee Era, but since the advent of the Greys they rarely put in an appearance. Perhaps they *are* the Greys, or something that has simply transmuted into them.

The MIB are also obsessed with time and life cycles, as can be seen from Peter's brush with the weird being who asked, *"Can you spare your life?"*, and the fascination with the carriage clock in Adele's story. Her MIB seemed to think that because the *clock* belonged to her father that somehow the *time* it showed was also his, which suggests that wherever the MIB come from, their idea of time – if they have one – is not the same as ours. Another episode illustrates this perfectly.

Mrs Ralph Butler and a female friend were watching the display of bright lights – which they called "little flashers" – that they saw in the field outside Owatonna, Minnesota, almost every night in the 1960s, when one of the lights began to swing back and forward over the ground. The women could see that this "little flasher" had become an object with a definite shape, with coloured lights along its saucer-like rim. Her female companion[12] suddenly collapsed onto her knees with a little gasp. She was very still, apparently in a trance. Then, 'A strange voice, stilted and metallic, came spasmodically from her lips. 'What . . . is . . . your . . . time . . . cycle?' The voice asked. Mrs Butler recovered from her surprise and tried to explain how we measured minutes, hours, and days. 'What . . . constitutes . . . a . . . day . . . and . . . what . . . constitutes . . . a . . . night?' The voice continued. 'A day is approximately twelve hours long – and a night is twelve hours long,' Mrs Butler replied. There were a few more innocuous questions, then the other woman came out of her trance. 'Boy, I'm glad that's over,' she remarked simply."[13]

Immediately after the incident both of the friends were on a "high", thrilled to have been telepathically in communication with UFOs, but as soon as they attempted to share their experience with others they became devastated with fierce headaches. The only time that Mrs Butler found she could discuss it and not suffer from intense pains in her head was when she contacted John A. Keel after she read an article by him in a magazine. He said: "I asked her all of my weird and seemingly silly questions, and she had all the right answers. She had been having unusual telephone problems and had also been receiving strange voices on her citizen's band (CB) radio."[14]

Mrs Butler suddenly asked John Keel whether anyone had "ever reported receiving visits from peculiar Air Force officers", to which he replied "cautiously" that he had "heard a few stories about them". Despite his caution, she was encouraged to go on, telling him a weird, but fairly typical, MIB encounter story. According to Mrs Butler, a man calling himself Major Richard French called at her home the previous May (1967), when both she and her husband were in, saying he was interested in CB radio and in UFOs. She described him as standing about 5ft 9in [1.7 m] with "a kind of olive complexion and pointed face". He had long dark hair that struck her immediately as being "too long for an Air Force officer." His English was that of a well-educated man (presumably with an American accent). He was smartly dressed in a white shirt and black tie – but his suit was not black, but *grey*. (Is this the only MI*G* on record?) Mrs Butler noticed that, "Everything he was wearing was *brand-new*". Eschewing the favoured MIB mode of transport – the black Cadillac – this man arrived in a white Mustang, whose registration number (which Mrs Butler's husband wrote down), was later discovered to belong to a real car, from the Minneapolis area. Then things became rather stranger.

Mrs Butler recalled that, "He said his stomach was bothering him. I told him that what he needed was some Jello [jelly]. He said if it kept bothering him, he would come back for some." That was already bizarre enough – why would an Air Force officer drop by at a strange civilian home just for a dish of jelly? If organizing a simple bowl of gelatine and water is beyond the skills of the USAF then there is not much hope for the world. But sure enough, the next morning, Major French showed up at the Butler residence and was duly presented with a bowl of Jello. What happened next was extraordinary. Mrs Butler asked Keel rhetorically, "Did you ever hear of anyone trying to *drink* Jello? Well, that's what he did. He acted like he had never seen any before. He picked the bowl up and tried to drink it. I had to show him how to eat it with a spoon." (Shades of Adele's MIB who politely *looked* at the glass of water.)

The same man later visited some friends of the Butlers in Forest City, Iowa, but unfortunately we have no details of his behaviour on that occasion. Perhaps he had learnt some table manners by then.

But did the weird Major French really exist? It transpired that while there *was* a Richard French in the USAF at Minnesota, he was nothing like the oddball who turned up at the Butlers' home.

Once again, the stranger seems to have had some connection – although it was never made explicit – with the sighting of the UFO and the friend's unheralded gift of mediumship. The paranormal elements in this story are strong: Mrs Butler and her friend regularly observed the "little flashers", and since the flap in their local area in 1966, the Butlers experienced all manner of poltergeist activity in their home. As John Keel noted: "Objects have been moving about of their own accord, glass objects have suddenly and visibly shattered without cause, and strange noises have resounded throughout the house."[15] (One is reminded of the Alex Birch hoax, with the underlying theme of paranormality running through it, and the story of Maggie Fisher, whose UFO encounter was just one of many unexplained events in her life.)

John Keel points out that the UFOnauts – and especially the MIB – seem to have an obsession with time: the voice that spoke through Mrs Butler's friend asked about her "time cycle", while the MIB that called on Adele in Scarborough seemed to react rather strongly to the idea that the clock belonged to her father, *as if he owned the time it represented.* Clocks and time appear time and time again as a theme in these encounters – but why? Could it be that whatever and whoever these aliens may be, they have no concept of time themselves? The MIB that caught up with the young man on Victoria Station asked him – with a sinister underlying threat – if he could "spare his life", as a beggar might ask a passer-by if they could spare some change.

Keel adds a little tale to this collection: in December 1967, a young student called Tom Montelecone, studying psychology in Adelphi, Maryland, claimed to have talked with the occupant of a grounded saucer, one "Vadig", who wore the standard shiny overalls. Some weeks afterwards, Tom was supplementing his allowance by working in a restaurant when Vadig turned up, dressed (fortunately) in a normal manner, with nothing unusual about him except for his "bulging thyroid eyes". They were to meet another three times, and the alien always ending their conversation by saying, "I'll see you in time".

As the ever-wise John Keel remarks: "Part of the answer to flying saucers may lie not in the stars but in the clock ticking on your mantlepiece.[16]

The accepted version of the MIB phenomenon has it that the strangers come to warn UFO witnesses off with threats – which many have taken to be proof that they are government agents, sinister *X-Files*-type individuals. Yet although the armed forces and intelligence agencies do produce some weird characters, surely none quite match the level of surreal "high strangeness" of cases such as Adele's and Mrs Butler's. And that type is considerably more common than might be imagined. Yet in a few cases it seems that the strangers may indeed have been government men, perhaps exploiting the prevailing paranoia about the sinister Men In Black for their own purposes. There is a case for the suggestion that the mysterious visitors who turned up on Crisman and Dahl's doorsteps during the Maury Island Hoax may have had the intention of covering up the illegal dumping of radioactive waste – which would explain the alleged physical effects, such as the burns, that the witnesses felt afterwards. Undoubtedly the authorities of any country would have no compunction in frightening off anyone who had seen something they shouldn't have – something highly classified, Above Top Secret, not for prying eyes or public consumption. How they choose to scare witnesses is up to them – perhaps some departments are more creative than others. One recalls how Ref Heflin's photographs of the UFO he saw near Santa Ana, California, in 1967 were taken away, never to be returned, by an alleged USAF Intelligence agent, who may or may not have been a MIB.

It is significant that Men In Black are not exclusively figures from the modern UFO era, although certain UFOlogists would have us believe so. In fact, there are many reports of strange beings dressed (for the most part) in black from all eras, where they are often associated with demonology. British author Patrick Harpur points out that: 'In the thirteenth century, Caesarius of Heisterbach was telling stories of demons in the guise of "big, ugly men dressed in black or, when set on seducing a woman, as a finely dressed fellow'[17] – or even as horses, dogs, cats, bears and other animals."

It is perhaps significant in this context that Albert Bender claimed that the dark-suited men he encountered turned into 'monstrous beings with hairy skin and glowing eyes.'[18]

However, sometimes the dark men are neither bad nor good, they just *are*. There is an interesting example of such an ambiguous quasi-MIB in the *Autobiography* of the Black Militant leader Malcolm X, where he describes such a being materializing in his prison cell:

> As I lay on my bed, I suddenly became aware of a man sitting beside me in my chair. He had on a dark suit, I remember. I could see him as plainly as I see anyone I look at. He wasn't black, and he wasn't white. He was light-brown-skinned, an Asiatic cast of countenance, and had oily black hair.
>
> I looked right into his face. I didn't get frightened. I knew I wasn't dreaming. I couldn't move, I didn't speak, and he didn't. I couldn't place him racially – other than I knew he was a non-European. I had no idea whatsoever who he was. He just sat there. Then, as suddenly as he had come, he was gone.

Is it significant that some witnesses to the 1890s' airship flap claimed that the aeronauts looked like "Japs"? Were they describing entities from the same race as the later MIBs?

Sometimes what appeared to be MIBs are later realized to be the aliens themselves. The case of abductee "Dave", which was investigated by Dr John Mack, involved an encounter when the witness was just three years old with three "black" men on motorcycles, in an episode that had certain surreal qualities. Later, under hypnosis, Dave realized that the motorcyclists transmuted into the more familiar aliens, which suggests that at least some other MIBs may be the UFOnauts themselves, coming to check up on, and apparently warn – although incredibly ineffectually – the witnesses to the phenomenon. If that is the case and there is any lesson to be learnt from the MIB experience, perhaps it is that we should not be afraid of the aliens, whoever or whatever they are, for even if they mean us harm, they seem incapable of actually harming us. On the other hand, perhaps the whole MIB thing is yet another experiment designed to test the witnesses's reactions to illogical and irrational behaviour, although the beings behind the experiment may be rather more terrestrial than is generally imagined, if certain conspiracy theories have an element of truth to them.

Notes

1. Jenny Randles, *Investigating the Truth Behind MIB: The Men In Black Phenomenon* (London, 1997) p. 7.
2. From Bender's account, quoted in Hilary Evans' *Visions*Apparitions* Alien Visitors* (London, 1984), p. 140.
3. *Ibid.*
4. *Flying Saucers and the Three Men* (Clarksburg, West Virginia, 1962).
5. One assumes the phrase "dark complexioned" or even "dark-skinned", common in the literature of that era, describes a swarthy, Mediterranean or Asian type rather than someone of Afro-Caribbean origin.
6. Widely acknowledged to be the worst screenwriter/director in the history of Hollywood. His greatest achievement was *Plan 9 From Outer Space*, during the making of which, movie legend Bela Lugosi, playing the lead role of a vampire/alien, died. Filming continued with Wood's dentist playing Lugosi's role – unfortunately he looked nothing like him, adding to the unwitting hilarity of the film. Truly a cult classic.
7. Evans, p. 141.
8. *Ibid*, p. 143.
9. *Ibid*, p. 144.
10. In several conversations with the author.
11. He wrote to *The Unexplained*, but unfortunately his letter was lost. The story is told here from memory.
12. This story is taken from John A. Keel's *UFOs: Operation Trojan Horse* (New York, 1971), p. 184. Mrs Ralph Butler's companion is not given a name.
13. *Ibid.*
14. *Ibid*, p. 185.
15. *Ibid*, p. 186.
16. *Ibid*, p. 187,
17. *De Divinatione daemonum*, lines 8-18, quoted in Harpur's *Daimonic Reality*, (London, 1994.)
18. See Randles, p. 49.

CHAPTER 5

THE INVADED

The phenomenon of alleged alien intervention in human lives, including abductions.

Why do you believe you were abducted?
You believe because you're crazy.
How do we know you're crazy?
Because you believe you were abducted.

Budd Hopkins

"UFO abductions are physically real events. But they are dramas materialised into three-dimensional space for us by the Phenomenon. They are dreams that the Phenomenon made come to life in very frightening vividness . . . Once someone has entered into physical contact with the Phenomenon the link may become permanent, and reactivate periodically."

D. Scott Rogo

(Sometimes the line between a mere "contact" and an abduction is a fairly thin one, and there is inevitably some overlap between this section and that of "Sightings and contact" cases. Unfortunately unexplained phenomena are rarely easy to categorize, despite the best efforts of the researchers.)

THE PHENOMENON

Not all abductions begin with a UFO sighting – causing some researchers to question whether there is always a connection between the two phenomena – but most are heralded with the witnessing of an unusually bright light. This can be accompanied by a structured craft, or the sighting of humanoid or Grey aliens, which can also be seen walking about inside the abductee's home, sometimes straight through solid walls. Often anomalous phenomena, such as vivid dreams or electrical malfunction also usher in the abduction proper, although the experiencer typically fails to recall the trauma until later, sometimes many years afterwards. Immediately before the onset of the experience, a buzzing sound is heard, or a noise like Velcro being ripped apart, (also a common feature of Near Death Experiences). Usually any non-abductee companion falls into a deep sleep or trance as if "switched off": spouses or fellow passengers in cars sleep like logs while the "chosen one" is taken.

Memories become muddled – although often there are none at all, simply a realization that many minutes, perhaps hours (or, in the case of Travis Walton, days) have been lost, for which there is no obvious explanation. After this, a host of physical and mental ailments, such as sinusitis, and anomalous markings like "scoop" marks on legs, gynaecological and obstetric problems (even, on one occasion, a virgin pregnancy), nightmares and an overwhelming, nameless dread surface, causing the abductee and their families untold anguish. They may also suffer from an unexplained loss of self esteem and uncharacteristic timidity, resulting in problems with relationships and in the work place. Sexual and marital trouble is common. In many respects, the reaction of the abductee – although they may not consciously recall their experience – is very like that of a victim of sexual abuse.

Later, perhaps recalling the events spontaneously – but more often than not through hypnotic regression – the full extent of the experience begins to dawn, usually accompanied by powerful sensations of disgust and hatred for the captors, the alien beings who kidnap, molest and even rape with such impunity. The abductee has suffered a crime with the same need for counselling and compassion – if not more so – as a more conventional sort of victim, yet all too often the very help they need so badly is denied them because, afraid of ridicule, they do not seek it, often for many years.

Even if the experiences never occurred in a literal way, exactly as recalled, from the cases given below it is only too easy to see that *something* happened to these people for which there is no precedent and no *context*: nothing like this has ever happened before to them, or to anyone they knew. Society as a whole is blind and deaf to their plight, but if they are lucky, they can find some rare help from the likes of Budd Hopkins or Dr John Mack, the best-known American abduction researchers whose intelligent and gently questioning approach has drawn some astonishing revelations from the experiencers.

Anyone, of any age, colour, religion or nationality can be "taken", although the interpretation of their experience may differ depending on their culture. No one knows how many abductees there are, and although they are scattered throughout the world, almost certainly the greatest concentration is in the United States, where some suggest a sinister military connection (for a detailed discussion of Military Abductions – MILABs – see Conspiracies).

It seems likely that the known cases are very much the tip of a massive iceberg, of which the following stories can only be a small sample. This, then, is a peep into a terrifying world where our cosy view of reality is turned inside out and the rules no longer apply . . .

1952, DESERT CENTER, CALIFORNIA

The early contactees, who although rarely coerced by the space people, paved the way for the later – and considerably nastier – abduction scenario. And some of them were more influential than others.

George Adamski, the most famous of the early contactees, met his first alien just after noon on 20 November 1952 in the middle of the desert in California, not far from the Arizona border, although allegedly he had seen spacecraft for several years before that.

Calling himself "Professor", he claimed to have a professional connection with the giant telescope on Mount Palomar, but in fact he only ran a hot-dog stand a few miles away with his wife Mary. But there were many (and possibly still are) who regard Adamski as a very important person, for he claimed to have met tall, blond entities from Venus, besides aliens from elsewhere in the galaxy, such as Saturn, and to have been taken on board their craft on a tour of the stars. His alien friends told him that the humanoid shape was found throughout the universe, imbuing him with a sense of brotherhood and love, which earned him a unique status in the early, uncritical, days of UFOlogy.

He also took photographs of the alien craft, which show bell-shaped, metallic UFOs tipped backwards to reveal what look like – possibly for obvious reasons – three white ping-pong balls on its underside. This image has had an astonishing hold on the imagination, becoming the standard "flying saucer" which is still accepted today, although flying triangles and the like are now in the ascendancy.

Adamski's 1953 book *Flying Saucers Have Landed*, co-authored with British writer Desmond Leslie, became one of the most influential of the day, effectively ushering in the "contactee" era, and paving the way for the darker, more disturbing nightmare of the modern abductees.

However, Adamski's character and background suggest strongly that there was at the least an element of fraud in his claims to have been chosen by the space brothers for initial contact. He was later to tell two followers[1] of his exploits in making and selling wine during Prohibition, by means of founding a monastery at Laguna Beach, California, called "The Royal Order of Tibet", and claiming to need the wine for religious purposes. He said: "I made enough wine for all of Southern California . . . I was making a fortune." Unfortunately Prohibition ended and the market was flooded with alcohol once more. If that had not happened, Adamski said, "I wouldn't have had to get into this saucer crap." But although his professorship was bogus, he was genuinely knowledgeable about Eastern religions, on which he gave public lectures.

It was on the night of 9 October 1946 when Adamski was watching a meteor shower above Mount Palomar that he saw "a large black object, similar in shape to a gigantic dirigible" hovering above him. The next year he witnessed no fewer than 184 "saucers" moving majestically through the sky in the same area. (Note that his initial sighting predated Kenneth Arnold's watershed encounter with the "saucers" by several months – according to Adamski's *later* account.)

Adamski claimed that in 1949 Joseph P. Maxfield and Gene L. Bloom from the San Diego Naval Electronics Laboratory arrived at Palomar Gardens and asked him to help photograph the saucers.

Using a camera mounted on a six-inch (15 cm) telescope, Adamski said he managed to get "two good pictures of an object moving through space", which he gave to Bloom.

By 1950 Adamski was famous in the local area, and in Spring of that year gave a lecture on his favourite subject, the saucers, where he described his encounter with Bloom and Maxfield and the photographs they had taken away. Unfortunately for his credibility, a reporter from the *San Diego Journal* checked up on his story. It transpired that the Naval Electronics Laboratory had never heard of the incident, Bloom and Maxfield had never sought Adamski's help, nor had they taken any photographs away. Bloom stated unequivocally: "Everything Adamski wrote about us was fiction, pure fiction."[2]

Adamski hoped he would make contact with the occupants of the saucers, continuing to try to take photographs of the craft, although few amounted to anything. Then on 20 November 1952, Adamski, his wife Mary, Al and Betty Bailey, Alice Wells, Lucy McGinnis and Dr George Hunt Williamson drove off in two cars into the Californian desert. Without warning, they saw a "gigantic cigar-shaped silvery ship" in the area of Desert Center. Not unnaturally, Adamski was almost beside himself with excitement, especially after his abortive attempts to photograph the saucers and his longing to meet their crew. Crying "That ship has come looking for me and I don't want to keep them waiting!", he got Al Bailey and Lucy McGinnis to drive him to a deserted spot, where he told them not to return unless he signalled for them to do so. All the others waited and watched.

They saw the cigar-shaped vessel depart from the area at the approach of several military craft, but within a few minutes a smaller "scout ship" appeared, which Adamski photographed. The military planes drove that one away, too.

However, all that was to pale into insignificance compared to what happened next. Adamski saw a long-haired blond man in a tightly-fitting brown "ski suit", and had no hesitation in declaring that he "was in the presence of a man from space – A HUMAN BEING FROM ANOTHER WORLD!" – in this case, apparently, the planet Venus. It turned out that communication was relatively easy – the alien understood some English and basic hand gestures, besides employing a form of telepathy. At one point he made a shape like a mushroom with his hands and said "boom! boom!", indicating, according to Adamski, that radiation from nuclear fall-out was a grave danger to the galaxy, and that he had come to warn about the dangers of the Bomb.

The blond Venusian said that his fellow aliens believed in a "Creator of All", about which Adamski later wrote: ". . . we on Earth know very little about this Creator . . . our understanding is shallow. Theirs is much broader, and they adhere to the Laws of the Creator instead of laws of materialism as Earth men do."

The alien told Adamski that on occasions humans had caused other spaceships to crash – which was perhaps why his compatriots did not intend to make a public landing. If they had, he said "there would be a tremendous amount of fear on the part of the people, and probably the visitors would be torn to pieces by the Earth people, if such landings were attempted."

When asked if the space-people had ever carried off humans, the Venusian smiled and nodded . . . then he pointed at the ground, where his shoes had made unusual marks on the desert floor. After the scout ship took the alien away again (with the film holder he requested), Dr Williamson took plaster casts of the footprints (although how he happened to be carrying the necessary kit is unclear), the group travelled into Phoenix to report their exciting news to the *Phoenix Gazette*, who published an article, together with a very muzzy photograph and some sketches by Adamski.

The excited contactee knew the Venusian would be back. Sure enough, on the morning of 13 December 1952, the scout ship flew slowly and soundlessly towards him. Hovering overhead, a porthole opened and a hand dropped the film holder out and waved goodbye, after which the saucer went on its way, flying over Palomar Gardens where Sergeant Jerrold E. Baker managed to capture it on film.

When the film that the Venusian had dropped was developed they saw it was covered in strange markings exactly like those on the alien's footprint in the desert.

It was at this point that Adamski approached Desmond Leslie, resulting in his sixty-page account being appended to the end of that author's newly-finished book, entitled *Flying Saucers Have Landed*. Unsurprisingly, it became a massive best-seller, and is still considered to be a classic contactee story. As Curtis Peebles writes: "It was Adamski's tale that created the 'Contactee Era' of the 1950s.

It supplied 'their' description (handsome), motivation (fear of nuclear tests), and most importantly of all, their message of love to stave off the abyss of nuclear war."[3]

However, Adamski did not have it all his own way. Criticism and accusations were not long in coming: a newspaper[4] described how he had claimed in a lecture that all his material had been "cleared by the FBI", but when three FBI agents put him on the spot about this, Adamski denied having said it. Then, after signing a statement to the effect that he fully understood that the FBI do not endorse individuals and the seriousness of making false claims, he used that document to another interviewer as proof that he had been "cleared" . . . After that, the FBI visited him to "read the riot act to him in no uncertain terms".

Worse was to come, when Al Bailey, who had been with Adamski on the memorable night of the first contact with the Venusian, denied having seen anything other than the mothership and a light in the distance, adding that none of the others could have seen any more than he did. This contradicted statements about what they allegedly saw, as detailed in *Flying Saucers Have Landed*. Then Jerrold Baker who was supposed to have taken the pictures of the scout ship as it passed over Palomar Gardens, added his voice to the growing clamour, alleging that the historic trip to the desert was not a spur of the moment event, but had been preplanned, and his wife said that Dr Williamson (whose title, like Adamski's, was "honorary") had not seen the spaceman either. Most damning of all, she reported that Adamski had said that "in order to get across to the public his teachings and philosophies, he couldn't be too 'mystical' . . . he must present all the happenings on a very material basis because that is how people want them." When she had objected that this "was as good as lying", he responded, "Sometimes to gain admittance, one has to go around by the back door."

The brotherly love preached by the space people failed miserably to percolate through to the Mount Palomar Adamski set. In 1961 Lucy McGinnis left her post as the "prophet's" secretary because he had taken up the dubious practice of trance mediumship,[5] but this was minor stuff compared to the damage he did to his own reputation when declaring he had personally visited Saturn, and met with the Pope and President Kennedy. As Curtis Peebles writes:[6] "By the spring of 1964, the situation at Mt. Palomar was an open palace revolt."

Yet even though it seems that Adamski certainly went off the rails with such wild claims, this may not be such an open-and-shut case of charlatanism as it first appears. While there is sufficient doubt about the story as Adamski told it, there is still the possibility that *something* happened in the desert, even if it did not involve tall blond Venusians with strange shoes. And unlike other contactee prophets, Adamski was neither particularly dangerous – except in promulgating fantastic tales – nor avaricious, and seemed quite genuine in his fascination with the *idea*, at least, of universal brotherhood, philosophy and mysticism.

His story unleashed a flood of contactee claims, some wilder and/or more amusing than others which raised the profile of the flying saucer phenomenon beyond the level of simple sightings. Now the occupants were beginning to impinge on mankind – usually, in theory at least, for the better. The really disturbing stories of alien intervention in human lives were still some years in the future.

SEPTEMBER 1955, AUSTRIA

What has been called – with some justification – the "single most bizarre encounter case"[7] took place in September 1955 when Josef Wanderka rode his moped along a road in his native Austria, and, through a momentary loss of concentration, found he had driven up a ramp straight into a UFO!

The aliens on board, who conveniently spoke German – and to whom he apologized for his intrusion – told him they had come from the star system of Cassiopeia. They were intrigued by his moped and questioned him about how it worked. It was all very amicable, but Wanderka had no wish to be taken away from his native Austria – or his native planet – so he insisted on giving them a lengthy anti-Nazi lecture. Apparently it bored them stiff, so they ejected him from the UFO and he was free to continue his journey on his trusty moped.

Certainly one of the more amusing and harmless alien contact stories, one wonders if the key was the witness's loss of concentration. It is known that extremely boring journeys can induce a form of self hypnosis,[8] which may have resulted in a realistic-seeming fantasy of meeting, and even boring, aliens. On the other hand, we know very little about this case: there is always the possibility that it may have happened exactly as the witness claimed.

5 OCTOBER 1957, SAO FRANCISCO DE SALLES, MINAS GERAIS, BRAZIL

The following is now widely regarded as a classic case, largely due to the meticulous investigation undertaken shortly after the event by the late Professor Olavo Fontes of the National School of Medicine in Rio de Janeiro, Brazil.

At around 11 p.m. on 5 October 1957 as twenty-three-year-old Antonio Villas-Boas and his brother were going to bed, they observed a strange light, like a searchlight, apparently coming from the adjacent corral, which then swept over the house before disappearing. That was only the beginning.

Nine days later, at about 10 p.m., Villas-Boas was on his tractor when a light, coming from the other end of the field, almost blinded him. Intrigued, he tried to get closer to it, but each time, although it seemed to "wait for him", it moved away. Witnessed by one of his brothers, Villas-Boas and the light played this game of "catch" about twenty times before he gave up. Finally, the light vanished. But at about 1 o'clock the next morning the witness observed a star-like object, reddish in colour, transmute into a much larger bright, egg-shaped thing with a rotating dome on top, changing from red to green, that hovered over him before landing quietly. By now rather frightened, the witness tried to make a getaway on his tractor, but the engine failed, so he jumped off. At this point he was seized, overpowered, and carried off into the object by four "men" wearing tightly-fitting greyish suits and helmets, who communicated with each other with peculiar slow growls, (which were however, "neither high-pitched nor too low").

Inside the craft, Villas-Boas fought his abductors, but they managed to strip him naked and cover him all over with some kind of thick liquid before leading him through a door, with strange hieroglyphics or letters on it, into a second room. After vomiting violently, he was alone for what seemed like ages, when suddenly a naked woman came in, giving him (not surprisingly) a "terrible shock". She was a blue-eyed blonde, with a straight nose, a face "wider than that of an Indio native", a mouth so small as to be almost invisible, and a sharp chin. She was very small – her head only reached his shoulder – although broad-hipped.

Very quickly the witness and the female humanoid were enthusiastically locked in primitive sexual intercourse, which was only slightly marred by her propensity to utter short barks like a dog when aroused. Afterwards, one of the male creatures entered the room and pointed at the female, who then gestured at her abdomen and then at the stars.

Shortly after this, they returned his clothes and took him to yet another room where the crew were sitting around, growling to each other. Convinced by now that they meant him no harm, Villas-Boas had a good look around, noticing what he described as "an alarm clock" – a glass-topped box – which had one hand and marks to indicate the various hours. But as time passed and the hand did not move, the witness decided the object could not be a clock after all. Although he had thought of trying to steal it as a souvenir, he was not given the chance to do so, being taken on a tour of the craft before finally being let out via a metal ladder. Once back on the ground, he watched the UFO take off, disappearing from sight within seconds. To his amazement, he discovered that four hours had elapsed since his capture.

For some weeks following the incident Villas-Boas noticed peculiar wounds on his limbs and, as in the case of Maurice Masse, suffered extreme drowsiness.

It was not until 1980 that the witness spoke publicly about his extraordinary experience, confirming his original account but adding that the female humanoid had taken a sperm sample away with her in a container. He died in the 1980s, without changing his story in the slightest.

At first glance one might be forgiven for being sceptical about this case: it sounds too much like erotic sci-fi fantasy. But in fact it contains many elements of a much older tradition – that of abducting fairies and demonic lovers – not to mention the modern concept of creating hybrid human/alien babies, suggesting that Villas-Boas was telling the truth as he recalled it. (It is also unlikely that a mere macho fantasy would have included the humiliating detail of vomiting before engaging in intercourse with the alien female.) Besides, Villas-Boas was no publicity-hungry sensationalist: he only told his story in response to a newspaper feature calling for reports of UFOs in the area, and even when the respected investigator Professor Fontes had won his trust, he was still extremely reluctant to describe the more intimate parts of the story.

Dr Jacques Vallée, in his seminal *Passport to Magonia* (1970), remarks of this case that, "The symbolism [of the clock] . . . is clear. We are reminded of the fairy . . . country where time does not pass . . . It is the poetic quality of such details in many UFO sightings that catches the attention – in spite of the irrational, or obviously absurd, character of the tale – and makes it so similar to a dream." The failure to make off with some evidence of the incident – in this case the clock – is, he comments, "a constant feature of fairy tales . . ."[9]

SEPTEMBER 1961, NEW HAMPSHIRE

The joint abduction of husband and wife Betty and Barney Hill in 1961 is still one of the most significant and controversial cases on record.

Although recounted in great detail elsewhere,[10] it is instructive to examine the official report of the incident, written for Strategic Air Command, Pease Air Force Base, New Hampshire by Major Paul W. Henderson:[11]

During a casual conversation on 22 September 61 between Major Gardiner B. Reynolds, 100th B W DC01 and Captain Robert O. Daughaday, Commander 1917-2 AACS DIT, Pease AFB, NH, it was revealed that a strange incident occurred at 0214 local on 20 September.

No importance was attached to the incident at the time. Subsequent interrogation failed to bring out any information in addition to the extract of the 'Daily Report of Controller' [. . .]

> On the night of 19-20 Sept. between 20/0001 and 20/0100 Mr & Mrs Hill were traveling south on route 3 near Lincoln, NH when they observed . . . a brightly lighted object ahead of their car at an angle of elevation of approximately 47 [degrees]. It appeared

strange to them because of its shape and the intensity of its lights compared to the stars in the sky. Weather and sky were clear. They continued to observe the object from their moving car for a few minutes then stopped. After stopping the car they used binoculars at times.

They report that the object was traveling north very fast. They report it changed direction rather abruptly and then headed South. Shortly thereafter it stopped and hovered in the air. There was no sound evident up to this time. Both observers used the binoculars at this point. While hovering objects began to appear from the body of the "object" which they describe as looking like wings which made a V shape when extended. The "wings" had red lights on the tips. At this point they observed it to appear to swoop down in the general direction of their auto. The object continued to descend until it appeared to be only a matter of "hundreds of feet" above their car. At this point they decided to get out of the area, and fast. Mr Hill was driving and Mrs Hill watched the object by sticking her head out of the window. It departed in a generally North-westerly direction but Mrs. Hill was prevented from observing its full departure by her position in the car.

They report that while the object was above them after it had "swooped down" they heard a series of short loud "buzzes" which they described as sounding like someone had dropped a tuning fork. No further visual observations were made of this object. They continued on their trip and when they arrived in the vicinity of Ashland, N.H., about thirty miles [48 km] from Lincoln, they again heard the 'buzzing sound' of the 'object'; however, they did not see it at this time.

Mrs Hill reported the flight pattern of the 'object' to be erratic, changed directions rapidly, that during its flight it ascended and descended numerous times very rapidly. Its flight was described as jerky and not smooth.

Mr Hill is a Civil Service employee in the Boston Post Office and doesn't possess any technical or scientific training. Neither does his wife.

During a later conversation with Mr Hill, he volunteered the observation that he did not originally intend to report this

> incident but inasmuch as he and his wife did in fact see this occurrence he decided to report it. He says that on looking back he feels that the whole thing is incredible and he feels someone foolish – he just cannot believe that such a thing could or did happen. He says, on the other hand, that they both saw what they reported and this fact gives it some degree of reality.
>
> Significantly, the Air Force Base's radar recorded an unidentified object at around the time that the Hills' experience was taking place.

The report quoted above contains only the bare outlines of the sighting itself, but omits almost everything that is now considered especially significant.

In the weeks following their sighting, both Betty and Barney began to suffer from nightmares. This prompted them, in desperation, to seek psychiatric help from a Boston therapist Dr Benjamin Simon who used hypnotic regression to recover the memories of what happened during the missing time. Gradually an extraordinary story emerged, which contained elements that were to become staples of the alien abduction scenario.

Under hypnosis, Betty recalled that, when they had stopped their car to get a better look at the bright light, some "men" had approached them, opening the car door and pointing a small thin object at her, which she said later "could have been a pencil".[12] She described the men as being not "as tall as Barney, so I would judge them to be 5' to 5' 4" [1.5 to 1.8 m]. Their chests are larger than ours, their noses were larger . . . like Jimmy Durante. Their complexions were of a gray tone, like a gray paint with a black base; their lips were of a bluish tint. Hair and eyes were very dark, possibly black . . ."[13]

Barney's report, made under hypnosis, while differing in some respects, more or less tallied with his wife's. He said: "The men had rather odd-shaped heads, with a large cranium, diminishing in size as it got toward the chin. And the eyes continued around to the sides of their heads, so that it appeared that they could see several degrees beyond the lateral extent of our vision. This was startling to me . . . [Their mouths were] much like when you draw one horizontal line with a short perpendicular line on each end. This horizontal line would represent the lips without the muscle that we have. And it would part slightly as they made this mumumumming sound. The texture of the skin, as I remember it from this quick glance, was grayish, almost metallic looking. I didn't notice any proboscis, there just seemed to be two slits that represented the nostrils."[14] His wife also noted that some of the aliens wore a "Nazi"-like insignia.

The entities spoke aloud to each other in an unintelligible language, although they communicated telepathically to Betty and Barney in English – which, according to Betty, somehow had an accent. Barney, however, said:

> I did not hear an actual voice. But in my mind, I knew what he was saying. It wasn't as if he were talking to me with my eyes open, and he was sitting across the room from me. It was more as if the words were there, a part of me, and he was outside the actual creation of the words themselves.[15]

Entering the craft, the couple were taken to different rooms for physical examination. Betty's "examiner" asked her several questions, but appeared not to fully comprehend her answers. After closely examining her throat, skin, ears and feet, needles (attached to a kind of EEG machine) were run over her body, her dress having been removed "as it was hindering the testing". The needle, the aliens said, was a "pregnancy test", which they promised would not hurt: however, when a long needle was inserted into her navel she felt an intense wave of pain, which seemed to surprise them. The chief examiner waved his hand in front of her eyes and the pain disappeared. This seemed to be a psychological and emotional turning point for Betty, who "became very grateful and appreciative . . . lost all fear of him, and felt as though he was a friend."

The examination ceased when the examiner was called out of the room. On his return he touched her teeth, wanting to know why they could not be removed, while Barney's could. Amused, she explained that her husband had dentures, adding that losing one's teeth was a feature of getting old. This meant nothing to the aliens, who had no concept of old age.

Remarkably calm and composed by now, Betty asked if she could take some proof of the experience away with her, to which her captors agreed, allowing her to select an imposing book, the pages of which were covered in "symbols . . . in long, narrow columns". She then asked the aliens where they came from, at which they showed her a "sky map", but refused to pinpoint their place of origin.

According to the story that surfaced in their dreams, when Barney was brought back, a row broke out among the aliens, resulting in Betty's "proof" – the book – being taken back. Angry, she asked why they had done so, to which they replied that they had decided the Hills would not remember the experience. They were then led back to their car.

Immediately after their abduction, both Betty and Barney, while having no conscious memory of the event, began to be plagued with nightmares about being abducted and examined by aliens, and suffered from recurring health problems.

A few weeks after what was still simply a UFO sighting to them, Betty and Barney spent six hours telling their story to Walter Webb of NICAP, followed a month later with a meeting with two more officers of NICAP and a retired Air Force major (also a friend of theirs). It was at this meeting that a pivotal question was asked – "What took you so long to get home?" It had never occurred to them that their journey of just 35 miles (56 km) had taken two hours too long. The sudden realization of the missing time came as a shock, making them wonder if their nightmares were about real events.

Barney was having serious problems with high blood pressure and ulcers, which his GP believed to be psychosomatic in origin, so he was referred to the respected Boston psychiatrist and neurologist Dr Benjamin Simon, who regularly used hypnosis as a therapeutic tool. Significantly, Dr Simon was quick to see that both husband and wife needed treatment, deciding that while they had both experienced the UFO sighting, *the abduction had only happened to Betty*. He came to believe that the whole thing was merely a dream, which had become a *folie a deux* by means of Betty's stronger personality "infecting" Barney's unconscious so that he came to believe he had also had the dream.

Dr Simon remained resolutely sceptical about the reality of the Hills' abduction. When asked in an interview with *Look* magazine's

Editor, "Do you believe the Hills were abducted by spacemen?" he replied, "Absolutely NOT".[16] To him, it was the symbolism of the "dream" that was important, which he interpreted as representing Betty's problems with her mother.[17] He also felt that there was too much absurdity in the experience – the aliens knew nothing of time, but at one point Betty claimed they told her to 'wait a minute', for example – for it to be real. However, Simon felt strongly that Betty was not consciously lying, she truly believed every word of it – as far as she was concerned both she and Barney had been taken on board a spaceship and had interacted with aliens.

Later, Betty was to draw the star map she had been shown by the aliens, inspiring researcher Marjorie Fish to reconstruct a three-dimensional representation of stars close to Earth, which seemed to show that the Hills' aliens hailed from Zeta Reticuli 1 and 2. Like everything else about this seminal case, the star map remains hugely controversial.

Since the Hills' case was made public through John G. Fuller's *Interrupted Journey* in 1966, it has attracted enormous attention, much of it increasingly critical. Yet the fact remains that their experience, be it a dream or a "real" event, cannot simply be explained by Dr Simon's idea of the resurfacing of Betty's unhappy family history, or social tension (Betty is white, Barney was black).

Certain elements in the story, such as the insertion of a long needle into Betty's navel and the "Nazi insignia" some of the aliens wore, are particularly suggestive of something deeper and more disturbing at work. In his 1991 book *The UFO Encyclopedia,* which he compiled and edited for the British UFO Research Association (BUFORA), John Spencer writes: ". . . This is significant because at the time there was no such recognised medical test [as the needle insertion], whereas in subsequent years such a test has become prominent"[18].

It may also be significant that a fifteenth-century French calendar, the *Kalendrier des Bergiers (The Shepherds' Calendar),* depicts demons thrusting long needles into the abdomens of the hapless people they had "taken" . . . Was this, even then, an image from the collective unconscious? Or was the medieval artist drawing on his own experience of being examined by aliens?

Whatever the origin of the experience, there is no doubt about the importance of the Hills' case in the annals of UFOlogy – not to mention twentieth-century mythology. As Curtis Peebles says: "The Hill abduction story would set the pattern for the future – a person sees a UFO, experiences 'missing time', then tells stories of a medical exam under hypnosis. But not just yet."[19]

Now a widow, Betty claims to see UFOs virtually every night, although some witnesses report that they are simply misidentified aircraft.

THE EARLY 1960s, LEEDS, NORTH OF ENGLAND

The difficulty in even beginning to be objective about alien abduction is illustrated by the multi-levelled complexity of the story of Maggie Fisher (a pseudonym), whose life, while rational and mundane enough on the surface, has always included a rich undercurrent of the paranormal and the mystical – something she is usually at pains either to hide or downplay. Now in her late forties, she works as a feature editor on her local newspaper in the north of England. This is her story, mostly in her own words.[20]

> As an only child I was always a loner, not because I had to be but because I chose to set myself apart. Somehow I felt puzzled by the rough and tumble of other kids. All I ever wanted to do was play quietly with figures cut out of magazines in my bedroom, although I did enjoy coming out of my self-imposed isolation to have a good laugh at comedy programmes on the radio or television. I've always had this strange double life in a way: I've always been deadly serious and a loner on the one hand, yet full of fun and loving to laugh – and make other people laugh – on the other. A very well-known stand up comic once told me that I had all the makings of being a successful comedian, but I have never tried to make my name in that way.
>
> One night when I was a child – curiously I can't remember exactly how old I was when it happened, perhaps about seven or eight – I woke up in the pitch dark, utterly terrified for some reason. I remember I crawled up onto my pillow in the corner of the room, clutching the bedclothes round me. I was too scared even to dash across the landing to wake my parents or climb in bed with them

for comfort, although I desperately wanted to. I knew something very scary was about to happen and that I had no choice. I had to experience it in some way. I didn't feel as if I was paralyzed with fear. It was more as if my bones had turned to jelly.

My eyes felt huge, as if compelled to strain into the dark but yet I really didn't want to see whatever was coming. Then suddenly the wall at the end of the bed seemed to mist over, and a woman's head appeared. I nearly fainted with terror. She had a dead white face, bright red hair parted in the middle and gigantic bright green eyes, almost "wraparound", that glittered horribly as she focused with enormous concentration down on me, this little vulnerable child crouching in the corner. I remember whimpering, and although I could see the woman's lips move, my heart was thumping so loudly that I couldn't hear very much at all. She seemed very annoyed that I wasn't getting her message and somehow I knew she kept repeating it. From the fragments that got through to me she seemed to be telling me that I would do something very important when I grew up, giving certain specific details, although she was at pains to point out that some of them were "symbolic". She repeated the word several times, although being so young I only had the haziest idea of what it meant. Obediently, though – for I was a very obedient child – I remembered her words, or at least those my thumping heart allowed me to hear.

I got the distinct impression she was not best pleased with me. She began to look sideways, as if consulting someone "offstage". She faltered and became so distracted that she stopped trying to communicate with me altogether. Finally, having listened intently to her unseen ally, the woman abruptly stared down at me again, her eyes seemingly even bigger and more terrible. Then she just said: "You're the wrong one" very accusingly – and vanished. What an anti-climax! I'd been through all that terror and it turned out she hadn't meant to appear to me at all!

The darkness seemed to crackle around me and the spell was broken. I leapt off my bed and dashed into my parents' room crying and babbling about a "funny lady on the wall". Of course they believed I'd just had a nightmare, and cuddled me until I fell asleep. But the next morning, and for the rest of my life, I knew it hadn't been a dream, but something very odd and special had happened. And although I had been terrified almost out of my wits – *and even though the woman told me I was the "wrong one"* – remembering the experience still makes me feel special, chosen, if you like, although I am very, very wary of using language like that. In all honesty, however, I can't think of any other way of putting it.

I don't discuss the details of the green-eyed woman's message. I've never described precisely what she said to me to anyone, although I will say that she predicted that I would do certain things that I realized quite recently I have actually accomplished, without any conscious effort on my part. But the vision – if that's what it was – proved to be only the start of a series of bizarre happenings.

Within days of the experience I went into my bedroom to collect something in the early evening and there was a thick, pungent smell in there – like incense I suppose, although as a child I had never been exposed to it and we certainly used nothing like it at home. Although I have a sensitive nose and some strong odours tend to sicken me, this one felt somehow friendly and inviting. I sat on my bed, and seemed to see the scent mistily moving round me. Gradually it became a ball of light, at first no bigger than my fist, but it soon grew into a large sphere almost the width of my little bedroom, completely enveloping me. I was inside the light, and felt radiantly warm and happy. I could see that there were silver and gold flecks of light moving around me, fizzing slightly like sparklers. This time, though, I wasn't afraid at all. This felt right. Then Mum shouted upstairs for me, and when I turned my head the light vanished and with it the perfume and the warm, happy feeling of being special. The spell was completely broken. I felt huge disappointment and was very sulky with Mum for days after that, which puzzled her.

Time passed, maybe a month or so. Then one night I woke up with a bad sore throat, feeling hot and thirsty. I jumped out of bed and ran to my parents' room – only to find that their door was closed and, apparently, locked. I can't exaggerate how bizarre this was. I had never known them to do this, ever. I knocked on the door but they didn't even answer. Panicking, feeling something was very wrong, I hammered on the door but there was still silence. I realize people might smile at this, thinking that obviously they must have been making love and didn't want to be disturbed, but I don't think so – they would certainly have opened the door to my desperate hammering – and in view of what happened next, I think it was all part of the same bizarre event. Having failed to rouse my parents, I decided that the only thing I could do to ease my throat was get some orange juice from the fridge downstairs. This was something of a first, too, because normally I slept right through and had never got up in the night. Feeling a bit scared in the darkened house, I slipped downstairs, guided by the light of the street lamp – and found the fridge door fast shut. Although it was normally very easy to open, it just wouldn't budge. By then my throat felt red raw and I was really hot, so I opened the backdoor and stood out in the yard to get some air. After the first blissful breeze I realized it was actually far too cold to be standing there in just my night-dress, but I soon forgot that because of what happened next. Coming down on top of the shed was what seemed to be a machine. It was a shiny, gun-metal grey contraption like one saucer upturned on another, with a couple of antennae that fizzed noisily. It's hard to calculate its size because I was so small myself and it was a long time ago – way back in the late fifties or early sixties – but this thing seemed to cover the shed top. I suppose it must have been about fifteen feet [4.5 m] wide and four or five feet [1.2 or 1.5 m] high. I stood rooted to the spot. I can't remember any feelings except awe.

The thing started to hum and seemed to change shape, but I can't tell you in what way because the next thing I knew I woke up on the sofa in the lounge with my father yelling to my mother that it was all right, he'd found me. Apparently they'd gone to my bedroom and found I wasn't there and panicked. The curious thing, though, was that I was wrapped up in one of Mum's coats, which had been hanging in the wardrobe in their bedroom, which I had failed to get into. I asked them why they'd locked me out and they denied it, being genuinely horrified at the very thought.

Assuming that I had walked in my sleep, which was a reasonable supposition, they took me to the doctor, who checked me out. I was sniffling a lot, but I'd been sickening for a cold for days, so no one took much notice. I got a bit embarrassed when my nose suddenly started bleeding – all over his clean shirt! – but no one thought much of it. The next day a fragment of what appeared to be clean bone fell out of my nose, although it didn't hurt at all and it seemed as if no harm was done. Years later when I had a medical for insurance purposes the doctor asked me when I'd broken my nose. It was news to me that I *had* ever broken my nose! I was quite shocked. I may be wrong, but I don't think it's the kind of thing that can just happen without you noticing it, but as soon as the words were out of his mouth my mind went back to that weird night when I saw the machine on top of our shed and woke up on the sofa. What made me associate a broken nose with seeing the UFO?

Shortly after seeing that experience I had an electric shock off a loose light fitting, which also seems to be part of the equation in some way, but I'm not sure how.

After that, I almost became used to weirdness. We had what I now realize could be called a poltergeist – objects thrown, although never very dangerously, by an invisible force, strange smells and pools of water, that kind of thing. Things would go missing and turn up again days later in the oddest places. I remember once my birth certificate was found in a bag of coal – the real mystery being that the coal had only just been delivered! After scratching our heads over it for a day or so, we just forgot about it. That was always the easiest option.

I never discussed any of the weirdness with anyone else for many years, even though at first, when I was tiny, I firmly believed that such things happened to everybody. Gradually when I realized that the rest of the world doesn't necessarily have the same sort of experiences and on the whole isn't equipped to deal with those of us who seem to step in and out of a very strange realm on a regular basis, I became even more guarded, hugging the memories to myself. Later, I did tell these stories – but only ever piecemeal – to very close friends, including my husband (I have since divorced). It's very strange

> looking back on such intense, otherworldly experiences, because I know it's a cliché, but really there are no words to describe them. And in an odd way they continue to exist, as discrete happenings, inside a deep and secluded part of my psyche.
>
> I wouldn't be human if I didn't have the occasional doubt about their reality, but then the memories come flooding back, and I realize I didn't dream them, and that weirdness really did happen. The woman with green eyes seemed to break the seal between this reality and something else, and the doorway she opened up always remained there for me.

However, this was by no means the sum total of Maggie Fisher's brushes with the Otherworld, nor was she to find it easy to forget the terrifying vision of the green-eyed woman who told her she was "the wrong one". Many years later, Maggie became friends with a famous British astrologer – I will call her Carla – who had just graduated from a hypnotherapy course and was keen to practise her new skills. Curious to try to retrieve, if possible, detailed memories of her childhood vision, Maggie asked Carla to put her into a trance and take her back to relive it. What she described shocked and disturbed her.

"I found myself back in my little bedroom in the dark, and the woman with green eyes was already talking to me," says Maggie.

> But as I looked at that terrible white face the whole thing began to bubble and peel off the wall as if a blow-torch had been applied to a painted picture. Suddenly there was no woman – and no wall or ceiling either. I looked up from my bed and there was a huge UFO blotting out the sky with a ladder extending down to a few inches in front of me. I heard a snickering, clicking sound and spun round to see a couple of what I now realize were "Greys" right there, very close to me. I could smell a faint tang of something like burnt paper. What was upsetting was the proximity of these grey aliens with their enormous black eyes. The creatures didn't blink at all, just turned their eyes on me and made this tiny snickering sound. It was surreal and somehow desolating because they weren't of my own kind. Being there with them made me feel completely isolated. Then they got hold

of me with their long thin fingers and lifted me up effortlessly and sort of floated me up into the spaceship.

There I was, this small child in my crumpled pyjamas in a big round clinical room with what seemed to be some kind of a computer in it. The aliens appeared to be discussing something earnestly – while I stood there, too bewildered even to be truly terrified. Then they turned and looked at me. Although they didn't utter a sound, they seemed to say – in a well-modulated, slightly posh English accent, probably male, although I can't really remember – "You must come with us, a long way off to the stars". I know, 'B' movie stuff, but that's what they said.

That seemed to break the spell and I started to cry. I was so desolate, so lost and lonely all of a sudden. I sobbed, "No, you can't take me away. How can I go without Mummy and Daddy? What would Mummy and Daddy do?" After staring at me for a few moments, they went into a huddle again and then, unbelievably, took me by the hand, out of the ship, down the ladder and back into bed. They hovered over me clicking something at me, and were about to go, when – and I smile every time I remember this – I abruptly changed my mind. I *wanted* to go with them, never mind poor Mummy and Daddy! I wasn't going to miss the biggest adventure of a lifetime! I grabbed one of their skinny little hands and begged, "Take me with you! Don't go without me!" They looked at each other and then back at me. One of them gently pushed me back into bed. Then suddenly I woke up. That was what I recalled under hypnosis.

Maggie admits that she has reservations about the regression, although she feels it would be wrong to discount it altogether. She says, "Whitley Strieber's first book *Communion* had just come out at the time of the regression. That extraordinarily haunting image of a 'Grey' on the cover seemed to be everywhere. Once seen never forgotten! Aliens were the hot topic among all sorts of people, especially those with an interest in mystical or New Age matters. Carla, a professional astrologer, was by no means a critic of the Strieber scenario – although at the time I hadn't got round to reading *Communion* myself. It seems quite feasible to me that my unconscious mind had confabulated this version of my vision to please her and fit into the prevailing mood of our circle. Yet I have no way of knowing if that's the case or whether it really happened literally as I described it under hypnosis.

"Since then I've read prolifically on the subject of UFOs and aliens and have never come across a story in which anyone successfully refuses to comply with the Greys' wishes! I'm rather proud that I wasn't a victim, even if it didn't really happen and all it means is that my unconscious mind refuses to think of me as one. That's healthy, although I admit that the scenario summoned up by the regression left me feeling somehow unclean, as if contaminated by something. Perhaps it was the underlying taint of confabulation, or a whiff of the alien abduction hysteria, then in its infancy. I don't know, but I certainly had no desire to do it again, or even to think about it."

There was an interesting sequel to her hypnotic experience. While on a weekend break in the spa city of Bath in May 1989, Maggie finally decided to read *Communion*, which she did just before going to sleep in her hotel room. Perhaps that was a bad move: she dreamt that Grey aliens came floating through the window and levitated her, still covered in the bedclothes. They moved her, while in the air, to a position at right angles to her bed, which was still in place on the floor. Communicating with each other with high-pitched clicking sounds, they seemed particularly interested in the tips of her fingers, flicking them with their bony grey hands and examining them closely. The next thing she knew, daylight was streaming through the window and she was awake in bed.

"I was rather shaken by the dream," she recalls,'because it was so vivid, although I can't pretend it was anything more. I was lying there thinking it over when I noticed that the tips of my fingers were bright red, where the creatures had been flicking them and – more alarmingly – my blanket and top sheet were lying across the bed at right angles, matching my position in the air when they levitated me . . ."

She felt it was odd for a mere dream, but made a point of giving *Communion* a wide berth after such a disturbing experience. Had reading that book somehow invited the aliens into her bedroom? Or had something in Strieber's tale struck such a chord in her unconscious mind that she herself created the dream, after effects and all?

To this day, Maggie experiences the occasional paranormal event such as mild poltergeist activity or waves of extraordinary synchronicities (meaningful coincidences), besides frequently waking to see anomalous lights darting about the sky.

Unlike the traditional simple shepherd girls' visions of the Virgin, her green-eyed lady did not declare her to be the Chosen One. On the contrary, after consulting the unseen adviser "offstage", she announced her to be *the wrong one*, not chosen or at all. Yet even so, it is interesting that, by Maggie's own admission, the vision's predictions – which she has always kept private – have come true. Although Maggie says, they were, exclusively personal, not global, one wonders how they compare with the prophecies of the more standard Lady-visions, even the much-vaunted three prophecies of Fatima.

In 1990, I had the opportunity to have a long discussion with Jacques Vallée, then in the UK on a promotional tour for his book, *Revelations*. Knowing Maggie's story well, I described it to him and asked what he made of the peculiar, "You're the wrong one" denouement. He said: "It would certainly help her to remember it. The absurd always makes you remember."

It is interesting too, that although the green-eyed woman took pains to let little Maggie know that she had failed the Chosen One test, the underlying "message" of the other experiences, such as the enveloping white sphere of light in the bedroom and the UFO on the shed, seemed to contradict this, making her feel special. And, as she says, "Let's be honest, the very fact of having a woman bother to materialize and tell you about your future seems pretty special to me."

But were the oddities in her experiences – the "wrong one" statement and her childish insubordination towards the Greys – merely unconscious embellishments of some archetypal fairy story? Or were her experiences no more than the literal truth? If so, how much faith can we place in the hypnotic regression when it was "revealed" that the vision was a "screen memory" to obscure the truth about an (attempted) alien abduction?

But if her hypnotic regression had been contaminated by the prevailing excitement about Strieber's *Communion*, it must be remembered that the core experiences happened many years previously, before she even knew of such things. Her childhood in that far-off era was UFO- and alien-free in ways inconceivable today, when both are so much part of our cultural baggage.

It seemed as if she was intended to go out into the yard to witness the UFO: she woke with a bad sore throat, not only failed to raise her parents but found their door locked, and couldn't even open the fridge door to find orange juice to ease the discomfort. She *had* to open the door to get some cool air.

Of course it is possible that she simply walked in her sleep, but then how do we explain her waking up wrapped in a coat that had been hanging up in the wardrobe in her parents' bedroom? And what about the strange bit of bone that fell from her nose, and the fact that her nose was later discovered to have been broken for years without her knowledge? Aliens, it is said, have a predilection for probing noses – implanting tracking devices, some believe: they may have been clumsy on this occasion.

It is possible that little Maggie may have had some kind of temporal lobe seizure which induced the bizarre mental images such as the UFO, after sleep-walking in the middle of the night, during which she fell heavily on her face, breaking her nose, after which she somehow dragged herself upstairs, sneaked into her parents' room (assuming that the door was now open as usual, or that it had never been "locked"), rummaging around in the wardrobe without waking up either of them to find her mother's coat, taking it back downstairs, and wrapping herself in it before falling asleep on the sofa. And all of this without having a nose bleed, switching on a light or making a sound. Sceptics would say either that she had invented the entire story or that certain events and images had become muddled in her mind – perhaps she had seen a UFO on television and dreamt she couldn't open her parents' door etc – merging into one bizarre, but basically false, memory.

It is interesting, too, that she should recall having an electric shock around the time of seeing the UFO (perhaps it was *before*, not after the experience?). Uri Geller recalls receiving a shock from his mother's sewing machine before discovering his well-known talents for psychokinesis and telepathy. As researcher Albert Budden has argued, electricity often plays an important role in *creating* such encounters.

Clearly deep down she has always felt she is *supposed* to be special, saying, "I chose to set myself apart". But did this sense of *otherness* really come from a conscious decision to be different, or had the aliens set her apart for ever?

1966–1980s, DETROIT, MICHIGAN, AND VARIOUS LOCATIONS INCLUDING "COPLEY WOODS", OUTSIDE INDIANA, INDIANAPOLIS

In September 1983 New York-based artist and alien abduction researcher Budd Hopkins received a letter from a woman in Indiana, whose privacy is still shielded behind the pseudonym "Kathie Davis". Her astonishing story – which was to prove even more amazing as the two-and-a-half year investigation progressed – has become a classic case.

Now in her early forties, the divorced mother of two boys recalls visiting friends in the Detroit area some time in 1966.[21] Playing outside, she came upon what appeared to be her friends' cabin, but when she entered she encountered what she described as a "little boy" who took her to his "playroom" – a windowless, round room – where he announced he was going to play a trick, which he accomplished by cutting her lower leg with a small machine. Briefly, the "boy" transmuted into a Grey alien, after which Kathie came upon her sister Laura, who appeared to be in a trance of some kind. Together they returned to their friends. Despite hypnotic regression, Kathie has failed to recall significantly more details of this encounter, although she bears a scar on her leg.

Then, when in July 1975 sixteen-year-old Kathie, together with her friend Nan and some other teenagers, visited Rough River State Park, Kentucky, another element of this increasingly complex scenario came into play. Kathie had mentioned to Budd Hopkins having had a recurring dream of being in a truck at night, talking with an unknown person over CB radio, and on further probing, realized that the "dream" was based on a real event, one that was also remembered (at least in part) by her friend Nan. It transpired that they had met four boys who claimed to be camping in the woods nearby – one of whom being the person to whom Kathie had been talking on the CB radio. Looking back, it occurred to her that there was something very wrong with the encounter. How had the boys found them, when the roads in that remote area had no names? And there were no campsites in the neighbourhood . . .

Kathie then remembered how it happened. She and Nan had been sitting in the truck talking to the boys on the radio, inviting them over to make a party of it, when Nan went indoors, leaving Kathie alone. Suddenly some lights – which Kathie naturally took to be those of another truck or car – approached slowly along the road, and at that moment the boy on the radio said, "I see you". The boys were invited into the cabin, where the strikingly "cute" blond one took charge. None of the other three, whom Kathie describes as "tall and skinny and . . . looked alike", spoke much or gave their names, which came to seem odd. The blond one told Kathie and her friends that they were part of a rock 'n' roll band , although they did not tell them its name, or invite them to any future gig, as one might expect.

Although they stayed for about two hours, it was the blond boy who did the talking, while the others just stood in the background. Kathie felt deeply attracted to him, feeling that he liked her too. Later, when Budd Hopkins checked the story with one of the others who had been present, it emerged that the skinny newcomers stood "sort of like guards while the blond guy ran things."[22]

Innocuous though the story may seem, it took on a darker hue when Kathie was regressed to the time of the event. She described sitting in the truck alone and seeing "real bright . . . pinwheels . . . four of them . . . something sparkly in the light . . . and I'm getting real scared . . ." It was immediately after that experience that the four boys turned up. When asked to describe the blond boy, Kathie described what, in Hopkins' words, "could be her brother, even her masculine twin", someone for whom she felt a strong empathy, saying, "I had feelings for this man like an obsession". In her hypnotised state, she noted that the "boys" "smiled with their eyes", a common occurrence in alien encounters. After coming round from her trance, Kathie remarked on how odd her friends had seemed when she took the boys into the cabin, as if they had been asleep.

Clearly, although this was by no means a "traditional" UFO encounter, the elements of high strangeness involved suggest that it was part of an alien grand plan, like the various paranormal events scattered throughout the life of Maggie Fisher. Kathie's life was punctuated in a similar way by odd happenings, culminating in her contacting Budd Hopkins, as if the pressure of the otherworldly contacts had reached a critical point.

It was in December 1977 that the story began to take on the by-now familiar outline of alien abduction. The adult Kathie consciously recalled being in a rural part of Indianapolis with her friends Roberta and Dorothy when they saw a strange strobe-like flashing light emitting from a hovering UFO. Terrified, the two other young women cowered in the car while Kathie alone was taken into the now landed UFO. It was there that her trauma really began.

Under hypnosis, Kathie was able to fill in the gaps in her memory of that night, describing in graphic detail a gynaecological examination that, "hurts . . . like toothache . . . a lotta pressure . . . wiggling and pushing, right in here . . . Kinda burning sensation from the waist down. Can't move".[23] After further questioning from Hopkins, she recalled being in a room with the same Grey alien she encountered before, who had appeared to her when she was a child as a "little boy".

It was at around the time of this abduction that Kathie, now in a relationship with her future husband, discovered that she was pregnant – a diagnosis confirmed by blood and urine tests and by her GP. So sure was she of her pregnancy that she brought forward the date of her wedding, yet within weeks she menstruated completely normally, and after another pregnancy test, discovered that she was not pregnant. She had not suffered a miscarriage or had an abortion. The foetus had simply vanished. Her doctor told her, "I don't know what happened. I think we'd best just forget about it." Kathie was distraught, crying over and over, "They took my baby . . . they took my baby".[24] But who were "they"?

Later she recalled being pregnant (with one of her boys) at her sister's house in 1978, watching television alone when she sensed a "prowler" in the house. As she lay on the couch, she felt someone touching her face and the small of her back gently, which turned out to be the prelude to another alien gynaecological examination. During the hypnosis session, Budd Hopkins was concerned to eliminate any element of sexual fantasy from the investigation, deliberately asking leading questions that could encourage her to reveal this undercurrent more obviously, but she resisted him. As he said, ". . . the experience seems . . . purely objective and clinical, a puzzling and real event rather than an erotic fantasy."[25]

The purpose of the examinations – and the missing foetus – was to become horrifyingly clear as the investigation proceeded. Kathie was to recall more abductions in which she was shown, and asked to hold, a tiny, listless, child, which she realized was hers – or rather, half hers. The ova taken from her had been fertilized by an *alien* male, and the child, a little girl, was one of a new hybrid human/alien race. Later, Kathie was to meet her hybrid daughter on a UFO again, by then a cute – if strangely weak – child "like a very timid bunny rabbit" who "was almost afraid of me". It seemed that the aliens needed Kathie to hold her in order to bestow some strength on the sickly little child. Whatever the cause of the recollection, Kathie was traumatized when the girl, for whom she felt intense love, was taken from her, but by then she knew that she was by no means alone in her distress. Her experience convinced her that the aliens had put in place a massive, long-term genetic engineering programme, in order to save their race.

Even that may not be the limit of the aliens' genetic interference. As Budd Hopkins says, "Perhaps the most unsettling idea of all is the possibility that a child born normally to an abductee may have been, prior to conception, subjected to some form of genetic tampering", noting that although he has no hard evidence for this, some other abductions he investigated happened on the night the abductees' children were conceived.

Certain sceptics have suggested that the abductees have unconsciously fabricated the story of being taken by aliens to cover up some form of traumatic sexual abuse, perhaps by members of their own family, although Budd Hopkins and Dr John Mack deny that there is any evidence for this supposition. Certainly, something appalling has happened to a great many innocent people – both men and women – which they interpret, or even recall, in all honesty as having being abducted and used by aliens.

25 JANUARY 1967, ASHBURNHAM, MASSACHUSETTS

It was ten years after her terrifying abduction that Betty Andreasson of Ashburnham, Massachusetts, fully recovered the memories of the experience. After writing to the tabloid, the *National Enquirer* in 1974, Betty's case was taken up by UFO investigator

Raymond Fowler on behalf of MUFON, who arranged for the witness to be hypnotically regressed. It was only then that an extraordinary story emerged.

On 25 January 1967, Betty, together with her parents and seven children, were at home in Ashburnham when, in the early evening, an anomalous light was seen outside the kitchen window. Later it emerged that Betty's father had encountered humanoids walking about the house – reminiscent of other abductions, such as those of Whitley Strieber and Kathie Davis – actually walking through the walls. Betty was taken up into a waiting UFO where she was subjected to the usual medical procedure in which a probe was inserted in her nose and navel (like Betty Hill, and possibly Maggie Fisher). She was also covered in a strange fluid, while tubes were inserted into her nose to enable her to breathe, calling to mind the early stages of Villas-Boas' abduction. Betty's experience included floating vast distances, over alien cities – one of which was made of crystal – and being surrounded by small reptilian creatures. She also reported that she heard "the voice of God".

It seems that the witness established a relatively cordial relationship with her abductors. The leader, very unusually, gave her his name – Quazgaa – and they *exchanged books,* Betty handing over a Bible, while he gave her some kind of slim religious work, which she lost. (The losing of alien artefacts, or their mysterious disappearance, is a common component of the classic contactee or abduction experience.)

Raymond Fowler found the case fascinating, writing three books on the subject,[26] questioning the objectivity of the abduction experience as a whole. Not only does Betty Andreasson's abduction include a great deal of apparently sexual and birth-trauma imagery, but certain elements – such as floating over crystal cities and hearing the voice of God – evokes other spiritual events, such as the Near Death Experience, or a shamanic initiation. Yet these experiences are intensely personal, and Betty's father was involved. Once again, the true nature of the experience remains tantalizingly elusive, yet continues to draw us ever deeper into the hidden recesses of the human psyche.

JUNE 1968, ARGENTINA

Seventy-year-old artist Bejamin Solari Parravicini was taking a walk when he encountered a "tall blond man with clear eyes who addressed him in an unknown language".[27] Believing him to be a little mad, Parravicini continued on his walk, but suddenly woke up inside an unknown craft, "where he was told, among other things, that the saucer people were keeping watch on the earth to avoid a catastrophe".

In those days, the aliens, although abducting the witness by causing him to lose consciousness, were not keen to subject him to painful and traumatic examinations. Although they had shown a harder side to their characters in the abduction of Betty and Barney Hill, such a modus operandi was only to become widespread later, after the publication of Whitley Streiber's *Communion* in 1987.

JUNE 1969, BANDJAR, WESTERN JAVA, INDONESIA

In June 1969, twenty-seven-year-old Machpud encountered a "beautiful but strange woman" who conducted him to her "house" in Bandjar, Western Java. This place comprised just one huge room, in which there was an unexplained "abundance of light", and in which the "woman" lost no time in making it obvious that she wanted sex – at which point Machpud passed out. When he came to he found himself wandering around as if in a trance in the depths of the Gunung Babakar Forest, totally naked. His clothes, it transpired, were up a tree. Discussing the case in her classic *The Complete Book of Aliens + Abductions* (1999), Jenny Randles says, "There is little doubt how many words in this story could be substituted in our language by 'UFO', 'alien', 'taking a sperm sample' [although the witness did not recollect any such procedure] and so on."[28]

The witness was totally unfamiliar with the kind of media exposure of UFOs and tales of abductions that have become so commonplace in the West, so his version of an otherworldly encounter had none of the expected interpretation, although it did feature several of the key elements.

1969, KANSAS, MISSOURI AND SEVERAL SUBSEQUENT TIMES OVER A PERIOD OF MANY YEARS

Although this case came to the notice of Dr John Mack, the Harvard University depth psychologist who has become one of the leading alien abduction investigators, in the early 1990s, the events it involved had been happening since at least 1969, when the experiencer, a woman called Jerry, was seven years old.

She recalled that she was living in a rural area, sharing a bedroom with her younger sister, and feeling anxious when an unexplained bright light filled the room, somehow making her go into the hall. She thought, "I shouldn't be afraid 'cause I know them",[29] but she felt *very* afraid, especially when she saw over twenty little entities coming towards her through the window. To her great astonishment, they picked her up and carried her through the solid glass, so high in the sky that she could look down on the roof of her home.

Once in a "big thing above me" – presumably a UFO – she began to weep, realizing that the aliens were also kidnapping her baby brother Mark, who was still fast asleep. Distressed and terrified, she wanted to run away, but felt as if she was paralyzed up to the waist, with a kind of "tremendous vibration" that seemed to shake her whole body. Recalling the trauma of the abduction under hypnosis many years later in the 1990s, Jerry said, "I'm worried about Mark . . . Okay, do it to me, but it's not fair to do it to him. He's a baby! I just hate them for it. . . . I thought at first they were all right. . . . I just thought they were cute and they wanted to come out to play." Clearly, Jerry had been repeatedly abducted since earliest childhood, and the pattern was being re-established with baby Mark.

She went on to describe a "real dark" being, and a lighter-toned taller entity she termed "the leader", who had a "nice face" and – unusually – a few strands of yellowish-white hair. He asked her "if the medication has been okay till now", which she found puzzling, before plunging a needle-like object into the side of her head, causing such acute agony that she began to sweat profusely and writhe as she recalled it. Screaming out loud, her legs in uncontrollable spasms, Jerry panted, "I can't stop it! Ahhhh! Ahhhh! I hate doing this! Stop it! Stop this!" After twisting the needle inside her head, they removed it, leaking blood or saliva into her throat. Jerry whispered, "They're awful. They're cruel. I thought they did something else. Oh, I didn't anticipate this." The alien told her that they had to monitor her, saying gnomically, "We have to do what we have to do" – which must have been little comfort. Afterwards, Jerry was drenched in sweat, and worried that whatever the needle had inserted into her head might still be there.

After this session Jerry told Dr Mack that the regression seem to have stimulated vivid dreams of UFOs, together with visions and out-of-the-body experiences, which argues strongly that the phenomenon of abduction has a psycho-spiritual component (or, more sceptically, that something triggers the part of the brain that can induce the *perception* of mystical phenomena).

However, it was in 1990 that she had by far the most traumatic of her many abduction experiences. By then she had married and lived in a flat in Plymouth, Massachusetts, generally happy with her lot except for a crippling aversion to sexual intimacy. Jerry has no memory of the lead-up to this particular experience, recalling only a strange tapping on the shoulder, before finding herself in a metallic-looking room where she was suspended in an upright position while an examination was conducted. During the procedure her necklace fell off, being retrieved by a small entity on the orders of a taller, blond humanoid, and put in a "plastic-looking pouch" because it was "contaminated". Later Jerry's mother found it in a box in Georgia – but was it the same one? Had she lost the necklace in some everyday way, merely hallucinating the events on the UFO? Or had it been returned to her, presumably after being decontaminated, in a manner similar to the materialization or dematerialization of objects in the seance room of Spiritualist mediums?[30]

During her subsequent abduction Jerry made the mistake of answering "fine" to a question about her "medication", and immediately she was flooded with such an excruciating, raw pain that was "even worse than childbirth", causing her whole body to jerk and ripple with uncontrollable spasms. She screamed, "Here I thought they were somehow perfect and loving beings. How could they have done that to me? I was so terrified. I blanked out after that." She woke up in bed.

During three abduction experiences in 1991, the Nordic-like

aliens transported Jerry to a very tall building near a seashore, where she was shown missiles and a triangular machine that became round when it was spun at high speed – probably the mysterious "flying triangles" that have perplexed UFOlogists and members of the public alike in recent years.

It seems that the abductions took on a more spiritual aspect after that, resulting in the witness finding a talent for something akin to automatic writing – producing over a hundred poems in a six-week period. Gradually, too, Jerry began to think of her captors as benevolent, loving and all-knowing beings, who came from far into the future.

Yet the fear was never far away, especially when she recalled the terrifying and humiliating deep internal examination of a teenage abduction, saying – with heart-rending pathos of her mother – "She wouldn't let them do this if she was here." Although the beings told her she would not remember the event, after the hypnosis, she lashed out at them verbally, saying, "They tried to make me think it was just a nightmare", adding, "What do they think, I'm just an animal, or something?" (Yet the apparently all-knowing aliens had failed to realize that one day she would recover the memories through hypnosis. This does not seem like omniscience.)

Jerry also described, very graphically, the cramped and "pinched" feeling of a medical instrument being inserted deep inside her, which felt as if it had deposited something beyond her cervix. Then, horrifically, she cried out, "I can't believe it. I'm too young for that. I'm only thirteen." What had caused this outburst was the sight of a "real tiny, skinny" baby about ten inches (22 cm) long that the aliens had placed in a transparent plastic cylinder full of fluid, like a futuristic incubator. Although her captors seemed to expect her to be proud of her child, she was shaken to the core and horrified, saying, "I didn't even know I had anything like that in me! . . . If they're gonna do this, they should at least tell you."

The taller being who appeared to be the leader explained telepathically to her that what they had done was "about creation", which she took to mean the creation of a new race . . . Unusually, the abductee had been told the leader's name, which after a tussle with herself Jerry told Dr Mack: it was something like "Moolana". After this session she was shocked, asking: ". . . don't they know us well enough to know that a thirteen-year-old doesn't do that?" She denounced them as being "pretty selfish" in their use of her as a "tool" for their own purposes. Later she had a brief vision of the letters "DNA" and heard the words "the marker trait" – a term used by geneticists but which was totally unknown to Jerry herself.

As in the cases of many other abductees, the experience also involved the next generation of the (Earth) family. Jerry's son Colin, who was three in 1993, was heard talking, apparently to himself, and rambling about owls with big eyes outside his window – recalling Whitley Strieber's experience of seeing owls with huge eyes as a "screen memory" for the big-eyed Greys. Colin spoke repeatedly of "scary owls with the big eyes" that "fall down [or 'float'] out of the sky", also evincing a great fascination with the planet Earth, which he said "go away" and "house go away", adding, "The owls with big eyes fall down and jump and I jump" . . . "there's a spaceship and I come out of the spaceship . . . my toe hurt". Indeed, on more than one occasion Jerry found he had a torn toenail. Later, when seeing a book with a Grey on the cover, Colin said immediately, "He's a Rocketeer. He goes up and comes down."[31]

Jerry seems to have remained unable to decide on her true feelings about the aliens: sometimes she regards them with loathing for the pain they cause, while at other times she sees that they do not have a true desire "to cause me fear and pain and agony", adding "deep down inside I think that what they're doing is somehow necessary", which seems to be connected with "races, beings or whatever, coming together to make another creation." She told herself it was an important thing, saying: "as a single person, compared to this big huge thing going on, I should look beyond myself and know it's for the greater good".

She later saw other hybrid children who were presented to her as hers and realized that the implantation and removal of foetuses had been going on for many years, each time reinforcing her terror of the gynaecological procedure, in itself like a rape. This would no doubt explain her adult horror of intimate touching. She says, "They've ruined that. I don't know what making love is because I'm always still too wound up in being tense and fearful of pain, and I associate sex with pain", although she admits she thinks it was never the intention of the aliens to ruin her love life. (Dr Mack delighted her by punning that what they had done was create an "alienation of affection".)

Further regressions elicited the horrifying details of an abduction when Jerry was just eight years old, during which an "ugly" alien quickly implanted something deep inside her vagina, besides probing her anally. Utterly disgusted and profoundly embarrassed, the memory of this made her say, "I wouldn't know the words like 'rape' [aged eight] . . . but it's like that".

Like many who have experienced either abduction or near death experiences, Jerry became interested in ecological and spiritual matters, besides writing a great deal of talented and incisive poetry – something she had never been interested in pursuing before. Much of her creative output appears to very similar to automatic writing, before which she seems – as John Mack notes – "to pause in awe". Jerry feels she must write a book about "Universe, Soul, God, and Eternity", writing in her diary,[32] "But why would you pick any ordinary housewife to do such important work as this?"

From obscurity, Jerry has moved into the realm of the Chosen Ones, and through personal pain and "alienation" to new levels of awareness and creativity. But all of that may be only the positive by-product of something much darker, some conspiracy against many more "ordinary housewives" – and thousands, perhaps millions, of others – for reasons we may guess at, but perhaps never know for certain.

12 OCTOBER 1973, PASCAGOULA, MISSISSIPPI

At about seven in the evening on 12 October 1973 Charles Hickson and Calvin Parker were fishing off the pier at Shaupeter Shipyards, Pascagoula, Mississippi, when they were startled by a "zipping" or buzzing sound. Then they saw a strange oval object, glowing a bluish-grey colour, hovering a few feet behind them. The witnesses were paralyzed with fear as a door opened in its side and three peculiar beings floated out. Moved forwards towards the terrified men "in a gliding motion", they were about five feet (1.5 m) tall, grey and wrinkled, with vestigial features, no neck and crab-like claws instead of hands. Hickson said later: "They didn't have clothes. But they had feet shape . . . it was more or less a round like thing on a leg, if you'd call it a leg . . . I was scared to death. And me with the spinning reel out there – it's all I had. I couldn't, well, I was so scared, well, you can't imagine. Calvin done went hysterical on me."[33] But worse was to come.

Parker said: "They were upon us before I knew it . . . I fainted as soon as they touched me". Hickson added, "The two things took me by the arms. I seemed to become weightless." Although Parker was unconscious throughout the following ordeal, Hickson remained awake and alert. They were floated into the UFO and suspended in mid-air where they were given a medical examination. Hickson said, "There was a large, optical-like device that came out of the wall . . . like a big eye". After about forty minutes, the creatures returned the men to the pier, then disappeared inside the UFO and flew away. Hickson managed to get to his car and gulp down some whisky before going with Parker to tell the *Mississippi Daily Press.* Finding the newspaper's office closed, the two abductees then reported the incident to the sheriff, who spent several hours going over their story with them.

Enter APRO in the form of one of their consultants, Dr James Harder, who interviewed the witnesses, wasting no time in concluding: "There is no reasonable doubt that the craft came from outer space . . . The experience of Hickson and Parker was a real one. It was not a hallucination".

Dr J. Allen Hynek, who joined Harder in the investigation, would only say cautiously, "There is no question in my mind that these two men have had a very terrifying experience"[34], while Harder went on elaborating his hypothesis in uncompromising terms: "My theory is that the Earth is a cosmic zoo. We are cut off from the rest of the universe. Every so often the keepers come in to make a random check of the inhabitants of that zoo . . ." but added, "There is nothing to suggest that the UFO occupants have harmful designs on humanity."[35]

After being tested for radioactivity – of which none was found – at Kessler AFB, the men were left high and dry by the Air Force. A PR officer explained the new official policy on such matters: "If anyone feels threatened, we send them to the local police. If they want a scientific investigation then we refer them to the nearest university." The Pentagon told the Pascagoula sheriff, in response to his request for help, "The Air Force will investigate only if there had been a direct threat to our national security. Nothing has taken place to jeopardize national security."[36]

A very different type of investigator took up the case, however. This was Philip J. Klass, editor of the highly conventional *Aviation Week & Space Technology*, whose long career as a high-profile UFO "Skeptic" has earned him the dubious honour as the man that mainstream UFO buffs love to hate. His delving on this occasion revealed that Hickson had been fired from the shipyard for "conduct unbecoming [to] a supervisor" – financial irregularities – the year before the incident, and clearly hoped the case would bring in some serious money. Shortly after Hynek and his APRO colleague had packed up for home, a local attorney called Joe Colingo became the witnesses' agent, handling media deals with his eye on millions. Klass also discovered that although it was claimed that the men passed a lie detector test, the polygraph operator had in fact flunked his course at college and was not certified to do the job. Also, inconsistencies began to appear in the men's story. Perhaps more damning was the fact that a 24-hour toll booth with a clear view of the pier had seen nothing on the night in question.

However, dubious though all of that may appear, it does not necessarily add up to a hoax. Jenny Randles points out: "When left alone in a police station (and secretly recorded), these men not only failed to show any sign that the story was untrue but one broke down and started praying into the empty room."[37] And, as we have seen, Dr Hynek said he believed that something had happened to terrify the men – but what it was will probably always remain unknown.

23 March 1974, GUSTAVLAND, STOCKHOLM, SWEDEN

On 23 March 1974, a young man, known as "Anders" (or "Harald" in some versions) was walking along a dark path in Gustavslund, near Stockholm, Sweden, when he heard a voice in his head instructing him to cross the road. At the same time a blinding light knocked him to the ground, where he lay unconscious, coming to on his own doorstep. His baffled wife, disturbed by his condition and a mysterious burn on the side of his face, sent him to the local

Danderyds Hospital where hypnotherapist Dr Ture Arvidsson put him into a hypnotic trance and regressed him to the time of the incident. He discovered how the beam of light had floated him up into the air, while tall hooded figures touched his head with an unknown device, saying they would meet again in the future.

After the abduction experience Anders became psychic, and at the same time both "electrical" himself and extremely sensitive to electricity. Interestingly, there was a witness to the flash of light, but not to the abduction. It seems that although the UFO "energy beam" was real, the aliens themselves may not have had an objective existence. Was Anders "taken away" or left prone on the ground, "dreaming" of contact later elaborated upon by his regression hypnosis?[38]

Or did the blinding flash cause an electrical storm in his brain, similar to that which produces temporal lobe epilepsy flooding his mind with vivid hallucinatory images of the UFO and the hooded figures? Perhaps the answer lies somewhere in between, in the strange unknown borderland between an electrical discharge and the opening of a stargate, through which beings of dreams and myth enter the human psyche.

31 MAY 1974, ZIMBABWE – SOUTH AFRICA

Driving across the scrubland between Zimbabwe (then Rhodesia) and South Africa, twenty-three-year-old Peter and his twenty-one-year-old wife Frances saw unexplained lights at 2.30 on the morning of 31 May 1974, which kept pace with their car around the Beitbridge area. As the light hit them, everything became unreal. Peter recalls, "It was like travelling in a dream. I lost trace of time. I felt as though in a coma."[39]

When they arrived at the border between Rhodesia and South Africa, the couple made an extraordinary discovery. Not only were there 200 miles (320km) less on the clock than there should have been, but they still had virtually all the petrol they started out with! Had the strange light somehow dematerialized them, rematerializing them at their destination? At the time both Peter and Frances had no conscious recall of anything other than feeling sleepy and weirdly unreal, but six months later when Peter was regressed by Dr Paul Obertik, the full story emerged.

Under hypnosis, Peter claimed that an alien "beamed down" into the back seat of the car, implanting images in his head of a clinical room where abductees were taken. Alleging he came from "12 planets of the Milky Way", the alien explained that although humans see them in the mould of stunted, hairless creatures, they could take any form – even passing as ordinary human beings, living in our midst. Their purpose, apparently, is to guide and help humanity, while maintaining a policy of non-intervention (although of course the later epidemic of agonizing experiments on abductees seems to indicate otherwise).

Jenny Randles notes:[40] "Hypnosis threw up one intriguing point. Peter commented that the aliens were 'about 2,000 years ahead of Earth' (in terms of their technology and progress). But Dr Obertik misunderstood this answer and said, 'So it would take 2,000 years for us to get there?" ' Peter immediately compounded the mistake by answering, "Two thousand light-years". Jenny Randles points out that, "This clearly shows how hypnotic testimony *can* be influenced unintentionally."

OCTOBER 1974, AVELEY, ESSEX, ENGLAND

The first known alien abduction in Britain took place in October 1974, when John and Sue Avis (pseudonyms) and their three young children were taken from their car near Aveley in Essex. It is particularly interesting in the ways that it diverges from the later – specifically American – cases.

The family spotted an odd blue light as they drove home from visiting relatives, and almost immediately afterwards found they had driven into a thick green mist, which made their car radio spark. Although they seemed to emerge from the mist within seconds, in fact they reached home with an hour and a half of missing time. Then their nightmare began, with broken nights punctuated with bad dreams of looming alien faces that lingered into their days. Finding this insufferable, they contacted a dentist and amateur parapsychologist who used hypnosis as a tool to recover lost detail of extraordinary events.

It was at this point that two local Essex researchers, Barry King and Andrew Collins (now the bestselling author of "alternative history" books such as *From the Ashes of Angels*) became involved, also hypnotizing the couple. As a matter of principle they did not attempt to hypnotize the children. UFOlogist Jenny Randles also sat in on one of the sessions, noting that this was when "my personal concerns about regression began".[41] She goes on: "This was notable when the wife, Sue, started to answer a rhetorical question I innocently posed. She did so as if she *was* the alien replying to me! In this way Sue was acting like a medium conveying tales from the other side, with the exception that her contact was not a dead person in some afterlife but a Nordic alien on board his UFO."[42]

The Avis family described being beamed up into a UFO, where they encountered the classic Nordic type of alien – tall, humanoid, blond – and smaller, furry-headed entities, a type of creature then very familiar from various sci-fi television programmes and movies.[43] The couple told Collins and King that when they were in the UFO they could see their bodies still sitting in the parked car below, indicating that the whole event was some kind of intensely vivid out-of-the-body experience, a psychic or spiritual happening rather than something wholly of the "real", tangible reality.

The beings conducted some kind of medical examination on the humans, but stopped short of inflicting the kind of pain and humiliation that characterizes the more modern, American experiences. After the examination, the family was split up to be taken on a tour of the UFO and given a lesson on the future of the Earth and their role in it. The aliens portrayed themselves as benevolent guardians of the human race, who nevertheless maintained a "hands off" approach as much as possible, even though humanity was in danger of destroying the planet. They conveyed these messages through a sort of holographic movie show that had a marked emotional and spiritual effect on the family. After their experience they found they could not eat meat and became deeply involved in ecological causes – quite a change from the ordinary western carnivores whose lives had been primarily concerned with everyday family matters.

The Avis case remains one of the most comprehensively investigated of all abductions, but there is something ultimately unsatisfactory about it, despite the total sincerity of the witnesses. As Jenny Randles says, ". . . they were describing images that were welling up inside their minds at the behest of the hypnotic state and urging of UFOlogists (myself included). I was not convinced that I was hearing about a real kidnap into a spaceship. Indeed, witnesses frequently react unconsciously as if they do not believe it either. If they *truly* did, then would they not demand round-the-clock police protection?"[44]

Jenny Randles ponders on the similarity between this case and that of the classic near death experience, which is triggered by a grave physical crisis (hence the term). Yet we know so very little about such matters – it was only in the late 1960s that Dr Elisabeth Kubler-Ross and a few other pioneers began to study the phenomenon. Could it be that there are other triggers that will produce similar experiences – some form of electrical or atmospheric phenomena, for example?

NOVEMBER 1975, SNOWFLAKE, ARIZONA

At the end of a long day wood cutting, a seven-man team were driving back to their base camp in Sitgraves National Park near Snowflake, Arizona, when they encountered a huge gold-coloured, domed UFO, hovering around the treetops.

Highly nervous, six of the wood cutters stayed in the truck, but the seventh – Travis Walton – got out and approached it, much to the horror of the others, who called him to come back. Suddenly the UFO emitted a blue ray, sending Walton cannoning back into the trees. At that, the others drove off in a panic to alert the police.

When the authorities checked the site there was no trace of either the UFO or of Walton. Indeed, he disappeared for five days, during which time, as it was suggested that the others in the woodcutting team may have murdered him, they were given polygraph (lie detector) tests, which established their innocence. Then Walton appeared, collapsed in a phone box some distance from home, telling how he had been abducted by aliens on board the UFO, whom he described as being hairless, with huge eyes and vestigial features. He was shocked to discover he had been away for so long – clearly this was the "missing time" syndrome, now so familiar in abduction cases, but a new feature in those days.

Walton recalled being abducted by aliens with spindly bodies and large heads, like "well-developed fetuses", about five feet (1.5 m) tall, with huge eyes. But there was another type of alien on board the UFO – more human-like, and dressed in blue.

One of the most controversial of all abduction stories, it has been claimed that Walton was in cahoots with the others in the tree-cutting gang, elaborating the tale in order to give themselves an excuse for being unable to meet a deadline. (Serious penalties would have been imposed, leaving them virtually bankrupt.) Certainly, Walton himself made a small fortune out of his story, having done an exclusive deal with the supermarket tabloid, the *National Enquirer,* who – together with certain UFO groups, "adopted" him.

Then in 1993 came the movie, *Fire in the Sky,* which confirmed the status of the story as a UFO classic, although the flashback scenes of being with the aliens is noticeably different from the witness's original account.[45] However, it seems that the change was at the behest of the producers, not Walton himself.

Over the years, although neither Walton nor any of his colleagues have ever admitted a hoax, there is good reason to question his story. One of the strangest things about the initial investigation – when Walton first went missing – is that neither his mother Mary Kellett, nor his brother Duane Walton (who was also in the wood-cutting team), expressed either shock or dismay at his absence. When Deputy Sheriff (of Heber County) Kenneth Coplan visited Mrs Kellett's ranch house near the UFO site, he was astonished by her reaction. She just said, "Well, that's the way these things happen."[46] What things? Mysterious disappearances – perhaps murder? When she told her daughter, she, too, was remarkably calm about the news.

While the authorities were looking for Walton, local UFOlogist, Fred Sylvanus of Phoenix, interviewed Duane Walton, who said that he was "having the experience of a lifetime!" – a remarkably upbeat and philosophical comment under the circumstances. Clearly, the Walton brothers had no fear of being abducted by aliens – quite the reverse. Duane went on:

> Travis and I discussed this many, many times at great length and we both said that [if either ever saw a UFO up close] we would immediately get as directly under the object as physically possible.

> We discussed this time and time again! The opportunity would be too great to pass up . . . and whoever happened to be left on the ground – if one of us didn't make the grade -to try to convince whoever was in the craft to come back and get the other one. But he [Travis] performed just as we said he would, and he got directly under the object. And he's received the benefits for it.[47]

Duane's words were particularly odd coming at a time when, as far as the police were concerned, this was a missing person case, possibly even a murder. Indeed, foul play was suspected to such an extent that the other wood-cutters were subjected to a lie detector test, carried out by C.E. Gibson of the Arizona Department of Public Safety, who asked the men three questions about whether they had killed or injured Travis Walton, and a last-minute question about whether they had seen a UFO when he had disappeared. Gibson stated:

> These polygraph examinations proved that these five men did see some object that they believe to be a UFO and that Travis Walton was not injured or murdered by any of these men, on that Wednesday. If an actual UFO did not exist and the UFO is a manmade hoax, five of these men had no prior knowledge of a hoax. No such determination can be made of the sixth man whose test results were inconclusive.[48]

Was the sixth man Duane Walton?

But almost immediately the men were off the list of murder suspects, for Travis turned up, with his story of alien abduction. Duane's phone became red-hot from all the calls he received from interested parties, including Coral Lorenzen of the influential UFO group APRO, who was in turn contacted by the *National Enquirer*. Ms Lorenzen suggested that the two Walton brothers be "sequestered" at the Scottsdale Sheraton Hotel, and the tabloid agreed to pick up the tab in exchange for an exclusive story, besides offering them $1,000 as an initial payment. The deal was that if Travis could pass a polygraph examination they would pay more, a five-figure sum. One of the *Enquirer's* reporters described his time with the Waltons at the hotel as "four days of chaos", remarking that although Duane was "one of the meanest and toughest-looking men I've ever seen" (he had kicked off the proceedings by saying belligerently, 'Nobody is going to laugh at my brother'), his first sight of Travis ". . . was a shock. He sat there mute, pale, twitching like a cornered animal." (Perhaps a reaction not out of keeping with having been abducted by aliens.)

As agreed, Travis underwent a lie detector test, carried out by John J. McCarthy, who asked him nine questions about his UFO encounter – and found that he was lying. In fact, he went further, stating unequivocally that the witness was guilty of "gross deception". With Duane baying for the expert's blood, Jeff Wells of the *Enquirer* said, "the office was yelling for another expert and a different result".

McCarthy's official report ended with the words: "Based on his [Walton's] reactions on all charts, it is the opinion of this examiner that Walton, in concert with others, is attempting to perpetrate a UFO hoax, and that he has not been on any spacecraft."[49]

Shortly after this devastating condemnation, APRO brought psychiatrist Dr Jean Rosenbaum and his psychoanalyst wife Beryl onto the scene. They spent several days in Walton's company, Dr Rosenbaum finally concluding that the story "was all in his own mind. I feel that he suffered from a combination of imagination and amnesia, a transitory psychosis – that he did *not* go on a UFO, but simply was wandering around during the period of his disappearance."

Perhaps even that was being charitable: McCarthy had called the abduction story a "gross deception", while there is Travis's mother and brother's curious reaction to his disappearance to take into account. Was the whole thing a put-up job by the Walton family? If so, where was Walton when the police were out looking for him? Could he have been in the last place they would have looked – *his own home*?

As Curtis Peebles, in his *Watch the Skies!* remarks, "This [the Rosenbaums' statement] should have been the end of the case. It was not."

The *National Enquirer* ran a story entitled "5 Witnesses Pass Lie Test While Claiming . . . Arizona Man Captured By UFO." But this was not the McCarthy test, but one carried out by a Dr Harder, whose conclusion could not have been more different. It stated that he was convinced Walton had indeed been abducted by a UFO. There was no mention of the flunked test.

Then over a year later,[50] APRO announced that Duane Walton had passed a polygraph examination carried out by George J. Pfiefer, which he had passed – as had the Waltons' mother, Mary Kellett, on a later occasion. These results featured in the *National Enquirer* of 6 July 1976, and the case won Travis and the other wood-cutters a $5000 prize for "1975's Most Extraordinary Encounter with a UFO" from the tabloid. APRO was fulsome in its praise, saying, "The Travis Walton case is one of the most important and intriguing in the history of the UFO phenomena", a conclusion endorsed by such luminaries of the UFOlogical community as Dr Leo Sprinkle and Dr J. Allen Hynek. Others were considerably less impressed, including arch-sceptic Philip J. Klass, who became suspicious when he read in the *APRO Bulletin* how Travis had answered "No" to the question, "Before November 5, 1975, were you a UFO buff?" in the Pfeifer test. Klass delved further, only to be told by Dr Rosenbaum, that "Everybody in the family had seen some [UFOS] and he's been preoccupied with this almost all of his life . . . Then he made the comment to his mother just prior to this incident that if he was ever abducted by a UFO she was not to worry because he'd be alright."[51] Clearly, judging by her reaction to her son's mysterious disappearance, Mrs Kellett had taken him at his word.

It is perhaps relevant that when John J. McCarthy carried out the "gross deception" lie detector test, he also conducted an in-depth interview with Travis, discovering that he had a criminal record – having been convicted of forging payroll cheques and theft in 1971. He had also used amphetamines, cannabis and LSD – which may, at least, cast some doubt on the witness's ability to think straight.

When Klass went public with his findings, all hell broke loose among UFOlogists – some of whom took his side, though most chose to ignore the sordid facts. APRO's reaction was to lash out at the polygraph examiner Dr McCarthy, calling him "unbelievably incompetent", while Dr Hynek said, "Walton's story seems more consistent than that of his detractors."

Perhaps no one will ever know for certain what really happened to Travis Walton. Whatever it was, judging by his appearance in the hotel immediately after his reappearance, it was clearly traumatic, although perhaps the stress was merely the result of maintaining a high-profile deception – and/or being afraid of his pugnacious and domineering brother Duane.

It could be that the Waltons' interest in UFOs simply crystallized in the forest that day in 1975, and the aliens, magnetically attracted by their eagerness to meet with them, came as if on cue.

18 MARCH 1978, CHARLESTON, SOUTH CAROLINA

Thirty-year-old Bill Herrmann, a truck driver from Charleston in South Carolina, USA, saw a "slick metal disc, about 60 feet [20 m] in diameter" swooping through the sky close to his house. He recalls that, "Suddenly it was right in front of me. I fell backward. The next thing I knew, there was light all around me, green and blue, and I felt myself being tugged upward."

It seemed at the time that it was only moments later that Herrmann discovered he was miles away, sitting in a field, with a shrinking circle of orange light around him. He sat and watched the UFO fly off in a peculiar three-cornered flight pattern that he was later told was designed to avoid being damaged by terrestrial radar. At the time, however, he was in the grip of trauma. He says:

> I couldn't remember anything. I didn't know where I was. A terrible fear came over me, and I stood there weeping for what seemed a long time. I felt dirty. I felt like . . . I can't describe it. I felt like I had been around something I shouldn't have been around.

When hypnotized later, he described having been on a table while being examined by 4½ foetus-like creatures dressed in rust-coloured jumpsuits with disproportionately large heads, huge eyes and "spongy" white skin. He remembers lights flashing around him.

A Fundamentalist Baptist, Herrmann was in no doubt that he had encountered Satanic agents, for when he invoked the name of Jesus, the phenomenon vanished. This technique also worked on subsequent occasions when the UFOs came to call.

The interpretation of the "saucers" as being agents of the Devil is widespread among Christian Fundamentalists, as may be imagined, but they are by no means alone in ascribing evil motives to the UFOnauts.

19 NOVEMBER 1980, LONGMONT, COLORADO

Driving to Longmont shortly before midnight on 19 November 1980 Mary and her husband Michael were startled by a beam of blue light that whistled and "whooshed" towards the car. As it did so the headlights dimmed and the radio emitted the sound of static. With that, the car levitated off the road.

The next thing Michael and Mary knew they were driving along the road with no UFO in sight and a period of 60 minutes missing time. When Michael stopped at a filling station he was so physically disoriented that the attendant suspected he was drunk. The couple reported their experience to social psychologist and UFOlogist Dr Richard Sigismonde of Boulder,[52] who initially wanted to arrange for the two witnesses to undergo separate sessions of hypnotic regressions, but was unable to do so because Mary had become very ill after the experience. Within hours of the encounter with the UFO she discovered an anomalous red patch on her abdomen, and was then rushed to hospital suffering from a serious infection. Once there, it was discovered that she was pregnant – the child, although born prematurely the next Spring, survived. After that, Mary refused to undergo hypnosis, preferring to forget all about the experience.

However, her husband agreed to be regressed, recalling that their car was enveloped in a strange mist and that there was a strong odour "like arcing electricity". Then a ladder came down and a small alien, humanoid in shape and with a "long head, bald with grey skin" and a pointed chin, waved them on board. Michael was spread-eagled, naked, on a long table and subjected to a physical examination, during which he experienced burning sensations in his legs and a feeling as if his mind was being scanned, then emptied, and something new being put in its place. Afterwards his bodily co-ordination was so poor that he banged into the wall at the filling station, much to his distress and embarrassment.

Although typical of modern abductions, this case is particularly interesting because while it happened three years after the release of Spielberg's *Close Encounters of the Third Kind*, in which the aliens are angelic beings who inspire rapturous loyalty, it predates Whitley Strieber's more ominous encounters. Clearly, whatever happened to Mary and Michael that night in 1980 was not an attempt, either consciously or unconsciously – perhaps in a *folie a deux* – to recreate the story of *Communion.*

Of course this, and many similar cases, prompts a multitude of questions. What was the smell like "arcing electricity"? Did some electrical phenomenon, either natural or induced, trigger an hallucination, as in Temporal Lobe Epilepsy? If not a natural phenomenon, who or what triggered the illusion? And if it was not some kind of hallucination, who or what abducted the couple – and why? And, perhaps most disturbing of all in its implications, did the aliens make Mary pregnant?

NOVEMBER 1980, TODMORDEN, WEST YORKSHIRE, ENGLAND

Driving his car at Todmorden, Lancashire, Police Constable Alan Godfrey saw a swirling light in front of him. As he approached, at first he thought it must be a bus lying sideways across the road, but soon realized it was like no bus he had ever seen, being a completely unknown diamond-shaped object, some 20 feet (6 m) wide and 14 feet (4.2 m) high, with a row of portholes and a rotating underpart. It seemed to be hovering about 5 feet (1.5 m) above the surface of the road. His police radio being dead, the quick-thinking officer sketched the object on his clipboard – which was the last thing he remembered before coming to some 100 yards (109 m) away. The anomalous object had disappeared, and were 15 minutes unaccounted for.

Later, during a session of hypnotic regression, he recalled what can be considered a standard abduction scenario: he found himself on an examining table with tiny creatures like robots ("the size of a five-year-old lad") examining him, making a "bid-de-bid-de-bid-de" sound reminiscent of the noise made by a character in *Buck Rogers in the 25th Century*, then a popular television series. But in charge of the examination was a tall, bearded man who introduced himself as "Josef" – and, bizarrely, at one point in the proceedings, a large black dog entered the room.

Godfrey exhibited a refreshing honesty in admitting that, between the experience and the regression, he had read extensively on UFOs, so his "recall" may have been contaminated. Since then he has kept an open mind about his experience, unable to decide whether it was real in the objective sense, or the result of some kind of internal, psychological process. Yet if it were a hallucination, how does one explain the fact that another local police officer saw something unusual that time – an anomalous bright light? Had some kind of electrical or light phenomenon triggered the abduction experience? But as Jenny Randles says: "It seems easier to accept that [the experience] was an hallucination than to accept that aliens fly through the stars with big, black dogs."[53]

15 AUGUST 1981, BORDER OF ZIMBABWE AND MOZAMBIQUE

On 15 August 1981 several native forest workers on the La Rochelle estate near the border between Mozambique and Zimbabwe saw an unexplained ball of light in front of them that drifted to the top of a disused fire tower. To most of the group, the light meant that evil spirits were present, and they all fled, except for the headman, Clifford Muchena, who was later interviewed by South African UFOlogist Cynthia Hind.

He told her that he watched the ball of light moved around the estate, becoming more disc-shaped. Believing it to be a fire hazard, he rang the fire alarm, but was stopped in his tracks by the appearance of three tall humanoids, who he took at first to be co-workers. However, as he drew close to them they shot a "bright flare of light" straight at him, hitting him in the eyes and knocking him backwards onto the ground, where he lay temporarily paralyzed, drifting in and out of consciousness. After some time, he came to his senses again, but the light and the beings had gone.

Cynthia Hind found it hard to communicate with Clifford because he only spoke the local dialect, which has no words that might be appropriate to describe UFOs and aliens. However, others came forward to talk of "spirits" carrying "torches" that threw beams of light much brighter than the strongest flashlight. When the investigator asked whether the beings could have been spacemen Clifford and his co-workers thought she was joking, and when she said that men had walked on the Moon, they "openly scoffed and said, 'Only God does that!' "[54] (It must be said that Cynthia Hind's questions about spacemen seem very much as if she were trying to lead the witness in the direction of a personally favoured hypothesis rather than simply report his story.)

Although the witnesses' culture has no framework or reference points that would allow them to interpret their experience as an encounter with a UFO – and very possibly at least one abduction – the story is still recognizable as a close encounter. But does the witnesses' readiness to describe the humanoids as bad "spirits" suggest that there have been other encounters in the past, which had been equally quickly ascribed to the depredations of evil spiritual forces?

19 JULY 1988, JOHANNESBURG, SOUTH AFRICA

Graphic designer Debra and her mother Pat stopped work on a rush-job for a television commercial artwork in the early hours of 19 July 1988, utterly exhausted. Piling into the car, Debra was driving her mother home when suddenly a bright light swooped out of the darkness at them.

At first Debra thought it was another car and that they were about to crash, but her mother locked the doors against possible attackers. Then the next thing they knew they had passed through a cloud of white mist and were inside a room where two beings, two female and one male, invited the women to lie down on a bed. Pat was too short to do so with ease, so the male touched her and she "floated"[55] up onto it. She was not happy with the experience, although compliant, while her mother put up more of a fight. The entities told them they had taken samples of DNA and blood.

One of the female aliens – called "Meleelah" turned out to be the commander of the UFO. (Jenny Randles notes[56] "how alien names seem earthly and can flow readily from our lips. Nobody ever seems to say – 'my name is Zzgyzptrxxcqy' ", although of course they could deliberately anglicise their impossibly alien names, or use a kind of "Babel fish"[57] with which to communicate with otherwise uncomprehending human beings.) The Commander had a peculiar high and sing-song voice. All of the entities were dressed like characters from *Star Trek.*

Debra and Pat came to with two hours missing time. They recalled the details spontaneously, without hypnotic regression, although they do not necessarily agree about what happened. As Jenny Randles reports, South African UFOlogist Cynthia Hind noted that: ". . . they responded differently. The mother recalls more, and was openly distrustful of the beings, but by the end of the experience found herself unaccountably swamped by a sense of 'love' [for them]."[58]

All the usual questions apply here. Why did the aliens bother to take the women as they were driving? Why not wait until they were asleep, then ensure – with their superior technology – that the witnesses had no memory of the event? It seems strange that so many alleged abductees either spontaneously recall their ordeal at the hands of the aliens, or find it relatively easy to do so under hypnosis. It is almost as if the aliens *wanted* them to remember what happened. Or perhaps the point of the exercise was not some kind of medical sampling, but to prompt the reliving of the humiliation – or other intense emotion, such as fear or even, as in Pat's case, overwhelming love. Do these creatures create the abduction scenario in order to generate extreme emotion in order to feed off it, quite literally? Mystics, psychics and occultists have always believed that emotion is a form of tangible energy that can be utilized, in healing, for example. Are the "aliens" actually vampires who live off human emotion – and are they becoming greedier?

NOVEMBER 1989, MANHATTAN, NEW YORK

Known in UFOlogical circles as "The Manhattan Transfer" this has become one of the most enduringly controversial cases.

Through a year-long series of sessions of hypnotic regression with famous abduction researcher Budd Hopkins, Linda Napolitano gradually recovered memories of a truly astonishing event. In the middle of one night in November 1989, she had been abducted from her apartment in Manhattan, and literally floated by Greys through the wall of the high-rise block into a hovering craft, where she was subjected to the usual medical examination. Meanwhile, her family slumbered on, unknowingly. What makes this case stand out from the mass of similar stories, however, is the claim that there were independent witnesses to the abduction.

Budd Hopkins had been contacted by a couple of secret service security men (bodyguards) who told him they were suffering severe mental health problems because of something they had witnessed in Manhattan while driving a "world leader" through the streets in the middle of the night. They said they had actually seen a woman in the hands of small grey men, floating out of the building, even offering to tell Hopkins the number of her apartment. At this point, however, Hopkins capped their story by telling them that he already knew, and that the woman had been returned safely to her bed.

Two years on, more witnesses came forward. They claimed to have been on the nearby bridge and they, too, had seen the woman levitate out of the high-rise block.

In many respects, this case seems too good to be true, and perhaps it is. For there are problems with the fact that the secret service bodyguards only communicated by mail and audio tape: Hopkins never met them face to face. The "world leader", revealed to be the former UN Secretary-General, Javier Perez de Cueller, denies any knowledge of the event, and night-workers at a nearby printing works saw nothing unusual.

While both Linda Napolitano and Budd Hopkins maintain that a *real, physical* abduction took place, other researchers are not so sure, suggesting that, like many similar cases, some psychological or parapsychological event caused the witness to believe in the literalness of her experience.

27 AUGUST 1992, TARBRAX, EDINBURGH, SCOTLAND

On the night of 27 August 1992 Gary Wood and his friend Colin Wright set out on the 15 mile (24 km) journey from south of Edinburgh to the village of Tarbrax in order to have a satellite dish mended by a friend. Driving the Vauxhall Astra at a steady 40 mph (64 km/h) they were chatting when suddenly Colin said: "*What the hell's that?*"[59] Following his gaze, his friend saw a 30-foot (9 m) wide UFO looming over the road. It was black, dark and shiny with no windows.

Gary put his foot down and they shot off at about 70 mph

(112 km/h). But in order to get away they had to pass under the UFO, and as they did so a wave of shining light enveloped them, to be followed immediately by total darkness . . . It was so total that Gary briefly thought he must be dead. Both men were unconscious for what seemed like a matter of some 15 seconds, before abruptly coming to. The car was out of control, swerving violently from side to side of the road. Greatly disturbed, they managed to bring the vehicle under control and continued on their journey.

Assuming that it would be about 10.40 p.m., they arrived at their friend's house in Tarbrax and unloaded the broken satellite dish from the car. They knocked on the door. No response. They knocked again and again. Several minutes passed. Then their friend put his head out of an upstairs window and demanded, not particularly courteously, just what they thought they were doing at 12.45 a.m.? Where had they been for the missing two hours?

Anticipating scepticism and ridicule if they reported the matter to the police, they turned instead to Malcolm Robinson of the Scottish branch of BUFORA, who suggested that they undergo hypnotic regression to attempt to uncover what had happened to them during the lost time. Both men had some misgivings about such a controversial procedure, but finally agreed to "sit" with professional hypnotist Helen Walters, who is also a psychic.

Gary's first session was not very successful because he became too emotional and could only recall dim shapes. But subsequent sessions elicited much more significant information: he remembered an alarming situation where he was lying on a table "and a long arm [was] extending over his chest towards his head . . . On another occasion, he remembers a hole in the floor, which seemed to be filled with a viscous liquid of some kind, like a gel. From this liquid a long, incredibly thin, frail-looking creature was emerging . . . like a skeleton covered in skin. He remembers that the skin over its ribs looked discoloured and bruised."[60]

There were about thirty of these skeletal entities, most of them tall and spindly, with washed-out grey colouring, although there was a smaller one with an "odd heart-shaped face. On its face were strangely familiar markings . . . facial stripes, three diagonally on each cheek . . . reminiscent of the tribal markings normally associated with natives of the American tribes."[61]

Even more disturbing was the presence of a naked young woman, a human being in distress. She looked at Gary, her face still streaked with tears, etching her likeness into his memory. When it was Colin's turn to be regressed he told of a similar scenario, but afterwards both men reacted in opposite ways: Gary became so deeply fascinated with UFOlogy that he became an investigator himself, while Colin never even wanted to think about the subject again, shutting off from it completely.

Brian Allan of the Scottish Strange Phenomena Investigations organization, who studied this case exhaustively, says: "Although I am not totally convinced about the effectiveness of regressive hypnotherapy, I think on the whole that Gary and Colin did see *something*. From their descriptions of the craft, there are more than a few similarities to the triangular vehicles seen from time to time."

Allan speculates as to whether it could "have been an example of *MILAB*, or military abduction, being perpetrated? There is increasing evidence that these events have [taken place] and still do . . . under the auspices of so-called 'Black Projects'. Experiments so covert that they are literally beyond top secret."[62] (For a discussion of the MILAB hypothesis see page 443.)

Colin and Gary's story offers a tantalizing glimpse of one of the relatively few British – and even fewer Scottish – cases of apparent alien abduction, prompting many more questions than answers. If they were spreadeagled on a table, does that mean they were subjected to medical examination, and had sperm removed for the aliens' genetic engineering programme? Was the young woman there for a similar reason? Are the aliens really making victims of us to create a half-human, half-extraterrestrial race?

AUGUST 1993, MELBOURNE, AUSTRALIA

On the night of 7/8 August 1993, twenty-seven-year-old mother of three Kelly Cahill left with her husband to drive to their friend Eva's house, where they were to celebrate her daughter's eighteenth birthday.

Dusk arrived as they passed through the outskirts of Belgrave South, a suburb of Melbourne, but Kelly could see clearly an unexpected object in a field, which although she only caught of glimpse of it for a matter of two or three seconds – due to the speed at which they were travelling – she later said, "I was almost certain I had seen something of an extremely unorthodox nature". Her husband, Andrew, seemed oblivious and at this stage Kelly decided not to tell him what she had seen for fear of ridicule. Later, however, she remarked: ". . . even if I had completely believed what I had seen a few moments before, never in my wildest dreams could I have anticipated the course of events which were to ensue later that evening."

As they continued to drive along, Kelly felt she could contain herself no longer and blurted out that she had seen a UFO – but Andrew, after suggesting it may have been a helicopter (as she wrote, "since when have you ever seen a round helicopter?"), lost interest.

Arriving at Eva's, Kelly found herself the butt of the inevitable "little green men" jokes, and was relieved to go off with her friend for an evening's bingo, arriving back at the house at around 11 p.m. Kelly and Andrew left for home at an estimated 11.45, arriving at 2.30 in the morning, although the journey usually takes just an hour and a half – leaving an hour unaccounted for. Although Kelly realized something must have happened during that time, her husband sought to rationalize it, assuming they must have left later than they thought.

Later that day, as the Cahills drove through the hills from Belgrave to Fountain Gate, they were startled to observe a UFO hovering about 400 metres (1,300 ft) in front of them. It displayed bright orange lights at the bottom, and a solid structure above with the suggestion of glass-like portholes. As Kelly put it, "Even Andrew could not deny what he saw with his own eyes". Excitedly throwing out ideas about what the object could be, they managed to drive on a few kilometres when suddenly a "brilliant light, like a shining sun" beamed straight into their eyes. It was so intense that Kelly had to shade her eyes to look through the windscreen.

Then, without warning, everything seemed to change. There was no UFO, no light, and no dizzying excitement – the adrenalin rush had been replaced by calmness, all apparently within the blink of an eye. All Andrew could think of was that they must have turned a corner, while Kelly began to worry that she must have suffered a blackout. Perhaps, she thought, she was epileptic.

Soon she realized that arguing about the apparent time loss with Andrew – "an extremely stubborn man" – was a complete waste of time, prefiguring many more tense conversations in the coming months about even weirder events. But they did agree, on that strange trip home, that there was a strong, but unexplained, smell of vomit in the car, and they were both suffering from stomach pains, which in Kelly's case extended from her lower abdomen to her upper shoulders "not unlike severe muscle fatigue after a day of strenuous weightlifting".

Still arguing about whether or not they had lost time on the journey, they arrived home, where Kelly discovered a triangular red mark under her navel, which looked like a burn "or as if the first layers of skin had been removed". More disturbingly to Kelly: ". . . was a small cut on my bikini line similar to a laparoscopy mark but finer, and with the appearance of being a few days old or partially healed. I knew what it was, as I had been beset with gynaecological problems since the age of sixteen and was more familiar with the incision mark left after exploratory surgery.

"Although not due for my menstruation, I was bleeding. This unexplained loss of blood continued for three and a half weeks until I became extremely ill and was hospitalized with an infection in the womb . . ."

As Kelly points out, such a thing is not an "everyday disorder. The usual cause is a pregnancy which has self-terminated and then festered in the womb, or an infection caused by non-sterile surgical procedure." Although Kelly knew she had not been pregnant – as indeed tests proved – her doctor insisted the infection must have come from a self-termination.

Later, in early October 1993 Kelly began to recall the events of the missing time, as a series of "flashbacks of definite conscious acts". Some of the memories were intensely personal, even humiliating, for Kelly, and she makes it clear that if she "could find an excuse not to [make them public] I would do so."

She remembered Andrew stopping the car, and a sense of wonder and awe – and of being somehow chosen – descending. As she said wryly, "Yes – I was indeed blessed". Bending down to retrieve her handbag from the floor was to prove an important trigger in recalling the events of the night, saying it was "one of the conscious acts that enabled me to retrieve the suppressed information". Having picked the handbag up, she looked behind and saw another car, perhaps white or light blue in colour, and felt relief that there would be other witnesses to the UFO. As Kelly and Andrew got out of their car and walked across the road she noticed that two occupants of the other vehicle had done likewise. She thought they were a man and a woman but "because of the incredibility of the object I did not spend a great deal of time observing the others".

Repeatedly reminding herself, "You are conscious, Kelly; this is real" she and Andrew looked towards the UFO through a fence, beyond which they could not go. The craft, now landed in a gully, lay before them – huge, majestic, with "a row of orange lights with what seemed to be solid rays of concentrated blue light beneath, arranged in a half-moon shape. The entire craft shone with a fluorescent glow".

They stood, just looking in awe at the spectacle, then became aware of a figure coming towards them – very tall, and seemingly black. Kelly was startled because "for some reason I had expected to see a human being, but this was not human – its shape was all wrong". Experimentally, Kelly tried communicating with the entity telepathically, but, as she wrote later this was a "BIG MISTAKE!", being immediately flooded with "the most horrifying fear I believe a human being is capable of knowing." The entity's eyes glowed a fiery red, as Kelly – perhaps only too tellingly – whispered to Andrew: "They've got no souls . . ."

Abruptly, the dark field in front of them was crowded with many of the entities, all with eyes of red fire, some of which came straight for the Cahills. Kelly recalls the maelstrom of emotions the sight of them stirred in her: "I was transfixed. I couldn't take my eyes off them. Their power or energy was unfathomable. My mind became a sea of intense confusion, a roaring wind inside my head, adding to my thoughts. I just couldn't seem to focus my faculties. It was as if something was interfering with the very way my brain functioned. I had to fight it. I was going to die if I didn't."

Screaming in terror, she yelled to the other people who were still further down the road: "They're evil! They're going to kill us!" As soon as the words were out of her mouth, there was a huge "*whoomph*" in her stomach and she found herself flying backwards. At the time she believed this to be the malevolent reaction of the aliens, but later, when she visited the site with Australian UFOlogist John Auchetti, he pointed out that there was an electrified fence immediately in her path, prompting her to admit that she could simply have bumped into it, although, as she later found out, "the official opinion is that the electrical current generated was not sufficient" to have such an explosive effect.

Sitting up, winded, in the road, Kelly was appalled to discover that she was apparently blind, and called to Andrew for help. But even more horrifying was the sound of his voice, coming from a distance, croaking with utter terror. He was saying "Let go of me!" to which the response, in a "clear, audible male voice" (and presumably in English) was the classic, "We mean you no harm". Bravely, Andrew came back with the question "Why did you hit Kelly then?"

It was the answer to this that affected Kelly profoundly, with a mixture of amazement and humiliation. The alien said "I wouldn't harm her. After all, *I am her father.*" Then the being chuckled sardonically.

Kelly cried out, "You're not my father. I'm not your daughter. You're evil. You're not my father. I hate you. You're evil, I hate you . . ."

Then, feeling desperately nauseous, she put her head between her knees and passed out. (Although she was unsure whether she had actually vomited, as we have seen, there was an unexplained smell of vomit in the car as they drove home.) Coming to, she was aware of the male voice continuing to speak "as if he were addressing a group". He uttered the ultimate cliché – and possibly the greatest lie – "We are a peaceful people", to which Kelly responded by becoming hysterical, sobbing uncontrollably. Turning to the other human witnesses, she yelled: "Don't believe them. They're not really peaceful. They're trying to trick you. They want your souls. They're trying to steal your souls." Then the male voice cut in calmly: "Will someone do something about her?"

Believing that they were faced with the embodiment of evil, malevolent tricksters, Kelly's utter disgust was intensified when the creature touched her lightly on the shoulder. Suddenly, her fear completely dissolved, and in its place was a towering rage. She shouted at the alien: "How dare you put fear into the hearts of these innocent people. How dare you. Get out of here. Do you hear me? Leave! In the name of God, go back where you came from."

That appeared to be the end of the experience. The next thing they knew they were back in the car, driving home.

Kelly says that she really believed she was "acting as a martyr for the whole human race", but looking back, is "still unsure of what I came up against, or of their intentions . . ." Since that time, she adds, "There were yet to be another four bedside manifestations by one of these creatures, the last occurring in January 1994." Perhaps tellingly, these experiences all took place "just after waking from a dream, although on each occasion I was perfectly conscious".

Yet there are many levels to this fascinating and complex case. Between October 1993 and January of the next year, Kelly had a series of extremely disturbing dreams – calling to mind those that so distressed Betty and Barney Hill in the 1960s. Kelly felt she was physically paralyzed, while being drawn up out of her bed – unsuccessfully endeavouring to rouse her husband. She described seeing shadowy, cloaked figures and later, suddenly feeling she had murdered one of them. She recalled another female abductee screaming "Murderer!" at her, but later one of the aliens explained that she had not killed anyone: the image had been deliberately implanted, he said, to control her anger – although exactly how this would work was not made clear.

So far this story is remarkable enough, but it was to become even more astonishing. Investigator John Auchetti actually managed to track down the occupants of the *other* car that Kelly saw stop at the scene of her abduction. Although they had no desire for publicity, these people co-operated fully in the investigation, adding such extraordinary corroboration of certain central details that this has been dubbed "the perfect case".[63]

These witnesses were a married couple and a female friend, and although the man remembered little (just like Andrew Cahill), the women – Glenda the friend, a nurse, and Jane the wife – were revealed to have played a key part in the abduction drama. They had seen the UFO, which made them feel queasy like Kelly, but managed to drive on until they had come across the UFO in the field beyond the electric fence. The women described the identical entities that had taken Kelly, even independently drawing them. Moreover, they, too, had been taken inside the craft and subjected to a medical examination that left them suffering from vaginal problems, besides noticing strange circular marks on their legs. Those were the women's conscious memories, but more was to come when they were hypnotically regressed.

They revealed that there had been a *third* car at the scene, containing a man on his own – whom the researcher also managed to locate. This witness, while unhappy about going public, did describe seeing a UFO at the same place that the women had their more intimate encounter with aliens.

The UFOlogists also tested the field where the drama took place, with interesting results – compasses fluctuated wildly, indicating a magnetic anomaly. Some might say this was the result of some unknown, but entirely natural, radiation, which somehow induced these shared hallucinations in the women, but such a theory is woefully short on evidence.

There are other possibilities, although once again they are not much more than colourful theories. Were the Cahills (and the other witnesses) subjected to some kind of military or mind control experiment? Perhaps it is significant that on first seeing the UFO, Kelly excitedly asked Andrew, "Was our own government more technologically advanced than we had ever thought possible?" But showing little faith in Australia's covert abilities, she answered her own question, perhaps with more accuracy than she knew, by saying, "Nah! If it had anything to do with governments then it was more likely the American military, not our own."

It is also interesting that the common theme of electricity featured in the story – Kelly thought she had been knocked over by the aliens' power, but the researcher pointed out that she could easily have bumped into the electrified fence at that point. Could some kind of electrical surge – perhaps triggered by unusual atmospheric conditions – have caused the couple to hallucinate? But in that case, where did the marks on Kelly's body come from? Even if the burn mark came from bumping into the fence, it would be highly coincidental for it to take on the same triangular shape as other abductees' "stigmata". And although a powerful electric shock could have caused prolonged anomalous bleeding, it seems unlikely that it would produce symptoms that the doctors confused with a self-termination. And how would *Kelly* having an electric shock make the other witnesses see the UFO?

Yet again, there is something very surreal about the encounter. Reading through the story of Kelly's ordeal, isn't the exasperated comment of the alien about her hysteria – "Will someone do something about her?" – somehow curiously human, just what stressed men would say with a screaming woman on their hands? Were there ordinary men hiding behind the "alien" theatrics? Was the "perfect alien abduction" actually a MILAB – a military abduction? The world of covert operations is much more like *The X-Files* than most people imagine, and sometimes even more disturbing.

However, perhaps Kelly herself should have the last word. Clearly seeing her experience as a form of spiritual enlightenment, she wrote: "Many followers of religion profess belief in an invisible spiritual world. Could UFOs be part of this? For one short moment had I been allowed a glimpse into another reality? Or were there really other civilizations somewhere out there in the universe? What if there was? Was God trying to tell me something?"

JUNE 1994, HARBIN, NORTH-EASTERN CHINA

UFOs in China seem to have an obsession with tree farms as we have seen in the Cases section. However, during one of their visits, to a tree farm close to Harbin, in north-eastern China in June 1994, their occupants did considerably more than clip the tops of trees and have a look around.

Spotting an unexplained object on nearby Mount Phoenix, three farm workers climbed up the mountain to investigate. The object turned out to be round and white with a scorpion-like tail, but further investigation was forestalled by "a very strong noise that produced unbearable pain." Driven away for the time being, one of them, Zhao Guo, decided to risk another attempt to get close to the UFO, and returned with a group of colleagues the next day. But when he was within a kilometre away, he saw (through binoculars) what Shi Li, of CURO describes as "an extraterrestrial with a raised arm emitting a beam which burned his forehead . . . Then he fainted, falling to the ground."[64]

During a train journey to hospital, the traumatized witness began to describe seeing – then and there – a female alien of weird appearance, although no one else could see her. He also claimed to have had sex with her, which the CURO UFOlogists found rather difficult to believe, recalling Villas-Boas' experience. However, when the witness had a photograph developed that he had taken outside his house on the night after his bizarre experience, there was an inexplicable white bar in one corner.

Unlike western abductees who are usually kept well away from politicians for fear of ridicule, Zhao Guo gave a talk at the International Space Research Congress in Beijing in October 1996, at which the President of China gave the welcoming address. Officials from the European Space Agency, NASA and the UN Outer Space Committee were also there in force, and although many of the questions aimed at Zhao Guo at the Congress were unashamedly sceptical and hostile, the fact that he was there at all, in that distinguished company, is in itself very significant.

Clearly although China is a late-comer to the UFOlogical scene, it has no intention of being left behind. Indeed, in certain respects it almost seems to lead the way already.

JULY 1994, MELBOURNE, AUSTRALIA

Driving home to Melbourne from a friend's house, Grace Kyriakidis, her niece Victoria McGinley, Tina Chatzibasile and Joy Bock noticed a bright orange light in the sky, moving parallel to the road. Then they saw a round object zipping about, flashing coloured lights "like a strobe". This appeared to be following the women's car, causing them to panic – especially when the whole road was lit up by an orange glow, emitted by a UFO just metres behind them.

Reaching a hotel, the women tried to get help, but none was forthcoming. When Joy – somewhat bravely – got out she saw high above them a massive diamond-shaped UFO emitting the orange light from its undercarriage and a "steady, droning sound".

After the UFO disappeared from sight behind some trees, the friends drove on, but the lights were still there, not far above them. Suddenly, Grace saw, next to a massive pine tree close to a house, what appeared to be a "shadowy figure . . . It was the shape of a human, but there was very little detail. That's when we knew we had to get out of there."

The orange UFO continued to follow them for a short while, then turned sharply across the road and disappeared. But then the "strobe" lights reappeared – six of them this time – encircling their car menacingly.

Somehow, the women got home, but for Grace at least it was not the end of the experience, for she had several more encounters with the humming UFO and the bright lights, on the last occasion immediately above her house. And all of them, after their terrifying night drive, suffered headaches, nosebleeds, memory loss and repeated electrostatic shocks.

Their journey, which should have taken them just 15 minutes, actually took an hour and a half – even given their stops to seek help, this was an inexplicably long time, suggesting that this was perhaps another abduction case, but that their memory of the experience had been wiped.

What were the bright flashing lights? What was the UFO? Was Victoria closer to the truth than she knew when she told reporters, "It reminded me of a stealth bomber: it was that big and the same shape"? Perhaps some secret craft was being tested over the outskirts of Melbourne in the early hours that day, and the women were inadvertently witnesses to it. Yet if they had stumbled upon the test flight of such a craft, why did it follow them? Surely the pilot would hardly have instructions to terrorize some women in the dark below. And what kind of terrestrial planes have the power to adversely affect the physiology of those who merely witness them passing over, not to mention playing havoc with linear time?

All the women could say was, "This story will sound too unbelievable for a lot of people, but we know what we saw."

FEBRUARY 1997, NORTH BERWICK, EAST LOTHIAN, SCOTLAND

Waking with a start in the early hours of the morning, Angie found that the only part of her body she could move were her eyes. Disbelievingly she stared into the gloom of her bedroom at a small hunched figure, about 3 foot (1 m) tall, that stood by her bed.

"I wasn't dreaming," she declares. "I was totally awake and alert.

When I saw the little figure so close to me in the semi-dark I was terrified, particularly as I couldn't move, not even a little bit."

As she continued to stare at the creature, it moved suddenly, emitting a strong smell of "burnt cardboard or paper". It leant over her while she lay helpless – apparently only to touch her pillow. But as it did so, she heard telepathically in her head mechanical sounding words: "Gran says do it in the morning". With that, the small entity vanished, leaving a heavy odour of burnt paper in the room, and Angie discovered she could move normally again.

After having slept like a log for several hours, she woke refreshed, recalling the strange event with great clarity. She was in no doubt about the meaning of those cryptic words. Her much-loved grandmother had died just two weeks previously, and Angie's grief had prevented her from applying for her "dream job" – as nanny for a family in Edinburgh.

"I felt as though my Gran had sent me a message," Angie says. "I knew I must pick up the phone and apply for that job straightaway." She did so, going for an interview the same day. It turned out that the couple were just about to hire one of the nannies they had already interviewed, but something in Angie's voice made them warm to her. She got the job, although she is in no doubt that she would never even have applied for it if she had not had the terrifying bedroom visitor.

Was Angie's experience simply a case of sleep paralysis and a classic bedroom visitor? Or did her dead grandmother really try to get a message to her, albeit in the most fantastical circumstances? However, another possibility should be considered. It could be that Angie's unconscious mind was looking for a way of encouraging her to apply for the dream job, knowing that she would only make the move if apparently urged to do so by her grandmother. That put the seal on it: effectively giving her permission to stop grieving and start living once again.

Yet there are elements in this story that are reminiscent of the classic alien abduction scenario – the smell of burnt paper, for example, which was particularly noticeable in Whitley Strieber's experiences with the Visitors. And the creature, although Angie failed to make out much detail, was tiny, just like the Greys.

This experience, relatively simple and short-lived though it was, is nevertheless one of those that appear to cross the boundary between the mythology of the alien and the bizarre world of the paranormal, and across the frontier between wakefulness and the hypnogogic, trance-like state experienced just before dropping off to sleep. Some researchers, such as Dr Karla Turner, tend to ascribe all peculiar events experienced during these "crossover" times, when traditionally magic is busiest, to the intervention of aliens, but humanity has been plagued by bedroom visitors when the vogue for explaining them was more uncompromisingly demonic, as we will see.

Whatever the explanation, Angie's brush with the unexplained changed her life for the better, and although she has not experienced anything paranormal since, she is still happy in her dream job. Today she believes her grandmother sent a messenger, firmly rejecting the idea that she may have been dreaming or fantasizing.

7 AUGUST 1997, SIOUX CITY, IOWA

When twenty-two-year-old Frank and eleven-year-old Justin were 45 minutes late in arriving at Grandma Brenda's home near Sioux City, Iowa on 7 August 1997, they brushed aside her scolding with an astonishing story.

They claimed that as they drove through the thickly-forested Stone Park ravine they had been pursued by a triangular UFO – and a bizarre entity with enormous eyes had actually collided with their car! Not surprisingly, Brenda wasn't too impressed with their excuse for taking an hour for what should have been a 15-minute journey, but as luck would have it, as they stood arguing on the doorstep, an orange UFO appeared in the sky, clearly visible to all of them, out of which smaller globes emerged. At that, the three of them jumped in the car to try to follow the UFO.

Although they drew a blank on that occasion, that night Justin had a terrible nightmare. He dreamt that he saw hands pawing him, and felt himself being put on a table and a medical needle appeared.

MUFON investigator Beverley Trout discovered that Justin had a background of paranormal experiences, including encounters with "angelic" beings that he did not altogether trust, in his bedroom. At one point his visionary experiences resulted in a referral to a psychiatrist, although he was not found to be suffering from any mental illness.

Beverley Trout says "we, as investigators, are left with more questions than answers."[65] It is interesting that in medieval demonology evil spirits were believed to manifest in the victim's bedroom, attempting to carry them off to hell, where needles were inserted into their navels. Are we looking at the modern manifestation of this? And if so, what was the UFO – a coincidence? Or is Justin some kind of medium, through whose strange talents UFOs can materialize in this reality?

JUNE 1990, SOUTH LONDON

Bob W. was idly thinking of "nothing much" when he took his rubbish out on the evening of 29 June 1990, placing the two black plastic bags in the bin at the back of his south London home. Out of the corner of his eye he noticed something move behind a pile of discarded plant pots, sneaking into his shed. Thinking it must be a rat, he called to his Border collie cross, but she refused to enter the shed, uncharacteristically whining and growling. When Bob made to go in himself she barked sharply, as if in warning.

Plucking up courage, he entered the shed, the dog following, tail between her legs and still whining. Nothing moved inside the dark coolness. Shrugging, he turned to shut the door when something moved again. Turning, he saw the vague outline of what appeared to be "a small thin child with a big head, with no clothes on". He blinked, and the "child" held up one "long bony hand, with four fingers and no thumb" as if to announce its alien credentials. The dog started to howl and Bob snapped out of the semi-trance like state into which he appeared to have fallen. Very scared, he tried to slam the shed door to prevent the creature from getting out, but there was a "puff of smoke or vapour" and a "bad smell of old eggs" and the creature simply disappeared into thin air.

Rushing into the house, both master and dog were violently sick, and the next day both of them were still suffering from nausea. The dog never went near the shed again. Bob is adamant that he saw a "small creature that couldn't have been human", and somehow scared it off.

Once again we find physical sickness a result of encountering such a being, and a sulphurous smell attached to the phenomenon, both of which suggest that this was not necessarily a beneficial encounter for either human or dog – perhaps it was never intended to be. Or perhaps the being had not intended to be seen.

It may be important that Bob's mind was idling, for this seems to be a factor in accidentally inducing paranormal experiences – an almost trance-like state often seems to open the doorway between this world and the magical realm.

MARCH 2000, GRIMSBY, LINCOLNSHIRE, ENGLAND

On the morning of 24 March, forty-year-old Elaine King was still in bed at her home in Tetney Lock outside Grimsby when, as she claimed,[66] "I was in bed and felt myself getting weaker and weaker like I was collapsing. Next thing I knew I was in a corridor." Apparently this was on board a large metal craft.

Assertively opening a door, she was horrified to find herself looking at some kind of surgical operation in which a human male lay on a table while aliens "peeled back his skin to look at his insides". Curiously one of the creatures, which appeared to be feminine because of its "milky white skin", seemed to be wearing a wig, perhaps to make it look more reassuringly human. The alien communicated with Mrs King telepathically, assuring her that her pets would be looked after during the experience – although, as it turned out, they had no need of alien kennel-care, for she was away for only about 10 minutes, suddenly finding herself back in bed.

Mrs King has a history of paranormal happenings, including out-of-the-body experiences, which may account, at least partly, for this "alien" encounter. The fact that she was in bed at the time, perhaps in the hypnopompic state between sleeping and waking, may have helped create the "vision", in which the corridor represented the divide between this reality and the Otherworld. The surgical operation could easily have come from any of hundreds of abduction stories. However, the Grimsby area was undergoing something of a UFO flap at the time, so Mrs King's experience may have been part of it – but whether that means that the phenomenon as a whole was created by freak electromagnetic or atmospheric conditions, or by genuine alien visitors in their craft must remain a matter for debate and speculation.

THE STARCHILD CONTROVERSY

The work of Dr John Mack, Budd Hopkins and others indicates that aliens are abducting humans in order to create a hybrid race, and that some people are, in fact, actually half alien themselves – the so-called Starchildren. Certainly there is an abundance of myth and legend that tells how the spacegods came to Earth and mated with the indigenous creatures – for example, in the Old Testament we learn how the sons of God (as fallen angels) mated with the "daughters of Man". In April 1999 at a UFO convention at Eureka Springs, Arkansas a fragment of a strangely-shaped skull was displayed to a rapturous gathering – but was it, as claimed, really that of a Starchild, a being whose ancestry lay in another galaxy?

Its custodian, forty-four-year-old American anthropologist, Lloyd Pye, claims that the skull was found in the 1930s in a cave in Mexico, and that it came from a child whose mother may have been a local, but its father came from much further away . . . The story goes that an American teenager on a family visit to a place near Copper Canyon in northern Mexico, was told to stay away from the nearby caves, which were traditionally taboo, but being a rebel, did the opposite – and found the skeletons of an adult "human" and a small deformed child at the bottom of the cave. These she put in an oversized basket she just happened to carry with her and hid them near the house of her relatives. However, when she went to retrieve them, all that was left were two skulls, a heavy downpour having apparently washed away the rest of the bones. She smuggled the skulls into the United States where she kept them safe until 1993, when, as an old woman, she willed them to a friend. This man, who has remained anonymous, passed them on to their present owners, who also are unidentified. Having seen a talk given by Lloyd Pye about his ideas of the origins of humanity (based in part on the work of Zecharia Sitchin – see Pye's book *Everything You Know Is Wrong*, 1998), they handed them over.

Most believers take the two skulls to belong to mother and child, which is the favoured opinion of the numerous psychics who have held or meditated over them. Pye believes that humanity was originally – about 250,000 years ago – a creation of the Anunnaki, an alien race, who needed a slave work force to mine gold for them. (Gold was apparently essential to maintaining an unpolluted atmosphere around their home planet.) Later, however, they returned to Earth to create human-alien hybrids in order to refresh the gene pool.

Of the two skulls, it is the child's that has had the most widespread publicity, perhaps for sentimental reasons. Some believe that the mother was hiding in the cave from the Star Beings, killing the child and then herself, rather than let it be taken away from her.

Certainly, the skull is very odd, apparently deformed. Although the bottom half of the face is missing, perhaps due to a massive blow (which may have been inflicted after death), the remainder is obviously very different from the normal human head: the cranium is disproportionately huge, and the eyes – which must have been enormous – are round and shallow, with a distinctive canal for the optic nerve at the bottom, unlike the normal human eye which has the optic nerve at the back. This indicates that the child had huge, fixed eyes, probably with panoramic vision. But it is the brain volume that is the most impressive – and controversial – feature of the skull. The normal human brain capacity is 1,400 cubic centimetres (85.4 cu in), but the volume of this child's brain was at least 14 per cent more. Lloyd Pye points out[67] that to anthropologists a 200 cc (12.2 cu in) increase in brain volume qualifies as a new species: had 'Starchild' lived to adulthood, he claims its brain capacity would have been 400 cc (24.4 cu in) greater than that of *homo sapiens*.

Although it is hard to discern the child's gender, it seems to have been a girl, about five, judging from the condition of her teeth. Strangely for one of her years, there are no sinuses, although they develop at around the age of five. Pye thinks their absence indicates an evolutionary advance that would provide more room for a bigger brain. However, one of the most exciting aspects of the Starchild enigma is the fact that the peoples of Central and South America have a common legend, as Pye says, ". . . where Star Beings supposedly came down from the heavens and impregnated local women. These women were allowed to raise these children for five to eight years. Then the Star Beings came back and took them away."[68] Did this skull belong to one who was somehow left behind?

In March 1999 a cover article of the *MUFON Journal* by that organization's international director, Walter Andrus, reported that a 40-inch (102-cm) tall skeleton, along with a breastplate and tiny sword and encased in an oval plaster "shell", had been found in the early twentieth century by Richard Wallace, a geologist, near the North Sulphur River, close to Ladonia, Texas. He had been assisted by Bob Slaughter, professor of palaeontology at Southern Methodist University, Dallas, whose widow was finally persuaded to reveal the extraordinary finds to the world. Andrus's article linked the tiny alien to the story of the airship crash at Aurora, Texas, in 1897 (see page 15), which immediately casts doubt on his claims as a whole, for it has long been exposed as a hoax. And is it simply coincidence that the late Professor Slaughter published a book entitled *Fossil Remains of Mythical Creatures* (1996) which included mermaids, leprechauns – and aliens – not to mention the fact that he is also known to have been a gifted amateur sculptor?

The Mexican Starchild, however, is harder to dismiss, although of course many have done so. Pye has had over fifty experts look at the skull fragment, but although almost all of them said words to the effect of "boy, this is weird"[69], "They will not accept that it might be alien, or an alien hybrid, because it is absolutely beyond their possibilities." The only way scientists would be persuaded to take the possibility of alien presence on Earth seriously, is if they found unknown DNA present in the child's skull. But to arrive at the position where they would be open to such a possibility they would first have to overcome the objections of sceptics, who cite the ancient South American practice of cradle-boarding – where infants had their heads tightly bound between boards in order to force them into unnaturally elongated shapes – as the explanation for the peculiarly misshapen skull. Physical anthropologist John Verano of Tulane University, who examined the skull in April 1999, had no doubts. He said: "There is nothing about his . . . child skull that would stand out if I were to mix them in with an archaeological collection of deformed skulls."

Other sceptics have argued that disease was the cause of the skull's deformity, such as Apert's Disease, Progeria or Crouzon's. But not all specialists who were contacted by the indefatigable Lloyd Pye have dismissed his ideas out of hand: Vancouver-based cranio-facial plastic surgeon Ted Robinson took an unusually long time to pore over the skull, and after some weeks contacted Pye to say that he "had been through every text book available and had proved to himself that nothing like the Starchild was recorded in the literature of human deformity".[70] In fact, Ted Robinson was so impressed by the skull that he agreed to take over Lloyd Pye's Starchild Project (together with Canadian researcher Chad Deetken) – their ultimate aim being to have the remains DNA tested. As Lloyd Pye says, "The beauty of such testing is that it cannot be disputed. It will say what it will say, then we will *all* have to deal with that in our own ways."

Unfortunately, this confident statement of impartiality evokes only too clearly the certainty of the "Shroudies" – those who believe the Shroud of Turin to be genuine – when they called for the cloth to be carbon-dated. So certain were they that the tests would prove it to be 2,000 years old (and therefore, to them at least, the cloth that wrapped Jesus's body, miraculously imprinted with his image), that they continually badgered the Vatican about the testing. However, when the result, announced on 13 October 1988, showed that the "relic" was in fact a medieval or early Renaissance fake, the very same people who had upheld carbon-dating as the ultimate proof, immediately sought to discredit it.[71] Similarly, it will be interesting to see what happens to the believers that the Starchild skull is part alien if DNA testing reveals it to be of terrestrial origin. Will it be the fault of DNA testing?

In his article on Starchild for *Fortean Times*[72], Max McCoy writes: ". . . doubts, however, rarely trouble the faithful. The Star-child skull strikes me as one of the first relics of a new religion and, as such, its significance cannot be measured empirically."

Recalling the scenes at the UFO convention where the skull was exhibited, he says: ". . . the UFO conventioneers who file past the skull don't need . . . confirmation. They are engaged in – well, there is no other word for it – adoration of the misshapen object. Already, stories of positive energies and spontaneous healings are sweeping the convention scene."

Notes

1. As quoted in Peebles, p. 113.
2. See Curran *In Advance of the Landing*, pp. 46–47, quoted in Peebles, who writes in the footnotes: "The similarity between Adamski's run-in with the three FBI agents and Bender's supposed encounter (q.v.), several months later, with the three men in black is suggestive."
3. Peebles, p. 117.
4. The Riverside *Enterprise.*
5. See Curran, *In Advance of the Landing,* p. 47.
6. p. 178.
7. See Spencer, *The UFO Encyclopedia,* p. 406.
8. Indeed, the authorities in Holland, where roads are extremely flat and featureless, have provided interesting treelines to break up the monotony in order to prevent accidents. The area around Dronten, just outside Amsterdam, is particularly deadly to drivers.
9. See p. 116 of Vallée's *Passport to Magonia.*
10. Notably in John Fuller's book *The Interrupted Journey* (New York, 1966).
11. Report No. 100–1–61.
12. As told to Jacques Vallée by Betty Hill.
13. From Fuller.
14. *Ibid.*
15. *Ibid.*
16. As told to Philip J. Klass, and quoted on pp. 198–9 of Curtis Peebles' *Watch the Skies!* (1994).
17. *Ibid,* p. 198.
18. Spencer, p. 185.
19. Peebles, p. 200.
20. "Maggie Fisher" is well known to the author. Her story as it appears here was pieced together from many long conversations over a period of years.
21. See *Intruders: The Incredible Visitations at Copley Woods* by Budd Hopkins (New York, 1987).
22. Hopkins, p. 110.
23. *Ibid,* p. 143.
24. *Ibid,* p. 147.
25. *Ibid,* p. 149.
26. Including *The Andreasson Affair – Phase Two* (New York, 1982).
27. See p. 147 of Jacques Vallée's *Passport to Magonia.*
28. Randles, p. 61.
29. See pp. 111–141 of John E. Mack's *Abducted* (New York, 1994).
30. Called "apports", these objects commonly include flowers and pieces of jewelry.

31. The author knows another little boy, who lives in Scotland near the UFO "hotspot" of Berwick Law in East Lothian, who has always pointed to representations of aliens, saying immediately, "Rocket man!" His mother has no idea where this comes from.
32. In a December 1991 entry: see Mack p. 140.
33. Spencer, p. 304.
34. See Philip J. Klass *UFOs Explained*, p. 298.
35. Warren Smith, "The behind-the-headlines story of the Pascagoula UFO kidnap", *SAGA's 1975 UFO Annual*, p. 81.
36. *Ibid.*
37. Pascagoula in *Flying Saucer Review* 20, p. 6 (1974), quoted in Jenny Randles' *The Complete Book of Aliens and Abductions* (London, 1999), p. 38.
38. Jenny Randles, *The Complete Book of Aliens and Abductions*, (London, 1999), p. 49.
39. *Ibid*, pp. 59–60.
40. *Ibid*, p. 60.
41. Randles, p. 40.
42. *Ibid.*
43. For example, the British cult classic series, *Dr Who*, then almost at its peak of popularity.
44. Randles, p. 41.
45. Jenny Randles, *The Complete Book of Aliens and Abductions* (London, 1999).
46. See Peebles, pp. 273–281.
47. *Ibid.*
48. *Ibid.*
49. *Ibid.*
50. On 7 February 1976.
51. *Ibid.*
52. According to Randles, p. 98, Dr Sigismonde "had become involved in UFOs by default, when between 1967 and 1969 his university was given a government grant to study the [UFO] mystery and – it would seem – dispose of the problem."
53. P. 139, Jenny Randles' *The Complete Book of Aliens and Abductions* (London, 1999).
54. *Ibid*, p. 60.
55. P. 58 of Randles.
56. *Ibid.*
57. The Babel fish is featured in Douglas Adams' seminal comedy-sci-fi novel *The Hitchhiker's Guide to the Galaxy* (1981). It is slipped into the ear of the intergalactic traveller, and although momentarily slippery, has the curious effect of rendering even the most outlandish alien language completely intelligible. Recently the name has been applied to the translation facility on some internet services.
58. Randles, p. 58.
59. From the files of Brian Allan of the Scottish organization Strange Phenomena Investigations, which he kindly shared with me.

60. *Ibid.*
61. *Ibid.*
62. *Ibid.*
63. By Jenny Randles, among others.
64. Source: *http://www.parascope.com/articles/0997/chinaufo.htm,* quoting original article by J. Antonio Hunees, International Co-ordinator of MUFON, in *Fate* magazine, vol. 50, No. 9, September 1997.
65. *Strange* Magazine No. 13.
66. In *The Grimsby Evening Telegraph.*
67. See *Fortean Times* No. 127, 'Starchild' by Max McCoy.
68. *Ibid.*
69. *Ibid.*
70. See *Fortean Times* No. 138, Forum article 'Starchild Still Mystifies' by Lloyd Pye.
71. See Picknett and Prince, *Turin Shroud: In Whose Image?* (2000).
72. No. 127.

THE VISITORS

Over the past 100 years over 7,000 encounters with aliens – or "Visitors" as many people call them – have been recorded, and presumably there are many more that were never made public, perhaps going to the grave with the experiencer.

The nineteenth century scareships were – on the whole – manned by *people*, usually bearded characters with a nautical air, as befitted the crewmen of air*ships*, although there were reports of more sinister beings on board, often of an "Oriental" appearance. Perhaps this was a reflection of the "Yellow Peril" paranoia that was prevalent at the time – an extension of a very widespread xenophobia. If strangers were generally suspected, what chance did the "scareship" crews have of being accepted by the remote rural communities over which they flew?

The Foo Fighters of the 1930s and 1940s were apparently simply machines – no pilot was ever reported – dodging teasingly around the Second World War Spitfires and Messerschmitts. Sometimes they appeared to be merely blobs of light like fairies or certain types of more modern UFOs.

Then came the flying saucers of Kenneth Arnold, and the suggestion that they may have an extraterrestrial origin – a rumour that actually began not with the media, as is generally believed, but with cutting-edge *scientists* . . . But whoever began the Extraterrestrial

Hypothesis it soon took hold in the mass imagination: thousands began to look up at the sky, enthralled with the idea of alien visitors, implicitly believing in their reality and expecting them to come calling at any minute. It was a powerful concept for a world that had suffered a tumultuous global conflict, spiritually decimated and badly needing signs and wonders – especially after entering the age of the Bomb, with the daily uncertainty and underlying terror that an enemy might just *use* it. Any day. Any day now . . .

The Space Brothers did come. A tall blond Venusian came to George Adamski, in the Californian Desert, speaking of the aliens' concern about the Bomb, and expressing his sorrow that humans had shot down some UFOs. Even if there was nothing but strife on Earth, there was hope from a superior, compassionate and intelligent race – if only we would listen, if only we honoured the Chosen Ones who had made contact with the Space Brothers.

Humanoid entities were very common in Adamski's day, especially in Europe, where they were often reported to materialize inside the contactees' rooms without necessarily arriving in a visible UFO. Usually good-looking and fair, occasionally they were also described as having Oriental-type slanting eyes and olive or swarthy skins. Sometimes the humanoids are noted in the background while the Greys do the dirty work in abductions; but whether they are superior to the smaller, less human-like, creatures is uncertain. Their humanoid appearance may enable them – worryingly – to blend in seamlessly with crowds of earthlings. They may also be the Men In Black (suitably attired in their traditionally sinister trappings), occasionally seeming to be unsure of acceptable human behaviour, or even how to dress.

A subdivision of the humanoids are the Nordics: although similar, they are very tall – 7 foot (2.1 m) is not unknown – with strikingly "Aryan" good looks: blue eyes and blond hair. Also commonly reported in the 1950s, they are keen on showing visions of the future to their human contacts, and encouraging them to develop their psychic abilities. Clearly, they are no friends of the Greys – unlike the humanoids, who work closely with them – and often vilify them as their inferiors. Billy Meier's Pleidean visitors were of this type. They often leave peculiarly pointless objects behind, such as pebbles, reminiscent of the "apports" that appear in Spiritualist seances.

Other entities from the early days of the modern contactee era include tiny tricksters – about 3 feet (900 m) tall, sometimes described as "children" – attired in tightly-fitting "ski suits" or uniforms. Seemingly obsessed with the concept of time, they are the most obviously similar of all the aliens to the Little People of fairy lore. Sometimes they become balls of light, like Peter Pan's friend, the fairy Tinkerbell.

Other entities from the 1950s and 1960s were the Mothmen of Virginia,[1] and assorted dwarfs, gnome-like creatures and goblins. Often described as being hairy, with red glowing eyes and pointed ears, they seem to have been resurrected as the dreaded Goatsuckers of South America, where the links with demonology and vampirism are marked.

Among the modern alien visitors, the most predominant type by far is the now-iconic "Grey", with the spindly body, huge head, massive, pupil-less eyes and vestigial nose and mouth, although there are occasionally others reported, including robot-like creatures and beings that perhaps belong more properly in the category of angels. Yet these creatures were barely mentioned in the 1950s, 60s and 70s,[2] so what caused the wave of Greys that has overwhelmed UFOlogy in more recent years?

American researcher Curtis Peebles believes he has the answer, writing in his *Watch the Skies!*[3] that it was one, magically seminal movie that changed the mass perception of aliens for ever: Steven Spielberg's *Close Encounters of the Third Kind* (1977), starring Richard Dreyfuss as Roy Neary, whose encounter with the aliens sends him off on a Pilgrim's Progress, ending with his voluntary "abduction". By far the most lyrical and tender movie of the genre (arguably before or since), it summons an irrational but benign force, imbuing it with more than enough magic to make us all believe . . . In *Close Encounters*, the camera dwells lovingly on the upturned faces of those witnessing marvels, in a kind of religious ecstasy. Watching the rapture, we all wanted to believe, but more than anything, to *see* these wonders for ourselves. The movie made millions *want to be abducted by aliens* . . . Who can forget the extraordinary experience of coming out of the movie back in 1977, and seeing hundreds of fellow-movie goers craning their necks to look at the skies, faces already illuminated by anticipation ? We *needed* the magic that would come from the skies, we *demanded* the rapture of the encounter with another world. It was not for nothing that Spielberg's mothership arrived in a mass of thunderously boiling clouds – a conscious evocation of the cinematic phenomenon that heralded God's presence in Cecil B. de Mille's classic *The Ten Commandments,* from the silent era of many years before. When the mothership approached, it was more than a tribute to one of Hollywood's great directors – it was meant to convey a subliminal message to the audience. These beings are not only superior, *they are divine.*

It was in that upsurge of desire for *contact*, that ushered in the reign of the Greys. As Curtis Peebles writes: "*Close Encounters of the Third Kind* defined the shape of the aliens. In the film, 'they' were short, with large heads, slanted dark eyes, and light gray skins. Their noses were small and their ears were only small holes. The aliens' bodies were elongated and very thin. The fingers were also long. Their overall appearance was that of a foetus. By the early 1980s, this 'shape' would come to dominate abduction descriptions."[4]

However, there was more than a physical outline involved. In Spielberg's masterpiece, we see crowds across the world gathering to honour – almost worship – the coming beings from the skies, while the plot implies that the hero is blessed by his removal from Earth, taken away from this vale of tears to a better life, just as all religions have taught that the believer will go to heaven when he dies. If the Greys did not exist, it was time to create them.

Even the disappointing whimsy of *E.T.* a few years later, in its own way, underlined the emerging myth of the Grey as friendly, even cute. The feeling was almost that every home should have one – and if one is to believe the extraordinary surveys about the frequency of alien abduction from the United States in the 1990s, virtually every home *does* have one. The era of the menacing, abducting and raping Grey began in the late 1980s, coincident with another artistic endeavour – the publication in 1987 of *Communion: A True Story: Encounters with the Unknown* by the American horror-writer Whitley Strieber, whose previous books had included *The Wolfen*, *Black Magic,* and *The Night Church.* Finally, after having been rejected by many publishers (who had their doubts about putting out such a book as a work of non-fiction), *Communion* suddenly erupted onto the bookstands, selling millions of copies. Perhaps even more significant, however, was the curiously powerful image of the front cover: simply the face, in close up, of a Grey alien. Everywhere one looked – from news-stands to bookstalls on railway stations, posters and advertisements in magazines – there was a truly remarkable image that had the power to seize and ravage the imagination. This was no wrinkled but cute "E.T." wrapped in its dressing-up clothes like a child, but an entity that could truly be called *alien*. Beginning by haunting the dark alleys and shady porches of the collective imagination, it was to emerge into full-blown nightmare, projected into all-too-often hellish reality.

However, it would be quite wrong to suggest that Steven Spielberg invented the Grey alien, or that Whitley Strieber was the first who had an experience at their hands. Science fiction magazines had been presenting the public with a variety of bug-eyed, hairless monsters for many years before either *Close Encounters of the Third Kind* or *Communion* were even thought of. Sinister otherworldly entities with more than a passing resemblance to today's Grey icons had long been known in occult circles, as Clive Prince and myself point out in *The Stargate Conspiracy*. Indeed, in 1904 the infamous ritual magician, Aleister Crowley (whom the newspapers called "The Wickedest Man in the World", much to Crowley's delight), undertook a magical "working", or rite, in Cairo with his wife Rose in order to conjure up his Holy Guardian Angel. Whether he managed this or not is open to question (certainly there was little that was holy, at least in the accepted sense, in his subsequent life), but the working saw the "coming through" of entities known as Aiwass and Lam[5] – and the latter, apart from the size of the eyes, is strikingly similar to the Greys. Was this an early contact by the same race of entities who are today's abductors, dressed up in magical terms to appeal to Crowley – or do we have Crowley to blame for the twentieth century wave of increasingly savage alien interaction with humanity? Did Aleister Crowley open the gateway and let the Grey aliens in? (If he did, it did him little good: always a larger than life character, he descended into heroin addiction, dying poverty-stricken in a prosaic English seaside boarding house, his last words being either "I am perplexed" or "Sometimes I really hate myself" depending on the source. Either way it was hardly a grand exit.)

A deeper approach to the aliens was suggested by one of John Mack's abductees, Eva, who said that the extraterrestrials "that we physically see" are "just a form they take when they enter this dimension . . . Wherever they come from, [they] don't live physically per se that way." She explained that their souls can take different visible forms, adding, "That's why we get different pictures . . . Some people call it reds, grays, browns, you know, with wrinkles, without wrinkles, whatever – it's a combination of their biochemical energetic makeup and our perceptive devices . . . But there will be some common ground."[6]

Perhaps even more common ground is being established due to the aliens' alleged genetic engineering programme, the attempt to create a human/alien hybrid race. "Kathie Davis", the central character in abduction researcher Budd Hopkins' book *Intruders* (1987), recalled, under hypnosis, having gynaecological samples taken from her on board a UFO by her alien captors, and later being shown a tiny listless baby, whom she realized was half hers – its father being one of the aliens. Later, a child of about four was brought to her, in what appeared to be a formal presentation of daughter to mother. Recalling the encounter with a great upsurge of emotion, Kathie said: "She was real pretty. She looked like an elf, or an . . . angel. She had really big blue eyes and a little teeny-weeny nose, just so perfect. And her mouth was just so perfect and tiny, and she was pale, except her lips were pink and her eyes were blue. And her hair was white and wispy and thin . . . fine . . . real thin and fine. Her head was a little larger than normal, 'specially in the forehead and back here . . ."

As a group of "little grey guys" stood watching, Kathie just looked at the little creature, "And I looked at her, and wanted to hold her. She was just so pretty, and I felt like I just wanted to hold her. And I started crying . . ." The Greys stood there holding the child's hands, "and she was almost like she was timid, like a very timid bunny rabbit, and she almost was afraid of me . . . I'm pretty sure somebody told me I should be proud. Her eyes were so blue and huge, and her pupils were so blue, and she blinked them at me . . . It was almost as if her eyes rolled up. Her skin was creamy . . . it wasn't grey. She was pale and soft and creamy . . ."[7]

Perhaps the hybrids will rapidly grow out of their bunny-rabbit timidity and turn into something quite different from pale, soft and creamy toddlers – more in keeping with entities that come to take our energy and rob our souls. Or perhaps they will prove to be benevolent guardians, improving on their human ancestry and representing the next stage in the evolution of the race.

Some "aliens", however, seem to have been singularly unsuccessful in using a mask-like persona while perpetrating crimes upon the persons of United States' citizens. Some abductees are beginning to recall not Greys, Nordics or any other kind of alien reported so far, but *men in uniforms or white coats*. Unless *that* is the latest screen memory – the way the real aliens are choosing to manifest these days – then the US military are up to something extremely nasty while hiding behind the more familiar myth of the alien abduction. The MILAB (military abduction) conspiracy theory is just beginning to take hold, but so far it seems to be horribly persuasive. Time, and evidence, will tell. (See Conspiracies.)

However, not all "aliens" are CIA, NSA or Pentagon lackeys in more or less effective disguise – although some of them may well be. As long as humanity has been conscious, there have always been encounters with shapeshifting entities, be they creatures from the deepest recesses of the collective psyche, or beings with a tangible reality, including fairies, demons, angels and vampires. Even when the day of the Grey is over and the horrors of MILABs are exposed, there will always be something that moves in the shadows of the mind, bringing new terrors, new excitement and opening new doors of perception, whether we welcome them in or not.

Notes

1. See *The Mothman Prophecies* by John A. Keel.
2. See p. 112 of Jenny Randles' *The Complete Book of Aliens and Abductions* (London, 1999).
3. *Ibid*, pp. 281–2.
4. Curtis Peebles, *Watch the Skies!* p. 282.
5. See Picknett and Prince, *The Stargate Conspiracy*, p. 313.
6. John Mack, *Abducted*, (New York, 1994), p. 258.
7. *Ibid*, p. 194.

CHAPTER 6

THE CHOSEN ONES

Contactee "prophets", Space Brother cults and channellers

"The journey will be almost inconceivably hard, but also rich with marvels and full of hope"
Whitley Strieber

". . . this is a complex phenomenon, we don't understand it. It is real, and the solution is not to worship it, or to blindly follow anybody who pretends to have an answer. Let's step back and look at what it's doing to us."
Dr Jacques Vallée

Alleged contact with extraterrestrials is by no means exclusively a phenomenon of the twentieth and twenty-first centuries. Way back in the 1850s, Victor Hugo, author of *The Hunchback of Notre Dame*[1] and *Les Misérables,* attended seances during the time of his political exile (from 1853 to 1855) on the Channel Island of Jersey, at which alien beings "came through". According to new information,[2] Hugo, together with his wife and friends, were present when entities, including one Tyatafia, from the planet Jupiter, tapped out messages on the table top (as in the early stages of the Philip Experiment – see page 478).

This fascinating contact told a rapt audience that life was tough for Jupiterians, for their home planet was a "prison world", inhabited by unhappy souls from elsewhere in the galaxy who were serving out life sentences for past misdeeds. On the other hand, the aliens from Mercury – who mysteriously insisted on communicating in Latin – were considerably jollier. They described themselves as being half-spirit and half-animal, with "two eyes which remained open all the time, a huge but very light head, and a long but very thin body"[3], and said that they floated around happily in Mercury's thin atmosphere, "suspended by six appendages ending in tiny 'suns' which they called their 'torches'."[4] They described their planet as a "reward world" – a sort of heavenly theme park, one might imagine – for souls that had proved themselves worthy in other incarnations elsewhere.

These beings called Mercury *petasus insani*, or "the insane messenger of the gods", which Victor Hugo believed might refer to the eccentric orbit of the planet.[5] The Mercurians also said that they were not immortal, although they never became ill unless they lost one of their "torches" that enabled them to exist.

Significantly, the Mercurian contacts were made possible, it was claimed, through the intercession of the ghostly Nicholas Flamel, the highly successful, but profoundly mysterious, medieval alchemist. This would no doubt have impressed Victor Hugo immensely, for their esoteric interests are said to have followed similar paths.[6]

Of course it is extremely unlikely that the spirits who communicated at the seances on Jersey really were aliens from Jupiter and Mercury (and elsewhere), although the communications are not without interest for other reasons. For example, the "suns" or "torches" from which the Mercurians were suspended are reminiscent of the stylized rays of the sun, each ending in a hand, that are found in depictions of the heretical Egyptian pharoah Akhanaten – something with which the esoterically-minded Hugo would have been familiar, and no doubt excited by.

Others in the pre-Contactee era claimed to communicate with alien beings, including the early twentieth-century Swiss medium known as Hélène Smith. Her claims attracted the attention of the respected academic Theodore Flournoy, professor of psychology at Geneva University, whose findings were eventually published in his *From India to the Planet Mars*. She claimed to recall past incarnations, including one as the tragic French queen, Marie Antoinette, and another as the wife of a fifteenth-century Hindu prince – whom she declared had been reborn as Professor Flournoy!

But much more significant (or so it seemed) was her claim of being in communication with the inhabitants of the planet Mars, for not only did she sketch[7] Martian landscapes, houses and people, but also gave many detailed examples of the Martian language, both in written and spoken form. This greatly intrigued Flournoy, who undertook a painstaking semantic analysis of her outpourings – which seemed so impressive on the surface. Yet it was revealed that the "Martian" was no more than a version of the French language: the sounds were French, the order of the words were French, and what little grammar was evident was also a crude version of the French language. Only the vocabulary was new, and that appeared to come from the unconscious mind of the medium. And as for the evidence of the Indian incarnation, it transpires that its colouring was taken from a rather bad history of India published by de Marles in 1823, although Hélène ". . . dug down to the very bottom of her memories without discovering the slightest traces of this work". Her mind had fabricated wonderful fantasies, based on books she had read some time before but forgotten at a conscious level. This phenomenon, known as *cryptomnesia,* is a common (if admittedly disappointing) explanation for many of even the most promising cases of past-life recall – although by no means all. The phenomenon may also help to explain other alleged communications from "aliens". In effect, these people may not be chosen by outside agencies such as beings from Mars, but by their own inner longing to be special.

Over the course of history many "alien" beings have appeared to humans with messages and prophecies, although some have been designated as angels, others demons, and more recently, as extraterrestrials come to save us from ourselves in their shiny spaceships. Whatever their message, the underlying implication is that the visionary or experiencer is *chosen*, special, set apart. In the past visionaries have claimed to be told great secrets by angels or spiritual figures, often founding cults or entire religions on the basis of this honour: the farm boy Joseph Smith, for example, founded the Church of Jesus Christ of Latter Day Saints (the Mormons) as a direct result of a series of visions in the early years of the nineteenth century. Now, although more orthodox religious visions are still reported, the beings that come to call tend to be space brothers – the aliens.

One successful space brother cult was founded by Claude Vorilhon whose book *Space Aliens Took Me To Their Planet* (subtitled:

The most important revelation in the history of mankind: the book which tells the truth) explains that the aliens chose his country of France as the seat of their endtimes operations because it is "where democracy was born", although this may be a sort of alien joke, because according to them the only way of surviving the "last days" is for humanity to abandon all the trappings of democracy, such as elections and the military. The aliens – which Vorilhon calls "the Elohim"[8] – will then deem it safe to return and rule the Earth. In the meantime, however, they honoured their Chosen One with a change of name – he is now the "Rael" after which the movement is named – and gave him a sacred insignia, a rather mixed symbol of a swastika set within a Star of David. The Raelian movement is now some 40,000 strong and showing no signs of diminishing. Recently it has announced plans to clone humans, against the possibility of mass extinction.

The One-World Family, founded by Californian Allen-Michael Noonan, also successfully attracted a huge following for many years. Allegedly, the founder was busy working on a billboard when he was abruptly translocated to another planet where he was surrounded by benign beings facing an empty fluorescent throne – which apparently was meant for him. A voice thundered: "Will you agree to be the Saviour of the World?" Unsurprisingly, Noonan accepted the offer. Since his initial – and sensational – contact, he remained in touch with an entity known as Ashtar (who is seemingly popular with many other groups). Noonan also claimed to have travelled to distant planets such as Venus, and believes himself to be a sort of intergalactic Messiah (one wonders what would have happened if he had refused to take on the responsibility), writing the "Everlasting Gospel" and enjoying the worshipful attentions of his cult members. The One World Family practise occult techniques and regularly use hallucinogenic drugs in their rituals.

One of the most famous, and indeed, enduring, of the "saucer cults" is the late George King's Aetherius Society, with its emphasis on inhabited planets and benign beings of the Intergalactic Parliament. The story of the Society is very much that of King himself (although since his death in the 1990s the cult has continued to thrive). Among other things he had been a London cab driver, although – like George Adamski – his true interests lay in Eastern mysticism, when in May 1954 he suddenly heard a voice saying: "Prepare yourself! You are to become the voice of the Interplanetary Parliament". Just over a week later "an Indian swami of world renown [who] had obviously walked straight through"[9] the locked door passed on information and teaching that was to change his life completely and result in the founding of the Aetherius Society. The swami explained that King was to drop his metaphysical research, even though: "we [it seems to be the royal 'we'] were on the verge of discovering a new method of cancer treatment which could cure certain forms of this malignant scourge. Nevertheless, this command came out of the blue in such a way that no receiver could do anything else but listen and obey . . . Quite soon after the deliverance of the Command, I was able to tune in and receive, telepathically, information which was replayed over millions of miles of etheric space. A message from Venus was recorded on our tape recorder . . ."[10]

Perhaps it would have been better if King had stuck to his cancer research and Eastern mysticism, for his pronouncements about life on planets such as Venus and Mars have been demonstrated to be completely absurd by the advances in cosmological knowledge. According to him, only Mercury is uninhabited, and the "Master Jesus" lives on Venus. King gleaned all this information through trance states – a form of mediumship – which can produce some astonishing gobbledygook. On one occasion a regular communicator known as "Mars Sector 6" solemnly instructed the faithful to: "Take those M-ions inside of yourself, then your brain cells will release an opposite female magnetic energy. This will counteract the hurricane-force".[11]

Named after King's main communicator, the Master Aetherius, his Society flourished, intent on fulfilling a number of arduous tasks, mainly to "recharge" the mystical batteries of allegedly sacred high places, including Mount Kilimanjaro in Tanzania and Ben Macdhui in Scotland, through missions with names such as Operation Starlight. These rituals are very exhausting and require great dedication and belief. For example, on 27 June 1981, 160 Aetherians travelled to Holdstone Down in south-west England to recharge batteries E-1 and E-3 with 219 hours of continuous prayer, which was augmented by a further 1,100 hours of alien recharging. On 23 April 1981 the society sent a "discharge" of prayer-power energy to Poland, in the hope of preventing a Russian invasion.[12] At the end of this big push for peace, Mars Sector 6 told George King approvingly that "there was a heavy resonance of Spiritual Energies over the whole of Poland".

From his base in California, the former London cabbie reigned – there is no other word for it – over the Aetherius Society, a flamboyant figure to the end, although by then he was no longer plain "George King". Over the years he had accrued an astonishing collection of titles, mainly from his own society (i.e. he had awarded them to himself), but others were bestowed on him by shadowy quasi-chivalric orders. By the time of his death, his titles included: Knight Commander; Metropolitan Archbishop of the Aetherius Churches; Count de Florina and Doctor of Sacred Humanities.

Many of the early contactees, such as George Adamski, believed they were chosen by the UFOnauts to be given information about other planets, and warnings about the way humanity was going – both technologically and spiritually. One incident, which took place on the same day – 24 April 1964 – as patrolman Lonnie Zamora encountered his controversial UFO in New Mexico, concerned dairy farmer Gary Wilcox of Newark Valley, Tioga County, New York . . .

At about 10 o'clock on a fine, sunny morning, Gary was spreading manure in one of his fields, when he saw something glinting in the sunlight at the top of the nearby hill. At first he thought that it was just the old abandoned refrigerator that had long languished there, but then realized that whatever was shining so brightly was actually between him and the fridge. He set out on his tractor to investigate, but walked the last 100 yards (90 m) when it occurred to him that it may be a fuel tank from an aircraft – and it might not be a good idea to bring his tractor near in case of explosions.

Then he saw that it was certainly not a fuel tank. The unknown object that had caught his eye as the sun bounced off it was "bigger than a car in length . . . shaped something like an egg . . . no seams or rivets . . ."[13] It was an estimated 20 feet (6 m) long, 15 feet (4.5 m) wide and 4 feet (1.2 m) high. With astonishing bravery (or foolhardiness), Gary hit and kicked the object, which "felt like metallic canvas".

No doubt curious about the racket, two 4-foot (1.2-m) tall humanoids suddenly appeared from under the UFO, carrying trays on which was a collection of "alfalfa, with roots, soil, leaves and brush". They wore white metallic-looking overalls, with no visible detail such as stitching or fastenings. More peculiarly, although they had humanoid legs and arms, no feet or hands were apparent, and when they raised their arms, instead of elbows, their sleeves simply wrinkled. Taken aback at first, Gary laughed, thinking it was "some sort of a trick . . . a sort of candid camera gag". But (perhaps unfortunately) there were no cameras, and his adventure was not destined for prime-time television. One of the entities said: "Don't be alarmed, we have spoken to people before." Later the witness found it impossible to describe the "man's" voice, because although he understood what was being said, something made him wonder if it was really speaking in English, and his voice seemed to come "from about them rather than from either of them".

Then the humanoid announced: "We are from what you know as the planet Mars", following this with a request for Gary to explain what he was doing. With that, he seems to have launched into something of a lecture about the subject of manure and fertilizer in general, which intrigued them so much that he offered to get them a bag of artificial fertilizer. The Martians told him they were "travelling this hemisphere", but refused his request to go back to their home planet with them because their atmosphere was too thin. They said that conditions were such that they could only get to Earth once every two years. Warning that man should not endeavour to travel into space, they predicted that two Soviet cosmonauts and US astronauts Virgil ('Gus') Grissom and John Glenn would die within a year because of the adverse effects of leaving Earth's atmosphere. They added that they needed to learn about terrestrial life because of the "rocky structure of Mars", but avoided Earth's big cities because the pollution adversely affected their spacecraft.

After sternly warning their human contactee not to mention the encounter, the aliens withdrew to their ship, which then took off towards the north, noiselessly. The effect on Gary Wilcox was interesting to say the least. Although the aliens had impressed on him the necessity not to tell others about his experience, the first thing he did on arriving home was telephone his mother and tell her everything – which ensured, as events were to prove, that the news would spread like wildfire. But after talking to his mother, Gary got on with life, milking the cows and doing odd jobs as if nothing had happened. However, he did keep his word to the humanoids on one matter: that night he left a bag of artificial fertilizer at the site of the landing. Next morning it had gone.

Soon the Tioga County Sheriff Paul J. Taylor heard of Gary's experience, asking him for a statement and carrying out an investigation into the subject on 1 May 1964. And there, for the time being, the matter rested. But in 1968 the respected psychiatrist Berthold Schwartz undertook a private investigation of the case, interviewing not only Gary himself, but also members of his family and many of the neighbours. After concluding his psychiatric evaluation of the prime witness, Dr Schwartz concluded that he was a "truthful person with no emotional illness, and that his experience was a 'real' event though the interpretation of the encounter is a complicated and uncertain matter".[14]

It might be thought that Gary had simply stumbled upon the UFO and its occupants accidentally, just as dozens of other witnesses interrupt aliens repairing their grounded saucers, but there is an underlying sense of *design* – rather than accident – about these encounters, as if they are part of a larger pattern. Certainly, the UFOnauts seemed happy enough to tell Gary interesting snippets about their alleged home planet of Mars, and vouchsafed the specific predictions about the deaths of the astronauts to him. Chillingly, although they did not come about in quite the time scale they outlined (perhaps Martians find our method of measuring time confusing), on 27 January 1967 Virgil Grissom *did* die – together with Roger Chaffee and Ed White, in a fire capsule at Cape Kennedy – and three years almost to the day after the encounter, Vladimir Komarov's capsule fatally plummeted to Earth after its parachute failed to open on re-entry into the atmosphere.

Superficially, there seems little reason to doubt that the story – surreal though it was – was simply a conversation between the denizens of two different planets during a chance encounter. Yet the chances are that if Gary Wilcox had encountered his alien friends many more times, they would have given him predictions that began by being convincing enough – many of them would have come true, after a fashion – but then, when the witness's reputation and livelihood were on the line, the last prophecy would have failed ignominiously, catastrophically. That is the way with the aliens, and that is also the way with channelled entities and spirits who "come through" Spiritualist mediums, for, arguably, the alien contactee experience is not an isolated one. It seems to be an intrinsic part of a much wider ploy to mislead, confuse and perhaps even destroy, both individuals and larger groups. Making the initial contactees feel special is all part of the softening up process, the building up of trust and massaging of ego that is all too often the road to ruin.

In the late 1960s, when John Keel travelled widely throughout the United States interviewing what he called "silent contactees" – those who do not write books or seek to become media stars – what he discovered was to change radically the way he thought for ever. Previously a sceptic about all matters unexplained, he became utterly convinced of the existence of a race of elemental entities, which he called "ultraterrestrials", who masquerade as spirits of the dead or aliens or whatever else is fashionable at the time. (See Explanations.)

It began, for Keel, when he arranged with the silent contactees that whenever the UFOnauts would land on their farms etc, the *aliens themselves* would talk to him on the telephone. Meticulously screening out as many hoaxes and time-wasters as possible, when the aliens talked he listened to them, following up their information – ending up by wasting months of his life by "searching for nonexistent UFO bases, trying to find ways of protecting witnesses from the 'men in black' "[15] and so on. But he also discovered another pattern: that of the increasingly valueless prophecy.

In May 1967 a number of the American silent contactees began to report that their alien friends predicted a massive power cut was about to happen – and, true enough, on the morning after the Arab-Israeli six-day war broke out,[16] the lights went out on a huge scale in four north-eastern states. Clearly, it seemed, the UFOnauts had a hotline to the future and we had better pay attention . . . Then the aliens warned that a *nationwide* power failure was about to happen, beginning on 2 July, which would last for three days, to be followed by natural calamities – including the destruction of New York City, which would be submerged beneath the waves. Not unnaturally, news of the forthcoming disaster spread like wildfire, and panic ensued. Hardware stores sold out of candles and torches, while a large number of people packed up and left New York. When the aliens said that Pope Paul would be assassinated during a visit to Turkey – and the Vatican announced that His Holiness was scheduled to visit that country – horror grew in contactee circles. Of course New York is still standing, and Pope Paul survived his Turkish visit. Yet, as John Keel points out,[17] other UFOnaut prophecies at that time did come true, including the plane crash that killed the newly-appointed US Secretary of the Navy, J. T. McNaughton. But there was another element to these prophecies, as Keel says: "What astonished me most was that these predictions were coming in from a wide variety of sources. Trance mediums and automatic writers in touch with the spirit world were coming up with the same things as the UFO contactees. Often the prophecies were phrased identically in different sections of the country. Even when they failed to come off, we still could not overlook the peculiar set of correlative factors."[18]

Yet even with so much apparent correlation, the *big* prophecies failed to come true, leaving the UFOnauts' contactees high and dry. As Keel says: "This is the tiger behind the door of prophecy. Some of the predictions are unerringly accurate[19], so precise that there are no factors of coincidence or lucky guesswork." He believes that what he calls the "ultraterrestrials" or elementals "are able to convince their friends (who sometimes also become their victims) that they have complete foreknowledge of all human events. Then, when these people are totally solid, the ultraterrestrials introduce a joker into the deck." But although he was caught up in the events surrounding the predictions of summer 1967, he remained untouched by the cultists. As he said: "I was lucky. I didn't cry their [the entities'] warning from the housetops. I didn't surround myself with a wild-eyed cult impressed with the accuracy of the previous predictions."

But he adds, ominously: "Others haven't been so lucky."[20]

There are many UFO and abductee cults that promise personal and global redemption in one form or another through devotion to the

"Space Brothers" or "ancient astronaut" gods. At one end of the scale there is the Raelian movement and the Aetherius Society, which may hold apparently bizarre views, but at least they are relatively harmless. At the other end of the scale are the likes of Heaven's Gate, a cult based largely on the television series *Star Trek,* which managed to persuade the cream of their membership to commit suicide in 1996 in order to be reborn on board a heavenly spaceship, and the sinister Solar Temple, who believed they were in touch with beings from Sirius, ending in fiery deaths in the early 1990s.[21] Whether it was better or worse that these disciples were not coerced into martyring themselves but did so with ecstatic joy is a matter for speculation, but the fact that it was so easy to convince them to do it at all is profoundly unsettling.[22]

The attraction of any cult is that it holds a certain *glamour* – not in the modern sense of being glitzy and fashionable, but in that its members are (often literally) *entranced by belonging.*[23] Frequently vulnerable members of society, with emotional, mental and physical problems, people are drawn to cults because they promise not only to make sense of their lives, but also to enable them to transcend their problems and reach a state of ecstacy. Perhaps even more significantly, they are no longer alone, but part of a magic circle of like-minded souls who *have the answer,* sharers in a great cosmic secret that marks them out as special, and sets them apart from the rest of the unenlightened world. The problem is that all too often the great cosmic secret proves to be fairy gold that tarnishes – or disappears – overnight, and their problems return, only this time much worse than before. Or the secret to end secrets involves the ruthless destruction of their personalities, perhaps even the demand for the ultimate personal sacrifice of their lives through suicide or murder. Sometimes, however, the end comes as a whimper, not a bang. Instead of death and destruction, the cultists "merely" become objects of ridicule – which in terms of their standing in the community, can be utterly crippling financially, emotionally and spiritually.

Sometimes, however, the entities get a helping hand from more terrestrial sources, as can be seen from the story of Chicago housewife Dorothy Martin, known previously under the pseudonym "Marion Keech",[24] who, in 1953, began to develop a talent for automatic writing.[25] At first these messages purported to come from her deceased father and other spirits of the dead, but soon they began to "come through" from sources claiming to be alien entities originating from a planet called "Clarion". Persuaded of their benign intentions, Dorothy called them the "Guardians",[26] although they were – all too predictably – to reveal utter contempt for her and let her down badly.

An enthusiastic circle of seekers formed around Dorothy Martin and her continuing contact with the Guardians of Clarion, including several housewives, a research scientist – and Dr Charles Laughead and his wife Lilian,[27] who were hardly newcomers to the world of mediumship and channelling. Originally both Protestant missionaries, Lilian had a crisis of faith due to a nervous breakdown she suffered in the years immediately following the Second World War, and after that she and her husband travelled widely seeking spiritual knowledge in many other belief systems. After a long conversation with George Adamski, they began to see that salvation might lie with the coming of the saucers, finally joining the Dorothy Martin circle in Chicago.

Excitement mounted as Dorothy's aliens prophesied in August 1954 that a huge calamity would befall the United States: the whole of the eastern seaboard, including New York, would slip under the waves on 21 December of that year. Britain and France would also suffer a similar fate. Charles and Lilian Laughead acted as spokespeople to the media, warning the readers of many national and local newspapers and magazines of the coming catastrophe, utterly certain of the truth of the prophecy despite the inevitable ridicule. With tongue firmly in cheek, the *Lake City Herald* carried this dire warning:

> Lake City will be destroyed by a flood from Great Lake just before dawn, December 21st, according to a suburban housewife. Mrs Marion Keech of 847 West School Street says the prophecy is not her own. It is the purport of many messages she has received by automatic writing, she says . . . The messages, according to Mrs Keech, are sent to her by superior beings from a planet called 'Clarion'. These beings have been visiting the Earth, she says, in what we call 'Flying Saucers'. During their visits, she says, they

> have observed fault lines in the Earth's crust that foretoken the deluge. Mrs Keech reports she was told the flood will spread to form an inland sea stretching from the Arctic Circle to the Gulf of Mexico.[28]

(Note the ironic repetition of "she says".) Perhaps it is significant that the *Clarion* was a local newspaper, besides being the name of the Guardians' home planet.

It was at this point that a group of sociologists from the University of Minnesota infiltrated the group with the specific aim of studying their reactions when the prophecy failed the materialize. Published in *When Prophecy Fails* in 1956, Leon Festiger, Henry W. Riecken and Stanley Schacter's findings make very interesting reading, and could apply to the awkward situations that also befall dozens of similar cults with predictable regularity.

Of course, just as in the case of John Keel's "silent contactees" (see above), New York remained intact – as did Britain and France. But what happened to the faith of the Dorothy Martin circle as a result of what might be imagined to be cruel disillusionment with their Guardians? Did they all simply leave at once, either angry or hurt (or both) without a backward glance? Did they denounce the aliens as impostors, and Dorothy Martin as a false prophet?

In fact, very few of the group even admitted to losing their faith: most simply came up with rationalizations for the debacle, some declaring it must have been a test of faith, while others clung to the hope that the disaster had actually been averted at the last minute by the power of their belief . . .[29] Ironically, when the group did disband, it was partly due to rows about which of their rationalizations was correct, and partly to the intensity of the ridicule that was heaped on them by the hostile outside world.

But was it all an invention of Dorothy Martin, or – more charitably – a fabrication of her unconscious mind? As with most such phenomena, it is tempting to think so, but there are disturbing elements to this story that suggest that the Guardians of Clarion had helpers somewhat closer to home. Although the sociologists declared that the entire phenomenon was a collective delusion,[30] this was not only rather unfair, but also revealed a tendency on their part to be over-selective in their data. For they had actually been present when Dorothy received not only telephone calls from the Guardians, but also a *visit* from five of them, including their leader, a young man named Sananda. If it was a hoax, it was an extraordinary one. Perhaps the callers were a variety of the Men In Black, but in this case it seems more likely that they were from "another group of people [who] were orchestrating both the events and the phenomenon of escalating belief".[31]

Who would bother to instigate and maintain such a ploy? Who but the people who have always shown an interest in the activities and beliefs of minority groups? The answer is, perhaps predictably, one of the usual suspects in conspiracy lore: an official (but largely secret) agency that has become something of a cliché in recent years, but whose activities have done little to endear it to the normal populace – the CIA. Although their involvement with UFOs and aliens will be discussed in the Conspiracies section, suffice it to say that wherever the voices of the aliens are heard, there the CIA – not to mention the Pentagon the NSA, and other intelligence agencies – will also be found. But whether they actually originated those voices or merely support the growth of cults around them, is another question altogether . . .

The Dorothy Martin story has an interesting postcript. She fled to become a leading light in a Dianetics (Scientologist) centre in Arizona, but the Laugheads – perhaps because of their turbulent history – were made of sterner stuff, sold up and moved on, still faithful to the Guardians. They were to become prime movers in the astonishing history of the ultimate cult, the Council of Nine, whose rise to power is told elsewhere (see Conspiracies).

Another channelling cult was the "Light Affiliates" of late 1960s-Burnaby, British Columbia. With considerably *chutzpah* they announced themselves with a flourish thus: "We wish to notify all those interested that a phenomenon has occurred here in Vancouver. A young girl, age 22, suddenly began channeling on 23.10.69. Her source is a being identifying himself as Ox-Ho, who is relaying transmissions from a galaxy close to our own . . . Her material is phenomenal in that she had been informed of the coming disasters, when to expect them, and what to do pertaining to the necessary evacuation of the danger areas and food supplies, etc, that will be needed."[32]

The entity had already made some changes. The channeller, whose real name was Robin McPherson, had become "Estelle", while her mother Aileen had been renamed (classically) "Magdalene" and a male groupie was to be known as "Truman Merit" – a positively Bunyanesque flourish.[33] (The name of the alien himself – Ox-Ho – will no doubt make British readers smile: *Oxo* is, famously, a beef stock cube, literally a household name.)

The alien informed the chosen one, Estelle, that the Day of Judgement was nigh. It would begin on 22 November 1969, and in the course of that last day humanity would be "given a last opportunity to repair his decadent house before the terminal series of disasters". If mortals rejected the last-minute chance to abjure their former wickedness, "the Space Brothers would remove the Chosen and return them to Earth after the planet had once again 'crystalized', and been spiritually, as well as physically, restructured." This rather sinister euphemism would mean that the Earth would be deliberately tilted on its axis so that continental land masses would disappear under massive tidal waves – with, of course, all the inhabitants of those doomed places. Unfortunately for the credibility of Estelle, but fortunately for mankind, these dire prophecies failed to materialize. Estelle faded from the picture, but her mother carried on the torch of the Light Affiliates, telling American writer Brad Steiger that:[34] "We misinterpreted them . . . because it all happened so suddenly. The first visions I was given of destruction were very upsetting. I can see things now in a much broader perspective . . . The thing is that it is the first ascension, and it is a *mental* ascension. The [Space] Brothers are trying to get as many people as possible into the Kingdom . . . You know, I've been told by the Brotherhood that Earth is like an encounter therapy centre for the psychotics of the Universe . . . I have been shown that the Earth is also wobbling very drastically on its axis."

As Kevin McClure says sagely:[35] "It is sometimes less painful to find ways of showing that your beliefs are fundamentally correct by means of some elaborate reinterpretation than to concede that they are simply mistaken . . ."

The idea of "ascension" was to figure in another saucer cult of the 1970s, one that was to find its destiny in bizarre suicides over twenty years later. The Human Individual Metamorphosis (HIM), which arose in California out of the burgeoning New Age, was led by a middle-aged couple known as Bo and Peep – real names Marshall Huff Applewhite and Bonnie Lu Trusoale Nettles. Known among their disciples simply as The Two, they preached that those who followed the true path would transcend bodily death by being transported physically to a realm beyond the environs of the Earth (to that, at least, Applewhite was to remain true, with deadly consequences). As Kevin McClure says, "Their teaching offered the advantages of life after death without the inconvenience of dying".[36] HIM's posters read:

> UFOs – why they are here? Who they have come for? When they will land? Two individuals say they were sent from the level above the human and will return to that level in a spaceship (UFO) within the next three months. This man and woman will discuss how the transition from the human level to the next level is accomplished, and when this may be done . . . If you have ever entertained the idea that there might be a real PHYSICAL level in space beyond the Earth's confines, you will want to attend this meeting.

At first The Two let it be known that they would be assassinated, miraculously rising again three days later, but later they quietly dropped this idea (probably because it was too dangerous to maintain). Disciples of the way of HIM had to follow rigorous lifestyle rules: even reading books was forbidden. The Two told Brad Steiger: "Husband and wife can take the trip [to the off-planet realm] at the same time – but not together. It would be impossible to become an individual if you went together on the trip . . . In order to leave this Earth's atmosphere, you must go alone and overcome whatever needs you have for any other individual or thing of the Earth. Anything for which you depend on another human being or any thing on this Earth must be overcome."

The disciples, of which there were to be many over the next two decades, lapped it up. One who rapidly saw through The Two, however, was Joan Culpepper, who acted as whistleblower, setting up a "halfway house" for cult members who wanted to escape.

Perhaps it came as a shock to them to discover that Bo and Peep had originally met in a psychiatric hospital (she was a nurse and he was a patient), and that they had been arrested for credit card fraud and car theft.

Over the years, HIM transmuted into Heaven's Gate, although its aims and beliefs remained more or less the same. However, the location of the heavenly realm beyond the Earth's atmosphere became a specified place – a spaceship that trailed behind the Hale-Bopp Comet in summer 1997. It was in order to be "beamed up" to this secret rendezvous with their destiny that over twenty members of the Heaven's Gate cult killed themselves, each previously confiding their rapture at the prospect to a video camera – which makes chilling and pathetically sad viewing. Yet the misguided cultists who followed Applewhite were not alone in believing that a massive spaceship trailed Hale-Bopp – among the others who used the media to disseminate their belief in this astonishing fable were science writer Richard Hoagland[37] and the most high profile alien abductee of them all, Whitley Strieber. Both of these men had their part to play in the unfolding drama that was late twentieth-century UFOlogy, wittingly or unwittingly helping to create a massive matrix of often inextricable myth, real experience and fantasy that was to colour, make sense of – and perhaps even distort – millions of lives.

On Boxing Day, 1985, well-known American horror writer Whitley Strieber felt distinctly uneasy. He and his wife Anne and young son Andrew were enjoying a Christmas break in their luxurious cabin in a wooded corner of Upstate New York, but he could not shake a feeling that something fearful was about to happen. His unease was soon to take tangible form. A noise, apparently coming from downstairs, woke him, but as the well-maintained burglar alarm had not gone off and the motion-sensitive intruder lights had not gone *on*, he tried to go back to sleep. Then, like something out of one of his own horror stories,[38] the bedroom door opened and out of the corner of his eye – perhaps the corner of his mind – he saw something that terrified his very soul. There, in his prosaic marital bedroom (Anne was sleeping soundly by his side), was a three-foot (1 m) tall being wearing helmet-like headgear with a brim and a tunic bearing a symbol of concentric circles on it. Was he dreaming? Although it later transpired that he was awake and alert at the time, immediately after posing the question to himself he lost consciousness, later recalling the subsequent events of that night only in confusing fragments.

He discovered himself in the forest with a small dark blue entity that seemed to be tinkering with his head, and an equally short, light grey being with huge eyes – which was to feature so sensationally on the cover of *Communion*. Later he recalled that he was taken to a small dark place where shadowy aliens poked a needle into his head, made a small incision in one of his fingers,[39] and inserted a probe into his anus. The morning after he had no recollection of his ordeal, but nevertheless became moody and nervy, unable to concentrate on his work.

Then it all came back to him, overwhelming him with a terrible sense of the surreal and impossible. It *couldn't* have happened as he remembered it. Yet deep down he knew it had – and suddenly a host of minor physical problems made sense in the light of the medical examination he had suffered at the hands of the aliens. Something directed him to read a book about UFOs, even though just picking it up imbued him with a phenomenal sense of dread, and when he read of an abduction, suddenly everything fell into place. Filled with a great wracking despair, he thought of suicide, but finally sought help from Budd Hopkins.

Whitley, Anne and Andrew all underwent hypnosis – and the stories they had to tell under trance were astounding. Although Whitley's only conscious memories of the night of 26 December were of seeing a large barn owl outside his bedroom window, the process of opening up under hypnosis revealed that the memory of the owl was in fact a "screen memory" – a device that the mind uses to shield the individual from memories that are too distressing. In his mind, the image of the large-eyed owl acted as a stand-in for the aliens with enormous black eyes who abducted him. (It transpired that screen memories of large-eyed animals such as deer and owls,[40] are extremely common among abductees.)

And while Whitley discovered he had been abducted many times since childhood, his wife revealed that she had witnessed the aliens' activities that centred mainly on the others. Young Andrew told of a weird dream in which "a bunch of little doctors took me out on the porch and put me on a cot. I got scared and they started saying 'We won't hurt you' over and over in my head." These revelations were traumatic for Strieber: the whole fabric of his relationship as protector of his family seemed to be torn to shreds. But this was only the beginning of his relationship with the aliens – or "Visitors" as he came to call them.

Partly as therapy, he wrote his story as the book *Communion* – originally entitled *Body Terror* – which was only accepted by a New York publisher after many others rejected it as unlikely *non*-fiction (the book's subtitle is, "A True Story: Encounters with the Unknown") – as being too fantastic altogether. To Strieber himself, however, everything in the book was literally true: he described it as "a chronicle not only of my discovery of a visitor's presence in the world but also one of how I have learned to fear them less".

The book became a phenomenal bestseller all over the world, and its cover, completely filled with a close-up face of a huge-eyed alien head, seemed to strike a chord in the collective unconscious. Other bestsellers may come and go, but this one *changed* the face of modern myth at some subliminal but pivotal level, where the soul touches the memory, and flinches with the strength of the remembrance. Soon there was a film, too, starring Christopher Walken as Strieber, which – arguably – failed to capture the essential quixotic and transcendental feel of the book, for whatever else may be true of Strieber, he is never less than a very accomplished, profoundly poetic, writer, handling complex ontological concepts with extraordinary deftness and supreme powers of evocation. Perhaps, though, he is still a poet in search of lost verses.

Any book that causes such a stir is going to generate criticism, and *Communion* received more than its fair share of harsh words, some questioning not only the sanity, but also the ethics, of the author. In order to clear his name – and put his own mind at rest – he voluntarily underwent a series of medical tests on his brain, including a CAT scan, an EEG and an MRI (an ultra-sensitive brain scan), none of which revealed any abnormality or malfunction whatsoever. This proved that, despite the suggestions from certain quarters, Strieber's experiences were not the result of Temporal Lobe Epilepsy (see page 464). Then, in order to quash the innuendoes to the effect that he might simply be a fantasist – or a *liar* – he took two lie detector tests, one of which was arranged by the BBC in London, and passed both with flying colours. Whatever had happened to him, at least he *perceived* it to be real.[41]

Although Strieber was to write other books on the same subject, including *Transformation: The Breakthrough* (1988); *Breakthrough: The Next Step* (1995); *The Secret School* (1997), and *Confirmation: The Hard Evidence of Aliens Among Us* (1998), it was *Communion* that provoked an unprecedented response from the public, largely because Strieber had appended an address at which he could be contacted. The landslide of letters from other abductees was extraordinary, and still continues. (The visitors themselves allegedly took a strong interest in the publication of the first book: strange small beings, incredibly well wrapped up in an assortment of clothes, were seen in a bookshop in New York, giggling over its contents, which they seemed to be speed-reading.)[42]

Yet it would be wrong to think that Whitley Strieber unquestioningly, even naively, accepted that he had been abducted by aliens who arrived in Upstate New York in their spaceships. In fact, his was a profoundly questioning attitude from the very first, a desperate search not only for some answers but also the right questions, as can be seen from his subsequent books. He put himself through the wringer, although in the end he seems to have become content with the concept that the visitors have been training him (and others) from childhood[43] – if you like, as Chosen Ones, although Strieber would no doubt be horrified at the thought of setting himself up as some sort of overt messiah.

The most intelligent and thought-provoking response to the furore surrounding *Communion* came from American journalist Ed Conroy, who painstakingly went over the whole story from Strieber's childhood to the first heady post-publication days, double-checking – as far as possible – all the details of that extraordinary tale. His findings, published as *Report on Communion: The Facts Behind The Most Controversial True Story of Our Time* (1989), make truly fascinating reading, for he has the wit to read between the lines, to puzzle, poke and prod into the often murky realm between hard fact, mystical subtext and the subtleties of myth-making. He does not come out and say emphatically that Strieber was lying or mistaken. Neither does he conclude that Strieber was literally abducted by beings from beyond the stars (at least not in so many words). Conroy's analysis is complex and often gritty, frequently finding the questions – if not the answers – in other directions entirely.

Conroy retraced Strieber's life, beginning with his widowed mother, Mary Strieber. Perhaps it is significant that she, the first person who ever knew him, said of his later work: "It's kind of wild, isn't it? . . . he had such a terrific imagination. . . . He was very imaginative, very . . ."[44] She also recalled that he was a "terrible practical joker, just awful".

A neighbour remembered the child as being so frightened of spacemen that he had to be taken home in the middle of the night, while his teacher, the seminarian Brother Martin "Mac" McMurtry recalled that the young Strieber was "always interested in the occult . . . when he talked about vampires he didn't talk about some movie, he talked about the historical vampire."[45]

As Conroy delved further, he began to see a pattern emerge, the early interest in the occult, the unexplained and the transcendent evolving into the storylines of his books, in which the wildest ideas from the fringes of possibility could be allowed free rein. Indeed, Strieber himself noted that his earlier work seemed like a preparation for the experiences that were to become *Communion*, writing: "The Wolfen were gray, hid in the cracks of life, and used their immense intelligence to hunt down human beings as their natural and proper prey."[46] He also cites *The Hunger* (1981), in which the vampire Miriam Blaycock "extracted the stuff of souls. . . . They were the source of her immortality . . . [she] describes herself as belonging to 'the justice of the earth'."[47] Conroy points out that *Black Magic* is about secret psychical research and mind control experiments – which may or may not be relevant to the case of Strieber (see the Conspiracies section).

(What may be more relevant, is the fact that he was fascinated by Arthur C. Clarke's *Childhood's End,* which Brother McMurtry said reminded him of Strieber's own *Nature's End* [1986], co-authored with James Kunetka], a pessimistic – not to say apocalyptic – view of the future that concludes that the only way of saving civilization is to reconstruct it completely.)

Although almost totally ignored by the media,[48] Strieber's first story, written after his unremembered experience on 26 December, was called *Pain*, which casts some intriguing light on his thought processes before, during and immediately after his pivotal first encounter. It centres on a man meeting a woman who is reminiscent of Miriam Blaycock, but not, as Strieber says, a vampire this time, but "a strange angelic demon whose purpose is to put people under so much pressure that they break through to a higher level of consciousness – and this, to me, is the essence of the visitor experience".[49] In other words, the process of being taken by the apparently predatory visitors is an experience of being broken down spiritually and emotionally in order to be rebuilt in a new, finer form. Significantly this process is the basis of two major examples of the rebirth of personality – the sinister technique of brainwashing,[50] and the traumatic ritual of the shaman.

Shamans are what used to be called "witch doctors", the highly-trained elders whose magical knowledge – gained painfully through feats of endurance and agonising tests of the body and spirit – heals and guides the tribe. Usually associated with undeveloped societies such as the Amazonian Indians, it seems that the highly sophisticated ancient Egyptians – the builders of the pyramids – may also have been a shamanic culture[51] (see also Explanations). The core of the shamanic experience is the opening up of the soul, what some call the "Third Eye",[52] and others, more recently, the "stargate", through incredible adventures of the soul, which are experienced as literal reality. During his trance, the shaman is seen as being physically dead, but his soul ascends to the realm of spirits where, after many trials and battles, in which he is torn limb from limb, he is taught great secrets, often of extremely practical use.[53] Although the shaman's training is incredibly excruciating on many levels, he ends up being revered by his tribe as a chosen one. The similarity between the classic shamanic experience and Whitley Strieber's encounters with the visitors was not lost on Ed Conroy, who likened them to Jean Cocteau's 1949 film *Orphée*, describing it as "the story of the poet's own love affair with death, taking place deliberately on the borderland between light and shadow, death and life".[54] If Strieber had been part of a more primitive culture, his experiences would have been understood as witness to the reality of the unseen world and its insistence on breaking through with its messages of power to its chosen.

The whole question of the "alien" experience as a form of magical initiation into a higher level of being involves other, but clearly interrelated, phenomena including that of the Near Death Experience (NDE) in which an individual is *literally* dead[55] while being taken off to another realm, commonly thought of as something akin to a day trip to the afterlife. Like the UFO experiencer, the NDEr sees a light, but this time with the inner eye – the vision of the mind, not the brain[56] – and enters into the dark night of the soul down a long black tunnel, at the end of which is comfort and even ecstasy. Sometimes, however, the experiences are hellish, profoundly disturbing visions of spiritual desolation and hopelessness, from which the NDEr is grateful to return to the pains and problems of life. Yet both types of NDE usually have a similar after-effect: the individual feels *reborn*, as if they have been taken apart and reassembled down to the very fibre of their being, just like the shamans on their magical trips to the otherworld.[57] In this context it is interesting that some of the most high-profile NDErs, such as the American Dannion Brinkley, co-author with Paul Perry of *Saved by the Light* (1994), claim to have been given shaman-style prophecies about the future – as was Whitley Strieber by the visitors. Brinkley wrote that he was taken, in spirit, to a "city of cathedrals . . . a monument to the glory of God" of which, as British author Marisa St Clair writes: "The very fabric of the buildings, which seemed to be somehow made of living glass, pulsed with the power of learning. There he encountered a panel of 13 beings of light, who answered his questions telepathically, filling him with a glow of knowledge. Later Dannion confided 117 predictions (concerning global events) to [NDE researcher] Dr Raymond Moody, and later it is claimed that 95 of them came true up until 1993 . . . However, some of Dannion's predictions have failed to materialize including a war between Russia and China, although the beings of light *impressed upon him that these events could be avoided if mankind had a change of heart . . .*"[58] [My italics.] Note how similar this is to other prophecies gleaned from supernatural or UFOlogical sources, as cited by John Keel, for example, in the discussion above. Some are startlingly accurate and encourage the growth of a discipleship, but then comes the crunch. Just as with Marion Keech's "Clarion" circle, the last and most important prophecy – usually about the date and time of the end of the world – is one huge, ruinous, damp squib.

Strieber, too, seems to have lost his way after a promising beginning. In 1997 he made much of a giant UFO trailing the Hale-Bopp comet,[59] implicitly allying himself with Courtney Brown (one of the more extreme "remote viewers")[60] and even with the demented cult leader of Heaven's Gate, who claimed that suicide would take his followers to immortality on board the Hale-Bopp UFO. Of course no one is suggesting that Strieber in any way endorsed Applewhite's position, but it is interesting that the visitors seem to have encouraged him to believe the same thing about the non-existent UFO.

Similarly, Strieber was vociferous in his support for the idea that there was a face on Mars – that the anomalous feature on the area of the red planet known as Cydonia was, as others suggested, actually the ruins of an *artefact*, a massive artificial construction depicting a "sphinx-like" lion's face. He wrote in *The Secret School* of having "dreamt" of visiting it – although the dream of landing on Mars in his pyjamas was very real – when he was a child. He later wrote that when he was shown the image of the Face on Mars by a scientist friend, John Gliedman, he recalled having seen it before – afterwards realizing that the visitors had shown it to him it during his long time at the "Secret School". Strieber said:

> No matter how I explained it away, seeing the face was still an enormous event in my life, far larger than I could ever have imagined or even – until recently – understood. It may well have been the trigger that caused the close encounter of December 26 1985 to take place. The mystery of Mars and the secret school, it would turn out, were deeply bound together.[61]

There is something very wrong here. Recently the Face on Mars has been revealed conclusively to be no more than a featureless rocky outcrop, despite all the excitement generated by the New Age. And after Courtney Brown talked of the Hale-Bopp UFO on the Art

Bell show, supporting it with the "evidence" of a photograph supposedly given to him by an anonymous astrophysicist, Strieber posted it up on his website, urging those who visited it to "meditate and try to establish a link" with the beings on board. Unfortunately, the photograph was exposed as nothing more than a doctored version of the comet taken by the University of Hawaii's observatory – a fake . . . One is left wondering that if the visitors are superior beings with a knowledge of the past, present and future, why they get so much *wrong*. Or if it is some kind of a test of their chosen ones, why do they permit them to make fools of themselves quite so publicly?

Traditional shamans know a great secret. They know that the beings they encounter when out of the body – in that magical realm – may appear to be all sweetness and light, offering them everything in the world, but many of them are in fact mischievous impostors, or perhaps downright evil spirits. The shamans' arduous training enables them to differentiate between the types of spirits they encounter, as did that of the old ritual magicians (and their modern counterparts), although many still succumbed to illusion, grasping at non-existent gold and insubstantial promises. If it is hard enough for *trained* shamans to know the difference between the good guys and the tricksters, what chance do the likes of Brinkley and Strieber have? Despite the stories of the lengthy training in the secret school, the latter seems to have no checks and balances in place. It is as if the visitors have told him to trust his intuition, which can be very much a double-edged sword in someone so gifted imaginatively – someone who can *create* reality for himself with his mind. And as for Dannion Brinkley, having an NDE as a result of *being struck by lightning,* is no guarantee that he will know how to sort out the true message of the experience from the false. There is a similar problem, on a massive scale, with those who – for all the most lofty reasons of self improvement – find themselves embroiled in the New Age, where dabbling in rites and techniques that are magical in all but name without a shred of training, discernment or questioning, is an "open sesame" to the masters of illusion, including those within their own minds. Streiber himself seems to have moved away from his original position of seeing his experience as a metaphor for spiritual rebirth, into becoming the voice of the abductees and visionaries – *their* Chosen One, at least.

One of the most vociferous advocates of the Face on Mars was (and still, unrepentantly, is[62]) Richard Hoagland, an American science writer who has made some astonishing claims for artefacts on Mars, drawing loaded inferences from the "fact" that the Face is a Sphinx and there is a Sphinx in Egypt . . . Could there be a connection? Could the same beings who built the Martian Sphinx have had a hand in the building of the great monument on the Giza Plateau on the outskirts of Cairo? While his arguments are dealt with in detail elsewhere,[63] suffice it to say that his ideas, while fervently held, do not seem to be his own. Perhaps a clue as to what is really going on is the fact that Hoagland, like many other prominent people, claim that the ancient Egyptian for "Sphinx" is *arq ur*. But it is not. *Arq ur* means "silver". But it is an easy mistake to make, for in Sir E.A. Wallis Budge's *An Egyptian Hieroglyphic Dictionary*, against the definition of the *arq ur* the word "Sphinx" does appear, together with the seemingly mysterious numbers "2, 8". In fact, the numbers are the giveaway: "2" is the volume number and "8" the page number of a French magazine called *Le Sphinx* – Budge's source. *This* is the "Sphinx", it is not the definition of *arq ur*. Yet in certain circles you will hear the mistake repeated with such authority as if the definition came from the gods themselves. And indeed, most of the people who propagate the error believe it *was* the gods who gave them the definition of *arq ur* – in fact, space gods from the distant star Sirius, called the Council of Nine (or just "the Nine"). And it is the Nine that is arguably the ultimate "UFO" channelling cult.

Incredibly, they have been going for over fifty years – in one form or another – making their debut through an entranced Indian mystic called Dr D.G. Vinod in a private paranormal research centre, the Round Table Foundation in Glen Cove, run by one Dr Andrija Puharich. The spirits who "came through" called themselves, "The Nine Principles", establishing their credibility by giving a variation of the Lorentz-Einstein Transformation equation (about energy, mass and the speed of light). Through Vinod, the Nine announced that they were nine aspects of God. The Nine, in fact, *were* God.

That was only the beginning. Since then they have been channelled by a large number of people, including (for a short time) Uri Geller, and after him, mostly by American medium Phyllis Schlemmer and hugely influential guru Dr James J. Hurtak.[64] Under the close supervision of Dr Puharich, the Nine became incredibly powerful – which perhaps is not surprising given their identity. Their spokesman, known familiarly as "Tom", revealed himself to be none other than Atum, the great creator god of the ancient Egyptians – the other eight members of the Nine being the rest of the old "Ennead", the nine major gods (including the famous Isis and Osiris). The Nine, who together make up One God, said that they come from Sirius and that they are heralding a time of great cleansing of the Earth, when the wicked shall be destroyed to make way for the good (very reminiscent of the Fundamentalist "rapture").

Four books have been written about (or by) the Nine: British writer Stuart Holroyd was commissioned in 1975 by Puharich to write *Prelude to the Landing on Planet Earth* (1977), renamed for the paperback as *Briefings for the Landing on Planet Earth*, and in the same year *The Book of Knowledge: The Keys of Enoch* by J.J. Hurtak, then in 1993 came *The Only Planet of Choice*, channelled through Phyllis Schlemmer, and finally the profoundly peculiar *Two-Thirds* by David Myers and David Percy, a novelized version of the Nine's teachings. To those not blinded by the glamour of hearing the words of the Egyptian gods, some of that material makes worrying reading indeed.

In among the usual outpouring of peace, love and light (and no normal person would seriously suggest there is anything wrong in that) are much darker hints. In *The Only Planet of Choice* we are told that all humanity was seeded from the gods of Sirius. Except, that is, for one group, the "indigenous" people of the Earth – the black race. Of course we are given stern instructions not to let that poison our attitude to black people, but of course the damage is insidious, and all the more dangerous because this form of racism kicks in at the powerful level of myth and spirit. How can it be wrong when it is the *gods* who give us the information?

No one should underestimate the power of the Nine – whoever or whatever they may be. In the 1970s they, through their channeller, were actually on the Board of a hugely influential think-tank, the Eselen Institute of California, wielding such power that they had the accountant fired! These days the Nine are still with us, hooking an enormous number of people – including top scientists and leading politicians, besides many in the intelligence agencies and the military – with their mixture of detailed knowledge, profound insights and apparent secrets of power. It is a heady mix. As Dick Farley, once close to the Nine's machinations[65] writes "[the Nine] maintain a working network of physicists and psychics, intelligence operatives and powerful billionaires, who are less concerned about their 'source' and its weirdness than they are about having *every* advantage and new data edge in what *they* believe is a battle for Earth itself."[66]

Although he has never made any obvious connection with the Nine, one wonders about Strieber's source. In *Communion* he describes how *nine* mysterious raps heralded the onset of his first encounter with the visitors, while in *The Secret School* he presents *nine* lessons, and has the same belief in the Martian face as Richard Hoagland and others who are connected with them. While Whitley Strieber's visitors may never actually mention the Nine, it seems that they may come from a similar background. If so, Strieber should beware. As with most other UFO cults, the history of the Nine is already littered with casualties.

One of the first of the Nine's channellers was a short-order cook from Daytona, Ohio, called pseudonymously "Bobby Horne", who came under such pressure from them that he was on the verge of suicide. Uri Geller, who got out of the scene as quickly as he could, later described the Nine as "civilization of clowns", while the famous British writer on the unexplained, Colin Wilson, called them "the crooks and conmen of the spirit world". Phyllis Schlemmer has suffered in both health and career. But why should the Nine worry, when they always have eager new channellers sitting at their feet?

The Nine, despite their hugely influential followers, are still a cult like any other – with all the attendant dangers. Yet even they talk sense from time to time, and urge the fostering of love and peace. It is when the entities are built up by their unquestioning human followers into God-like figures that it all comes tumbling down around their ears, aided and abetted by the weakness of the human ego. When the longing to be one of the chosen becomes too great, the entities begin to take advantage. Perhaps it is not good to forget that human beings are blessed with something that precious few, if any, of the aliens seem to have – the power of discernment, and the ability to say "thanks, but no more".

It is interesting to note a possible connection with the Nine in a case history cited by Nick Pope in his *Open Skies, Closed Minds.*[67] As he says, "The case is possibly unique among the files of Secretariat (Air Staff)." One would hope so, for the letter purported to be written by a female alien, whose benevolent race of "Nordic" sounding beings were locked in conflict with the Greys. As Nick notes, "They came from a planet or organization – the exact nature of the organism is unclear – called Spectra", which immediately evokes the Council of Nine, for in the early days Uri Geller claimed to be in touch with them via an intelligent computer called Spectra which journeyed through space. Nick comments that the letter reads like the fantasy it undoubtedly – or at least hopefully – was, noting that it has distinct elements of a *Star Trek* story. Again, Gene Roddenberry was an early acquaintance of the Nine, and although it is not known what his views were on them at the time of his death, his widow discovered among his papers the outline for the future television series *Earth: Final Conflict*, in which morally ambiguous aliens known as the "Companions" have taken over the world, ruling by a committee known as the "Synod" – a synonym for *Council.* Nick Pope muses on the fact that bizarre though the letter may seem, it was written before accounts of Nordics and Greys were well known, wondering, "Is it possible that some elements of this story are true?" Or is it possible that the Council of Nine was endeavouring – apparently unsuccessfully – to influence the Ministry of Defence?

People mostly live such mundane lives that any contact with the numinous or the paranormal – visions, dramatic conversions to a religion or cult, or abductions – sends shock waves through their whole systems. Consensus reality is cracked, if only briefly, and through it can be seen another world, perhaps another dimension or level of awareness. That is special. The problem is that too often the experiencer believes that he or she is inherently special simply as a result of having the experience, beginning a downward spiral that can take many others with it. (Where are all the UFO cults that believed the world was going to end at the Millennium? In the event, 31 December 1999/1 January 2000 turned out to be one of the most peaceable and joyous nights in history.)

Both Budd Hopkins and John Mack have written of the pain of the abductees in recalling their experiences, and their fervent hope that what they remembered was *not true*.[68] They counter arguments that the experiencers enjoy the attention by saying that they seek anonymity and constantly try to deny the more outrageous aspects of their stories – the removal of sperm and ova to make hybrid babies, for example. There is no reason to doubt that Hopkins and Mack were sincerely reporting the state of affairs at the time of writing – the late 1980s and early 1990s – but things have changed. These days dozens of books, television programmes and videos present abductees who positively boast about their status as chosen ones and "support groups" often turn into little more than clashes of personality, where each abductee tries to outdo the others with increasingly bizarre stories.[69]

Phrases such as "I realized I was very special" and "I now know I am chosen" are scattered throughout the literature, while the evidence for that chosen-ness is often conspicuous by its absence, on any level.

Perhaps the phenomenon is about to transmute into something else, some other form of picking on, and abusing, individuals while persuading them that it is all for their own good. Or perhaps the abductees are right, and it is the rest of us who are too blinded by cynicism to recognize the truth. Yet in the end, there is only one test, and that is encompassed in the Biblical advice: "By their fruits ye shall know them." In the meantime, it may be a good idea to beware visions and aliens bearing information. Maggie Fisher, who was told she was the "wrong one" may, in the end, have had a lucky escape.

Notes

1. Victor Marie Hugo, 1802-85, French poet and author, and leader of the Romantic Movement in France. He wrote the novel *Notre Dame de Paris*,

wrongly but popularly known as *The Hunchback of Notre Dame*, in 1831 – now a hugely successful Disney animation and musical play – and *Les Misèrables*, his greatest novel, in 1862. However, his esoteric interests and involvement is considerably less well known. Along with other historical luminaries such as Leonardo da Vinci and Sir Isaac Newton, his name appears in the list of Grand Masters of the secret society Le Prieuré de Sion – the Priory of Sion, which if true marks him out as a heretic, a Johannite and a goddess worshipper (for more details of the Priory's beliefs, see Picknett and Prince, *The Templar Revelation*, London, 1997).

2. See *Nexus* magazine, April-May 1999 and *Conversations with Eternity: The Forgotten Masterpiece of Victor Hugo* (Australia, 1999).
3. *Nexus*.
4. *Ibid*.
5. Which takes eighty-eight days to orbit the Sun, but only fifty-nine days to revolve on its axis. This was understood in Hugo's day.
6. Nicholas Flamel, believed to have accomplished the elusive alchemical "Great Work", on 17 January 1382 at his home in Paris – a mysterious rite, one side effect of which is (allegedly) to be transmuted physically into an immortal. He is said to have been Grand Master of the Priory of Sion – like Victor Hugo in his day – between 1398 and 1418. See Picknett and Prince, *The Templar Revelation*.
7. Using a form of "automatic drawing".
8. A term found in the Old Testament, meaning "gods".
9. See *You Are Responsible*, by George King. (London, 19??).
10. *Ibid*.
11. *Ibid*.
12. They did not succeed.
13. From Charles Bowen's files.
14. Dr Schwartz's report was prepared exclusively for *Flying Saucer Review*.
15. John A. Keel, *UFOs: Operation Trojan Horse*, (New York, 1971), p. 274.
16. The conflict began on 4 June 1967.
17. See p. 275 of *Ibid*.
18. p. 276 of *Ibid*.
19. On 9 April 1986 Whitley Strieber had a terrifying dream in which he appeared to be in a marshy plain close to a complex of low buildings. He said: "I was aware that the building was some sort of nuclear installation . . . There were masses of pipes running along the wall . . . Suddenly a big pipe fell apart and a great deal of water gushed out. Moments later the whole place started to explode."

    Although the dream was unusually vivid, he thought nothing more of it until on 25 April 1986 the Soviet nuclear plant at Chernobyl exploded. Strieber came to believe that the dream had been inducted into his mind by the "Visitors" as proof of their power and foreknowledge.
20. *Ibid*.
21. Whether they committed suicide or were murdered remains an open question, although the evidence points to the latter.
22. Although of course enthusiasm for martyrdom is nothing new among

fanatics. Early Christians were renowned for seeking their own death – something that even the Romans found distasteful – and often the grislier the better, for they believed it was through their suffering and death that they would go to heaven and be with Jesus.

23. "Glamour" is an old word for being under the spell of the fairies.
24. In the classic academic study *When Prophecy Fails* by Minnesota University sociolgists Leon Festiger, Henry W. Riecken and Stanley Schachter (1956).
25. Automatic writing, most often associated with psychic activities, is actually a common psychiatric tool. Basically, you just rest a pencil lightly against a piece of paper and wait. Often a few squiggles will appear immediately – random letters etc – but the material will make little sense, being a form of unconscious doodling. Sometimes, however, the result is more impressive: messages appear that purport to come from other entities – spirits of the dead or beings from other planets, for example – and perhaps predictions. Sometimes these prove to be at least partly veridical (checkable), giving rise to cults based around them. However, almost all such automatic scripts prove to be outpourings of the scribe's unconscious mind, although a very small percentage remain unexplained.
26. See *The Stargate Conspiracy* by the author and Clive Prince, (London, 1999), Chapter 5: 'Behind the Mask'.
27. Who appear pseudonymously in *When Prophecy Fails* – see (12) above.
28. Quoted in Kevin McClure's article 'Apocalypse Now?' in *The Unexplained*, No. 87.
29. This is very similar to the reaction to prophecies of doom given in visions of the Virgin. In many cases she has predicted that the world would end on a certain day, even at a specific time, because of widespread evil, but when it does not happen, either she or her followers declare that the prayers of the faithful averted the calamity. This suggests a common source for the apparently diverse phenomena.
30. See *Ibid.*
31. Picknett and Prince, p. 224.
32. Quoted in McClure.
33. John Bunyan, 1628-88, English writer and Puritan preacher, author of *Pilgrim's Progress*. The names of his characters have a similarly worthy ring to them.
34. In an interview in the mid-1970s.
35. *The Unexplained*, p. 1728.
36. *Ibid.*
37. See the Conspiracies Chapter.
38. Which included *The Hunger* (which was made into a movie), *The Wolfen*, *Catmagic* and *Black Magic*. Most of his novels have leading characters who are not human.
39. Perhaps reminiscent of the aliens who took such an interest in Maggie Fisher's fingertips in the extremely vivid dream that happened after she had read Strieber's *Communion* for the first time.

40. It may be significant that owls have traditionally been thought of as bringers of wisdom. The Greek goddess of wisdom, Athene, was often portrayed as an owl.
41. During a meeting in the late 1980s in London, I asked Jacques Vallée – a close friend of Whitley Strieber – what he thought had happened to him. He replied cautiously, "Something did happen to Whitley, something that changed his life. But what it was I can't say."
42. The incident was reported by former journalist Bruce Lee, then an Editor for William Morrow and Company, the publishers of *Communion,* and was described in that book. Ed Conroy, who double-checked the Strieber story for his book *Report on Communion* (1989), interviewed Lee on this subject (p. 18) and was satisfied that the incident was more than a publicity stunt. Lee also passed a lie detector test.
43. See *The Secret School: Preparation for Contact,* (1997).
44. See *Report on Communion.*
45. *Ibid,* p. 68.
46. *Transformation* (New York, 1988) pp. 125–126.
47. *Ibid.*
48. Except by Ed Conroy, see p. 85 of *Ibid.*
49. See *Ibid.*
50. For a discussion of the use of brainwashing by the British and American military, see *Double Standards: The Rudolf Hess Cover-up* (London, 2001) by Lynn Picknett, Clive Prince and Stephen Prior, with additional research by Robert Brydon.
51. See the Epilogue of *The Stargate Conspiracy* (London, 1999) by Lynn Picknett and Clive Prince.
52. Interestingly, John Keel tells of Men In Black who claimed to be from the "Nation of the Third Eye". See *UFOs: Operation Trojan Horse,* pp. 267-268. While the symbol of the "eye in the triangle" is found in the Great Seal of the United States, the reason for this is that it is a Masonic symbol, and most of the men involved with the approval of the Great Seal were high-ranking Freemasons. But the Masons took the symbol from the ancient Egyptian "eye of Horus", a magical glyph commonly believed to bring luck and protection. See also *The Stargate Conspiracy.*
53. Such as what combination of plants to use in order to heal specific illnesses – and perhaps how to build a pyramid . . .
54. See Conroy, p. 7. *Orphée* also features some Men In Black in the courtroom scene.
55. In most cases pronounced brain dead by doctors.
56. See Explanations section for a discussion of the mind *v.* the brain.
57. The literature on the Near Death Experience is huge. A recommended primer is Marisa St Clair's *Beyond the Light* (London, 1997).
58. *Ibid,* p. 80.
59. On his website, for example.
60. Remote viewing, which used to be known as "travelling clairvoyance", is the technique of leaving the body and visiting other locations distant in place (or even time). Many of the world's military organisations em-

ployed *and trained* remote viewers in what became known as "ESPionage". See *Remote Viewers: The Secret History of America's Psychic Spies* by Jim Schnabel (New York, 1997). For the links between remote viewing and other psychic experiences see St Clair's *Beyond the Light.*

61. Strieber's *The Secret School,* pxix.
62. With characteristic bluntness he dismissed the evidence that the Face is merely a rocky outcrop as "crap".
63. See Picknett and Prince, *The Stargate Conspiracy.*
64. See *Ibid.*
65. He was director of programme development at the Human Potential Foundation, partly funded by Laurence Rockefeller, resigning in 1994, in part because of Dr Hurtak's – and therefore the Nine's – close involvement with the Foundation.
66. Dick Farley, "The Council of Nine: A Perspective on 'Briefings from Deep Space' ", Brother Blue website (*http://www.brotherblue,org),* 1998.
67. See Pope, pp. 124–125.
68. In *Intruders* and *Abducted* respectively.
69. Sitting in on a British support group in the mid-1990s was quite an eye-opener.

CHAPTER 7

CONSPIRACIES AND COVER-UPS

The idea that world governments – in particular that of the United States – are engaged in a conspiracy to suppress the truth about UFOs is nearly as old as the phenomenon itself. The existence of such a conspiracy is an essential part of many key cases, such as the Roswell Incident and the Rendlesham Forest landing. And belief in the official cover-up is not confined to UFO enthusiasts: a survey by Ohio State University in 1995 found that 50 per cent of Americans believed that it was either "somewhat" or "highly likely" that their government is withholding information about UFOs.

The idea was initially popularized by Major Donald E. Keyhoe, the first writer on UFOs to reach bestseller status, and who produced a series of books in the 1950s and 1960s beginning with *The Flying Saucers Are Real* (1950). Keyhoe was a former Marine Corps pilot – and one-time aide to pioneer aviator Charles Lindbergh – who turned to writing after the Second World War. Keyhoe was first asked to look into the UFO mystery by the editor of *True* magazine, Ken Purdy, in 1949. Interestingly, Purdy's original brief, based on his own investigations, was that flying saucers were "a gigantic hoax to cover up [an] official secret."[1]

As Keyhoe dug deeper into the mystery, using his contacts in the USAF, he became convinced that UFOs were extraterrestrial – and that the authorities knew it. In his next two books, *Flying Saucers from Outer Space* (1953) and *The Flying Saucer Conspiracy* (1955), he developed the twin themes that UFOs were alien spacecraft and that a "silence group" within the USAF was preventing this information from reaching the public.

Keyhoe went further in his self-imposed mission to expose the cover-up. In 1956 he founded the National Investigations Committee on Aerial Phenomenon (NICAP), which became the foremost of the civilian UFO research organizations. However, from the beginning an important part of NICAP's mission was to lobby Congress to open up the top secret files that Keyhoe and his associates believed were held by government organizations, in particular the USAF and the CIA.

An essential part of Keyhoe and NICAP's argument was that official studies into flying saucers – which had all concluded that there was nothing to the phenomenon – were all exercises in misinformation designed to hide the truth from the public.

The flying saucer craze of the summer of 1947 had led to calls for an official investigation. The government charged the USAF with this task, and the first study, Project Sign, was launched in September 1947. Two years later it was redesignated Project Grudge, and in 1951 Project Blue Book, the name it retained until its closure in 1969. (In one of the frequent Kafkaesque episodes created by official secrecy, as the designations of military projects were themselves classified, ludicrously Project Sign used the "cover" name of Project Saucer when making public statements.) The team was based at Wright-Patterson Air Force Base at Dayton, Ohio.

The USAF studies, undertaken by a small team of under-funded, low-ranking officers, looked into reports by members of the public and issued periodic Press Releases on their findings. Perhaps predictably, while admitting that a small percentage of reports could not be explained by misidentification of conventional aircraft or natural phenomena, they concluded that there was no substance to the rumours of UFOs. Flying saucers didn't exist: it was official. But to researchers such as Keyhoe, whose own investigations on the ground had convinced them of the reality of UFOs, this had to be a smokescreen.

What Keyhoe and other researchers suspected was that the real work was being done by other, more highly classified teams. We now know that he was correct: since the advent of the US Freedom of Information Act (FOIA) in 1974, previously classified documents have been unearthed that showed that other, secret studies *had* been undertaken, in particular by the CIA. UFOlogists were quick to use the FOIA to find out what the US military and intelligence agencies really knew about UFOs, the lead being taken by Ground Saucer Watch. One organization, Citizens Against UFO Secrecy (CAUS) was specifically created with this aim in mind.

One of the earliest official pronouncements on the flying saucer phenomenon was written in September 1947 – although not made public until 1969. In a memo to his superior in USAF intelligence at the Pentagon, Lieutenant General Nathan F. Twining, after consulting with experts at the Air Technical Intelligence Center (ATIC) at Wright Field (later the Wright-Patterson Air Force Base), offered the conclusion that, "The phenomenon reported is something real and not visionary or fictitious." While this is an impressive endorsement – by USAF scientists, no less – that there was something to the UFO phenomenon, Twining's memo carried a sting in the tail for advocates of events such as the Roswell crash. It lamented the "lack of physical evidence in the shape of crash recovered exhibits which would undeniably prove the existence of these objects."[2] As, according to the generally accepted reconstruction of the Roswell Incident, the crashed UFO and the bodies of its crew were sent to Wright Field for examination by ATIC, Twining's memo presents a serious difficulty.

In mid-1948 Project Sign sent a report – known as the *Estimate of the Situation*[3] – to the USAF Chief of Staff Lieutenant General Hoyt S. Vandenberg, which concluded that UFOs were real and that an extraterrestrial explanation was the most likely. Vandenberg disagreed, rejecting the report on the grounds that the evidence it gave did not support this conclusion. The report was destroyed, and the exact contents remain unknown. However, its existence was made known by Captain Edward J. Ruppelt, the Director of Blue Book between 1951 and 1953.

These two documents show that very soon after the first flying saucer reports, USAF scientists were not only taking the phenomenon seriously, but also considering an extraterrestrial explanation. Interestingly, it appears that they were the first to look to outer space for a solution, as the extraterrestrial hypothesis (ETH) did not become popular with the general public until the early 1950s.

Perhaps the most significant of the secret studies was a CIA-sponsored group known as the Robertson Panel, after its chairman, the physicist Dr H.P. Robertson, who was then Director of the

Defense Department's Weapons System Evaluation Group. Convened in January 1953, this was made up of five very eminent scientists, including the Nobel prize-winning physicist Dr Luis Alvarez (who later became famous as the originator of the "deep impact" theory of dinosaur extinction). The primary purpose of the Panel was not to pronounce on the reality or otherwise of UFOs – although it scathingly rejected any suggestion that they might be real – but to look at the implications of the *belief* in them.

It took five years for a summarized version of the Panel's report to be made public, and it was top of the list for UFOlogists when the FOIA became law in 1974, when the full report was finally obtainable.

The report concluded that UFO reports constituted "a threat to the orderly functioning of protective organs of the body politic" and recommended:

a. That the national security agencies take immediate steps to strip the Unidentified Flying Objects of the special status they have been given and the aura of mystery they have unfortunately acquired.

b. That the national security agencies institute policies on intelligence, training and public education designed to prepare the material defenses and the morale of the country to recognize most promptly and to react most effectively to true indications of hostile intent or action.[4]

The report openly advocated the debunking of UFO reports in order to reduce public interest in the phenomenon.

However, matters are rarely clear-cut in the vexed world of UFO conspiracies. Opinions are divided about what the Robertson Panel reveals about the CIA's interest in, and knowledge of, UFOs. It confirmed – as Keyhoe and others suspected – that it was official policy to down-play and discredit UFO reports, but was this in order to suppress the truth about UFOs? In fact, this report – along with other declassified documents – shows that the CIA's main concern was in the social and psychological aspects of public belief in UFOs. The Panel had no hesitation in rejecting the reality of flying saucers

– most damning of all was its statement that there was no physical evidence of their existence, again arguing against the Roswell Incident.

The Panel's greatest concern lay mostly with the consequences of the public's fascination with UFOs. In those nervy days of the early Cold War, there was a fear that the Soviet Union might flood America with bogus UFO reports, creating a "flap" that would tie up military communications and resources in advance of an attack. For this reason, a high degree of public interest in UFOs was evaluated as at best a nuisance, at worst a danger, and had to be discouraged.

For the same reasons, the CIA were interested in the mass psychology of UFO flaps. (There is an undeniable social aspect to waves of reports, whatever really lies behind them.) How did such waves start, and how could they be stopped? There was also the possibility that the phenomenon could be used in reverse – against an enemy.

A few months before the Robertson Panel study, in September 1952, the Director of the CIA's Office of Special Intelligence wrote a memo to Director Allen W. Dulles on the security implications of flying saucers, posing the question of "whether or not these sightings: (1) could be controlled, (2) could be predicted, and (3) could be used from a psychological point of view, either offensively or defensively."[5]

A month earlier a secret CIA briefing paper instructed that the Agency should prevent the public from finding out about its interest in UFOs, as otherwise it might appear to give credibility to the subject – proof that the Agency *did* keep its investigations secret (although not for the reasons given by advocates of the "UFO cover-up" theory).[6]

As a result of the Robertson Panel's recommendations, the CIA effectively took charge of the USAF projects' policy. Not long after the Panel meeting, Captain Ruppelt and other USAF officers planned to go public with the Air Force data on UFOs – but the CIA stepped in and prevented them. Instead Blue Book was ordered to debunk the whole subject, if necessary by publicly ridiculing witnesses.

Three years after leaving Blue Book, Captain Ruppelt went into print, writing *The Report on Unidentified Flying Objects* (1956), which took an open-minded attitude to the phenomenon and even seriously considered an extraterrestrial explanation. However, he revised the book in 1959, explicitly rejecting the ETH. Many believe that the change was due to pressure from above. Ruppelt died of a heart attack a year later.

More suspicion centred on an Air Force Regulation (200-2) issued in August 1953 – again probably as a result of the Robertson Panel recommendations – which categorized UFO reports by USAF personnel under intelligence activities and forbade them from talking about their experience to the Press or public. It also instructed that, at the discretion of the base commander, details of a sighting could be made public if it could be identified as "a familiar or known object".[7] In other words, the public were to be given a distorted picture, and would only be told about reports that could be explained away.

The documents unearthed using the Freedom of Information Act generally showed that the CIA were open to the possibility that UFOs were extraterrestrial, but did not consider it likely. For example, a report by the CIA's Office of Special Intelligence in March 1949 referred to studies by several scientists and laboratories and noted:

> That the objects are from outer space or are an advanced aircraft of a foreign power is a possibility, but the above group have concluded that it is highly improbable . . . However, since there is even a remote possibility that they might be interplanetary or foreign aircraft, it is necessary to investigate each sighting.[8]

In a similar vein, the head of the CIA's Weapons and Equipment Division wrote in a memo after the July 1952 Washington wave that, while he believed that all UFO reports could be explained conventionally if enough data was available, caution dictated that the Agency should continue to investigate reports in case they turned out to be of "alien origin".[9]

So the declassified documents brought good and bad news for UFOlogists. They confirmed that the CIA – despite everything it had claimed previously – had been secretly interested in UFOs, had concealed that interest from the public, and that the public were being misled by the USAF investigations such as Blue Book. However, the reason for this was not to cover up what they really knew about UFOs. If anything, the documents reveal that the CIA were as puzzled as everybody else.[10]

British researcher Nicholas Redfern also used the American FOIA to search the records of another US security organization, the FBI. Despite the Bureau's claims that it had never been interested in UFOs, Redfern was able to show that they *had* investigated sightings – although, once again, none of the documents suggested that the FBI had any greater knowledge of what lay behind them than anyone else.

Since 1974, UFO researchers have been able to obtain several thousand pages of previously classified documents from the CIA, FBI, Defense Intelligence Agency (DIA) and all three armed forces. However, some FOIA requests have been refused on the grounds of national security. The biggest problem came with the National Security Agency (NSA), the organization whose task is to monitor communications throughout the world for information that is useful to the US Government. Although CAUS were able to obtain some documents from that agency, it withheld many on the grounds of national security. When CAUS petitioned the Supreme Court for their release, NSA Director of Policy Eugene P. Yeates produced an affidavit giving the reasons that the documents should not be made public. The affidavit was shown only to the Supreme Court judges – CAUS's lawyers were not allowed to see it – and, as a result, they decided not even to hear the case. Refusing to be beaten, CAUS filed an FOIA request to see Yeates' affidavit. It was released to them – but only after some 70 per cent, including entire pages, had been blacked out by the censor.[11]

In a public statement, the NSA explained that the documents were based on the interception of foreign communications, and that releasing them would compromise their methods by alerting foreign governments that they were being monitored. However, the suspicion remains in many UFOlogists' minds that this is not the real reason.

Leading American UFOlogist Dr Bruce Maccabee – a former US Navy physicist – claims to have inside information that reveals that the CIA have thousands of documents relating to UFOs that it refuses to acknowledge publicly.

During the 1950s claims began circulating within the fledgling UFO research community in the USA and other countries that researchers who were getting "too close to the truth" were being visited and warned off by representatives of shadowy government agencies. These bizarre events were described by theatre owner and flying saucer investigator Gray Barker in his 1956 book *They Knew Too Much About Flying Saucers.*

Barker became intrigued by the UFO mystery in the early 1950s, and joined the International Flying Saucer Bureau (IFSB) founded by Albert K. Bender in 1952, becoming its representative and principal investigator for his home state of West Virginia. But he soon found himself under surveillance by the FBI as, one by one, his best contacts fell silent after visits by mysterious individuals who revealed the "secret" of UFOs to them – a dangerous secret that their victims dare not pass on.

Albert Bender was the first. In September 1953 he claimed he had been visited by three men dressed in dark suits and black hats, who warned him to discontinue his investigations. He told Barker and other colleagues that he could not say what the trio had revealed to him, other than that it was dangerous knowledge. When quizzed Bender said that the men were not from the FBI but "another branch". In one of his few specific statements, Bender said that they had told him that the government had known the secret of UFOs for two years, and that they had said, "In our government we have the smartest men in the country. They can't find a defence for it. How can *you* do anything about it?" To a list of questions sent by another colleague, Bender gave a one-sentence reply: "The above questions are not to be answered due to security reasons." (For other aspects of the Men In Black phenomenon, see page 263.)[12]

Shortly afterwards the IFSB's Australian representative, Edgar R. Jarrold, told a similar story, and he too dropped out of the UFO field. Then so did another foreign contact of Barker's – he does not reveal his name or country – after a visit from a man from "a certain government agency which I will not name" who warned him to give up his research and "forget that flying saucers exist". After more UFOlogists fell silent or began avoiding certain issues, Barker became convinced that some agency was systematically disrupting the amateur UFO research groups.

However, despite his warnings and injunctions to Barker and others to give up their pursuit of the truth about flying saucers, Bender did eventually write a book about his experiences, *Flying Saucers and the Three Men* (1963). In this, unlike the human, if overly dramatic, description of his visitors that he gave to colleagues at the time, the "three men" had become ghostly visitors who appeared in his bedroom with glowing eyes that "seemed to burn into my very soul" and who communicated with him telepathically. Bender went on to describe the "secret" that the men had imparted, which centred on the existence of alien bases in the Antarctic. Given Bender's earlier fears and statements that it would be dangerous for him to speak, many believe his book to have been a fable designed to conceal what the three men had really told him – after all, if it contained the truth would he have been allowed to publish it? (It is perhaps significant that Bender was warned off just a few months after the CIA adopted its policy of discrediting the UFO phenomenon.)

The Bender affair is one of the classic Men in Black cases, leading many to the conclusion that MIB are government agents, out to warn UFO researchers and witnesses off.

Other UFO investigators were being visited by government agents in the 1950s, although they did not always bring threats. One of the most important such episodes involved Brazilian medical doctor and UFOlogist Dr Olavo T. Fontes in February 1957.

Fontes had been interested in UFOs since a major wave of sightings in Brazil in late 1954. Many of the reports came from military personnel, as the UFOs initially seemed to be interested in Brazil's Army and Air Force bases. On 24 October 1954 a UFO hovering above an air base near Porto Allegre had been seen by over 100 members of the base personnel. On 20 November a dome-shaped object surrounded by a glowing yellow halo was spotted by the crew of a military mail plane and landed on an island in the Rio Grande river. As the plane flew over to observe the strange object, it took off, its halo becoming brighter, and in seconds had disappeared from view. As the wave progressed, the sightings shifted to cities, and there were several mass sightings. What interested Fontes was that there was a discernible pattern to the flap: first over military bases and then cities and large towns, and moving gradually from the south of the country to the north – as if Brazil were being surveyed.[13]

The Brazilian authorities were initially open about the fact that they were investigating these sightings – declaring that "there is no ground for alarm, but much for keen interest" – but Fontes later noticed a clampdown by the military to prevent details of the more interesting cases reaching the public. He strongly disagreed with such secrecy, as he believed that the public had a right to know the truth.

Fontes became an enthusiastic UFO researcher, and Brazil's representative of APRO. He investigated such classic cases as the Trindade Island photograph and the Ubatuba debris. He is, perhaps, most famous as the investigator of the Antonio Villas Boas case – which was to lead directly to the incident in question.

Boas's encounter had happened in October 1957, but it was not until 22 February 1958 that he was interviewed by Fontes in his Rio de Janeiro office, accompanied by journalist Joao Martin. Four days later Fontes – who had made many contacts in the Brazilian Government and military – took a report on the meeting to Brazil's Minister of the Navy. That evening he was visited by two officers from Brazilian Naval Intelligence.[14]

After showing him their credentials, the pair opened by telling him that, "You know things that you have no right to know". This angered the doctor, as he assumed that the agents had come to threaten him, but their attitude changed to one of openness and frankness. Over the next two hours they told him a number of extraordinary things – some of which, they said, were not even known by the Brazilian President.

All of the governments of the world, they said, knew that UFOs existed, and had "absolute proof" they came from another planet. Six flying saucers had crashed in different places around the world – three in the United States, one in the Sahara Desert, one in Scandinavia and one in Britain. All were fairly small – between 30 and 100 feet (9 and 30 m) in diameter – and all had contained the bodies of their crews. The aliens were humanoid but small, between three and four feet (900 cm and 1.2 m) in height. All were dead. The recovered craft were being keenly studied, but the secret of their propulsion system had not yet been determined.

Of the aliens' intent, the officers said that this was the subject of intense concern by world governments. They had not shown any interest in contacting us, and when attacked or approached had no compunction in destroying military aircraft, although whether this was out of self defence was unclear. It was feared that they were preparing for invasion, although some believed that they were a kind of "police force" who wanted to stop mankind venturing into space and from continuing the development of atomic weapons. All this was being withheld from the public to prevent mass panic, and the agreed policy of world governments was to debunk the subject of UFOs.

Fontes wrote a letter to APRO in America about the experience the next day.[15]

There seems little doubt that the visit took place. Fontes was a greatly respected figure in UFOlogy, and he had no reason to lie. Coming within hours of his delivering a report on the Boas incident to a government minister, it does seem that it was as a result of official concern that Fontes was getting too close to the truth. But, unlike the bullying tactics employed against American UFO enthusiasts such as Albert Bender, it had apparently been decided to adopt a different, more reasonable, approach to the eminent Brazilian citizen.

However, the visit did not discourage Fontes from his UFOlogical activities, and the letter he wrote about the incident to APRO was published, without any repercussions for the doctor.

What of the UFO crashes that Fontes had been told about? Is there any independent corroboration of them? One of the three crashes in the United States might be the Roswell incident. The Scandinavian case referred to appears to be the Spitzbergen event of 1952, although this is now widely regarded as a hoax by a West German newspaper. Nothing of the Sahara crash is known.

The British case, however, may be one that had been reported on two years earlier by American journalist Dorothy Kilgallen, and which supposedly happened in 1945. Kilgallen said that she had been told about the crash by "a British official of Cabinet rank", who had told her that the craft had contained the bodies of "small men . . . under four feet [1.2 m] tall." Kilgallen also claimed that the British Government were withholding the official report on this incident.[16]

Recently, a very illustrious name has been alleged as Kilgallen's informant. According to the veteran British UFOlogist – and former intelligence officer – Gordon Creighton, he was none other than Earl Louis Mountbatten of Burma.[17] While Mountbatten and Dorothy Kilgallen are both dead, and so can neither confirm nor deny the claim, it is a fact that Mountbatten had a deep interest in the UFO phenomenon. When the contactee George Adamski lectured in London in early 1960s, Mountbatten and Air Marshall Lord Dowding – wartime head of Fighter Command – were among the audience.

THE NAZIS

Albert Bender's three visitors had apparently told him that part of the secret was that the aliens had established bases in the Antarctic. This links with another conspiracy claim concerning UFOs – one that links them with the Nazis.

There is a theory that the early flying saucer sightings were due to sightings of secret German flying craft that were captured by the Allies at the end of the Second World War. The idea was toyed with by Donald Keyhoe, but its main modern proponent is W.A. Harbinson, who used the concept in his 1980 novel, *Genesis*, but who later wrote a non-fiction book on the evidence that had inspired him, *Projekt UFO* (1995).

There is good evidence that the Germans were experimenting with circular flying craft. Under the Nazi regime, German science had set off on its own path, one that diverged from the rest of the west, and there was a greater openness to exploring ideas considered exotic, or indeed very much on the fringe of conventional research.

Captain Ruppelt wrote in his 1956 book, speaking of the new aircraft in development in Germany at the end of the war: "The majority of these were in the most preliminary stages, but they were the only known craft that could even approach the performances of the objects reported by UFO observers."

A recently declassified CIA report from August 1953 states: "'Flying saucers' have been known to be an actuality since the possibility of their construction was proven in plans drawn up by German engineers towards the end of World War II."[18]

And certainly, in the early 1950s, the USAF and CIA were concerned that flying saucers might be Russian inventions based on the work of German scientists they had captured at the end of the war.

The theory also offered an explanation for the "Foo Fighters" seen by Allied fighter and bomber crews over Germany in the closing months of the war, and which were reported in the British Press as a Nazi secret weapon. Adherents of the theory also cite the comments made by Sir Roy Feddon, who headed a team of British scientists who went in to Germany on behalf of the Ministry of Aircraft Production to examine German aircraft designs and technology. In 1945 Feddon stated:

> I have seen enough of their designs and production plans to realise that if they had managed to prolong the war some months longer, we would have been confronted with a set of entirely new and deadly developments in air warfare.[19]

In the mid-1950s, German newspapers published claims that between 1941 and 1944, *Flugkapitan* Rudolf Shriever had designed and built prototype disc-shaped flying machines at the BMW factory near Prague, Czechoslovakia. It is claimed that these had a diameter of over 100 feet (30 m) and could reach speeds in excess of 1,000 miles per hour (1,600 km/h). Schriever himself said that, although a pilotless version had been tested in July 1942, the larger manned version had never flown; it was ready for testing but had to be destroyed, and the plans abandoned, as Russian troops advanced through Czechoslovakia.[20]

In 1968 Italian aircraft engineer Renato Vesco claimed that the Germans had developed two circular aircraft in the closing stages of the war, and that these were responsible for the Foo Fighter reports.

The first was the *Feuerball* (Fireball), an unmanned, radio-controlled jet-powered craft, and the much larger piloted *Kugelblitz* (Ball Lightning), also jet-powered, which allegedly successfully flew – reaching a speed of 1,250 miles per hour (2,000 km/h) – in February 1945.[21]

To link these experimental craft with the first wave of flying saucer reports in America two years after the end of the war, it is pointed out that many German scientists -the most famous being Werner von Braun – were brought to America as part of a programme called Operation Paperclip. Significantly, many of those concerned with aircraft and rocket design were stationed at Wright Field.

So could the flying saucers of the 1940s owe nothing to extraterrestrials, but have been captured German craft, or American devices made according the designs captured at the end of the war? Against this, it has been pointed out that even though they would have been kept secret at the time, by now such craft should have been made public. Although there were several attempts to construct circular flying machines in the late 1940s and 1950s – most notably the decidedly flying-saucer-shaped "Avrocar", built by the A.V. Roe company in Canada for the US Army Air Force – none of these seemed to have resulted in a viable military aircraft.

However, in recent years a new set of theories has emerged that reintroduces the extraterrestrial dimension – not to say a distinctly dark aura of occultism – to the "Nazi UFO" theory. According to these claims – as promoted, for example, in the widely-available video *UFO: Secrets of the Third Reich* (Royal Atlantis Productions, 1995)- the story begins in the 1920s with a group of occultists who were allegedly inspirations to Hitler and the Nazis. They include well-known mystical and esoteric groups such as the Munich-based Thule Society and the more obscure Vril Society. It is true that several of those who were close to Hitler in the early days of the Nazi movement – such as Professor General Karl Haushofer, originator of the geopolitical ideas in *Mein Kampf*, and the racist poet Dietrich Eckhart – were deeply interested in mystical and "occult" ideas, as were some of the later Nazi leaders, most notably SS leader Heinrich Himmler. But what has this to do with UFOs?

According to the recent claims, members of these groups experimented with what is now known as channelling – mediumistic contact with higher intelligences. It is said that they made mental contact with extraterrestrials who had visited earth many thousands of years before, and had sown the seeds of cizilisation in the Middle East. These beings were able to give instructions for the new energy sources that would power circular flying craft, and, in a curious blend of technology and the paranormal, it was this that led to the development of the Nazi UFOs, which were called "Vril" or "Haunebu". Unlike the Nazi saucers of other theories, which used conventional propulsion methods, the Vril and Haunebu craft used novel forms of energy and were capable not just of extraordinary speed and performance within Earth's atmosphere, but even space travel.

There have been recent claims that the Vril Society's work was supplemented by the recovery of a crashed UFO, which they were able to study and combine with their channelled instructions. There are two versions of the story. One is that the crash occurred in the Black Forest in 1936, the other that it was in Czernica in Poland in the summer of 1938, and that the craft was captured when the Nazis invaded Poland a year later. In both versions, the crashed UFO was sent to the Vril Society to assist them in their work on developing the Haunebu craft.

Leading promoters of these claims include Vladimir Terziski (Bulgarian-born founder of the American Academy of Dissident Scientists), retired USAD Colonel Wendelle Stevens and James Hurtak, now best known as a New Age "prophet".

According to Terziski, an alien "tutor race" made contact with German scientists in the late 1920s, and introduced to them their "concepts of philosophical, cultural and technological progress"[22] – concepts that were adopted by the Nazis, which should tell us something about the tutor race's moral and ethical status. Hurtak agrees with the scenario, although placing the contact in the Nazi era of the 1930s, also arguing that the extraterrestrials chose that place and time for their first contact with humankind because it was the most advanced nation on Earth.[23] The most worrying thing about sources such as the *Secrets of the Third Reich* video is its marked lack of criticism for the Nazi regime.[24] As British researcher Kevin McClure writes: "It is clear that the concept of vastly superior

German technology demonstrated while Hitler was still alive and the Third Reich still unbeaten, is one issue of real importance to some people."[25]

The claims go on that, as it became obvious that Germany was losing the war, the Haunebu craft, along with other advanced technology given by the "tutor race", were smuggled out by U-Boats to underground bases in the Antarctic – and continued to operate from there. (It is true that the Nazis did have a great interest in Antarctica, sending a large-scale scientific expedition there in early 1939.)

In 1946-7 a huge US expedition – named Operation Highjump and comprising a number of US Navy aircraft carriers, aircraft and some 4,000 military personnel – was sent to the Antarctic under the command of Rear-Admiral Richard E. Byrd. Although the official reason for this was to undertake a detailed geographical, geological and scientific study of the frozen continent, proponents of the "Nazi base" theory argue that this was a cover for what was really an attempted military invasion to try to defeat the last stronghold of the Third Reich. The story goes that the Americans were beaten back by the superior weaponry, and the Nazis have since operated unhindered from Antarctica, and it is they who are responsible for the UFO phenomenon.

Among Vladimir Terziski's more extreme claims is that a Haunebu-3 craft reached the Moon in 1942, that a Nazi base still exists there, and that their underground Antarctic city (called New Berlin) has a population of 2 million "pure-bred Aryan SS" who still use slave labour.

As McClure comments: "This material amounts to another form of Revisionism, in which the Nazis won in the end."[26]

As Jocelyn Godwin has shown in his book *Arktos*, the belief in secret Nazi bases in the Antarctic is a common feature of modern neo-Nazi mythology, and seems to be a development – a kind of scientific updating – of the "polar mythology" that was at the heart of much of the original Nazis' mystical ideas.

Whether or not Nazi flying discs were responsible for some or all of the early UFO sightings, the more extreme form of the theory, with the alien alliance and the survival of the Nazis in Antarctic bases is more the stuff of myth. Why, if the Nazis had access to such advanced technology and such powerful allies, did they lose the war? And why have they continued to live on in their wasteland bases for half a century? Surely, if they were building their forces for a second attempt at world domination, wouldn't they have made their move by now, sixty years after they went to ground?

Ridiculous though all this may seem, there is, however, a very sinister aspect to it. The books and videos promoting this theory focus attention on a secret society called the Order of the Black Sun, which is said to have been a secret order within the SS. The Black Sun, it is said, was the most powerful of the Nazi occult orders, and it was they who controlled the means of contact – psychic or physical – with the aliens. And the suggestion is that, unlike the other orders, the Black Sun survives to this day, still possessing the secret of contact with these god-like extraterrestrial beings.

None of the experts who have studied the occult and esoteric groups allied to the Nazis has ever heard of this Order of the Black Sun. It appears to be a modern invention – but one that is trying to convince people of its pedigree, presumably in order to foster the idea that it is the heir to the true power of the Nazis. As to the question why any group would want to do this, there are several possible answers – all of them unsavoury, not to say profoundly disturbing.

That there is a strong right-wing extremist, if not fascistic, element to many extraterrestrial contact groups and cults is undeniable, but this important point has received very little attention. Among eminent UFOlogists, only Jacques Vallee and Jerome Clark have explored this sinister social dimension to the UFO phenomenon. It is a fact that most of the early contactees of the 1950s had right-wing views and connections. The most obvious example is William Dudley Pelley, pre-war supporter of Hitler and founder of the fascist Silver Shirts of America, who was interned as a security threat during the Second World War. In 1950 Pelley produced one of the first books based on channelled communications with extraterrestrials, *Star Guests*, and was one of the key figures in the early contactee movement. Even today, many of the more extreme forms of the "UFO cover-up" scenario are bound up with a right-wing agenda.

THE CONDON COMMITTEE

The USAF's Project Blue Book continued until the late 1960s, although its activities were much scaled down early in the decade. In 1967 it was decided to set up a committee under Dr Edward U. Condon of the University of Colorado to review its work and make a recommendation as to whether the project should continue. The Condon Committee delivered its report in 1969, concluding that UFOs did not warrant further study from either a military or scientific point of view and recommending that Blue Book be closed. That marked the official end of the US Government's interest in UFOs, although many UFOlogists maintain that secret investigations continue to this day.

Confidence in the Condon Committee evaporated with Condon's statement at the outset that, "My attitude right now is that there's nothing to it . . . but I'm not supposed to reach a conclusion for another year." Two committee members were sacked after leaking a memo from the project's co-ordinator, Robert Low, that declared, "The trick would be, I think, to describe the project so that to the public, it would appear a totally objective study, but, to the scientific community, would present the image of a group of nonbelievers trying their best to be objective, but having an almost zero expectation of finding a saucer."[27]

Critics of the Condon Report were quick to point out that the committee's findings actually clashed with its own data, as they were unable to provide explanations for over 25 per cent of the cases studied – a far greater proportion than was even claimed by civilian UFO organizations. The report actually converted many who were previously sceptical to a belief in the reality of UFOs. The most influential such figure was the astronomer Dr J. Allen Hynek, a committee member who had been involved with UFOs from the very beginning, when he was asked by the USAAF to look into Kenneth Arnold's sighting. He was also an associate member of the CIA's Robertson Panel. Throughout, Hynek had been sceptical, but the evidence in the Condon Report convinced him that there *was* some substance to the UFO phenomemon. From 1969 until his death in 1986, Hynek became the foremost advocate of the scientific study of UFOs, establishing the Centre for UFO Studies (CUFOS) in Chicago and devising the famous classification of UFO reports into Close Encounters of the First, Second and Third Kind (and, later, more "kinds"). He was the consultant to Steven Speilberg's *Close Encounters of the Third Kind*, and appears briefly in the final scenes of the movie, looking with keen interest at the alien spaceship.

The French rocket scientist Dr Claude Poher was another such "convert". When asked in the early 1970s how he became seriously interested in the UFO mystery, he replied, "I read the Condon Report."[28]

Following the closure of Blue Book, some very odd things began to happen in UFOlogy. During the 1970s, the conviction grew that the US Government had known for many years not only of the reality of UFOs, but had entered into ongoing contact with the extraterrestrials. This notion had not featured prominently in earlier "cover-up" scenarios, such as those of Donald Keyhoe, or of Olavo Fontes, although such claims had occasionally surfaced.

One of the most intriguing of these allegations dates from 1954, and involves several visits of UFOs to Muroc (now Edwards) Air Force Base near Los Angeles. The rumour goes that a number of eminent people visited the base to witness the craft and meet with the aliens – including President Dwight D. Eisenhower.

One of those who claimed to have witnessed one of the alien visits was Gerald Light, a member of the Borderland Science Research Foundation. This was one of the first groups to combine mediumistic communications with UFOs – and was run by a medium named Meade Layne. Light wrote to Layne describing the experience in detail, and referring to a visit by Eisenhower in February. Intriguingly, there is a mystery over Eisenhower's whereabouts on 20 February 1954, when he was on a golfing holiday in Palm Springs, California – not far from Muroc. There was some speculation in the Press when the President apparently disappeared for several hours, although this was later explained as an emergency visit to the dentist.

Light wrote to Layne, ". . . it is my conviction that he [Eisenhower] will ignore the terrific conflict between the various 'authorities' and go directly to the people via radio and television."[29]

Two leading British UFOlogists, Desmond Leslie and the Earl of

Clancarty (better known as Brinsley le Poer Trench), have said that they had spoken to former officers from Muroc (both of whom, unfortunately, wished to remain anonymous) who confirmed both the UFO landings and the President's visit.[30]

Light's belief in an imminent Presidential disclosure about UFOs was not the last time that such hopes were to be raised. As we will see shortly, in 1974 there was widespread anticipation that Gerald Ford was about to come clean about the US Government's knowledge of UFOs, and during his successful election campaign in 1976, Jimmy Carter (a UFO witness himself) declared that, "If I become President I'll make every piece of information this country has about UFOs available to the public and the scientists." Carter's successor, Ronald Reagan, was also known to be a believer in UFOs. However, the world is still waiting for such a revelation . . .

Since Keyhoe's first books on the subject, UFOlogists had generally believed that the reason that the US Government was covering up its knowledge of UFOs was to prevent mass panic – of the kind that is believed to have been generated by Orson Welles' radio dramatization of *War of the Worlds* in 1938. (Clearly Keyhoe and NICAP disagreed with the policy and believed that the American public was now mature enough to deal with the revelation that "we are not alone".) Others who agreed that there was an official cover-up, such as John A. Keel, took a different line, believing that the military authorities were hiding their *ignorance* about UFOs. On this reasoning, the government knew that UFOs existed, but didn't want to admit to the public that they didn't know what they were – or how to deal with them.

However, from the 1970s a more sinister explanation for the government conspiracy has steadily grown in popularity (and been taken up enthusiastically by movies and TV series such as *The X-Files* and *Dark Skies*). This is that government is suppressing information about alien contact because it has made its own unholy alliance with the extraterrestrials. This is either because it has done a deal in order to increase its own power – for example allowing the aliens to operate unhindered in return for alien technology – or that it has been forced to submit to the aliens because of their superior firepower.

In 1972, documentary producers Robert Emenegger and Allan

Sandler were contacted by two USAF officers who had been involved with Blue Book, Colonel William Coleman and Colonel George Weinbrenner, and, at a meeting in the Pentagon, asked whether they would like to make a film about UFOs – using USAF footage. To their utter astonishment, Emenegger and Sandler were shown films and still photographs of UFOs and grey-skinned aliens, both dead and alive. They were told that the living alien, which appeared on 16 mm film footage, had survived a crash in New Mexico in 1949, and had lived for three years.[31]

At a further meeting at Norton AFB in California, the two producers met the head of the USAF counter-intelligence agency, the Air Force Office of Special Investigations (AFOSI) – the department responsible for Blue Book – and were told that in 1964 a UFO had landed, by prior arrangement, at Holloman AFB at Alamogordo, New Mexico, where a meeting between the occupants and military officers and scientists had taken place. All this, they were told, had been filmed, and Emenegger and Sandler were promised access to this film for their documentary, *UFOs: It Has Begun.* Shortly afterwards Colonel Coleman withdrew permission, saying that the Watergate scandal had made it inadvisable for such revelations at that time.

Two years later a film researcher and UFOlogist, Robert Carr, announced at a Press Conference that, according to his inside informants in the USAF and Pentagon, the US Government had the bodies of dead aliens, which were preserved in Hangar 18 at Wright-Patterson AFB. Carr claimed in October 1974 that, "Five weeks ago I heard from the highest authority in Washington that before Christmas the whole UFO cover-up will be ended."[32]

Unfortunately, Carr linked his claims with the crash in Aztec, New Mexico, in 1948 that was the subject of Frank Scully's 1950 bestseller *Behind the Flying Saucers.* Scully's book was soon exposed as a hoax, his principal informants being convicted tricksters.

Another UFOlogist, the veteran investigator Leonard Stringfield, also used his contacts in the USAF and other agencies to find out what the government really knew about UFOs. In 1978 he claimed that he had information from "twenty-five unimpeachable sources" that the government had the remains of UFOs and bodies from several crashes at Wright-Patterson. None of his witnesses allowed their names to be used, but Stringfield – one of the most respected investigators in UFOlogy and certainly no crank – was convinced of their integrity. However, critics again pointed out that there was a strong connection with Scully's long-discredited tale: the account by one of Stringfield's informants, who was given the pseudonym of "Fritz Werner" virtually quotes word-for-word from Scully's book.[33]

But the claims continued. William H. Spaulding, Director of Ground Saucer Watch – an organization respected in UFOlogy for its careful scientific approach – obtained affidavits from several former officers in the US intelligence community attesting that a crashed UFO and its occupants had been transported to Wright-Patterson AFB after first being examined at the CIA's headquarters in Langley, Virginia.

Gradually a consistent scenario took shape. It began with the "revival" of the long-forgotten Roswell case, as described in a previous chapter. In 1978, Stanton Friedman and William L. Moore first tracked down the retired USAAF intelligence officer at the centre of the events of 1947, Major Jesse A. Marcel. As described earlier, initially the case only involved the recovery of strange wreckage from a ranch in New Mexico, but gradually witnesses came forward with claims that alien bodies had been found at a second crash site, and that there had been a massive military cover-up.

During his research into the Roswell case, William Moore had made contacts in the military and intelligence community, in particular AFOSI. He claimed to be in contact with a "dissident" group that opposed the government's policy of silence, a group he called the Aviary. Moore's main AFOSI contact was to play a key role in the revelation of alien contact: Sergeant Richard C. Doty, who was stationed at Kirtland AFB near Albuquerque, New Mexico. Sergeant Doty first appeared on the UFO scene in 1980, in one of the most bizarre episodes in recent UFO history: the Bennewitz affair.

Paul Bennewitz was an electronics expert and president of his own scientific instrument company, based in Albuquerque, New Mexico. His company supplied humidity control equipment to

Kirtland AFB. He was also a UFO investigator for the Aerial Phenomena Research Organization (APRO). In early 1980, Bennewitz accompanied Leonard Springfield on the investigation of the abduction of Myrna Hansen, and her six-year-old son.[34] (Abductions were then a relatively new phenomenon on the UFO scene.)

Initially, Myrna Hansen only remembered seeing five UFOs descend in a cow pasture near Cimarron, New Mexico – and nothing at all about the next four hours. Stringfield and Bennewitz had her hypnotically regressed, and she recalled seeing two figures, dressed in white suits, emerge from one of the craft and begin mutilating one of the cows. When she tried to intervene, she and her son were grabbed and taken on board separate craft. Hansen was subjected to a physical examination, including a vaginal probe, as reported in many abduction cases. However, the procedure was interrupted by the appearance of a *human* scientist, who apologized to her and ordered that she be freed. The man took her on a tour of the UFO, and also allowed her outside: the craft had apparently flown to a new location in the New Mexico desert. She and her son were then subjected to a process in which bright lights and loud noises battered their sense, before being returned to the field where their adventure had started.

Bennewitz believed that Myrna Hanssen had been implanted with a device to allow the aliens to communicate with – perhaps even control – her. He was also disturbed by her testimony that humans were co-operating with the aliens in abductions and cattle mutilation. He became convinced that the aliens were beaming signals into Kirtland AFB and other nearby installations, and used his expertise to construct equipment to monitor them. Bennewitz was able to detect extremely low frequency (ELF) pulses emanating from the base, and claimed to be able to decode them – and even that he had entered into a dialogue with the aliens. As a result, he began to tell fellow UFOlogists that there was a huge alien base, covering some 20 square miles (518 sq km), beneath the New Mexico desert near the town of Dulce. From this the aliens – with the knowledge of the government – were abducting humans on a massive scale and implanting them with devices that allowed the aliens to communicate with them. Their aim, he said, was to produce a race of human-alien hybrids.

Bennewitz's colleagues began to notice signs of instability in him, and his mental health went into a rapid decline. He became paranoid, delusional and violent, eventually suffering a breakdown. Most of his colleagues in the UFO community regarded his wild claims as the result of his delusions.

Oddly, however, Bennewitz's claims attracted the attention of AFOSI, who sent Sergeant Doty to interview him in October 1980. Intriguingly, Doty took with him Jerry Miller, the Chief Scientific Advisor to Kirtland AFB and a former Project Blue Book member. As the sceptical British UFO writer (and former Editor of *The Unexplained*) Peter Brookesmith has pointed out, the fact that a former Blue Book member should still be on call for UFO-related investigations some ten years after the project was disbanded suggests that there was an ongoing interest in the subject on the part of USAF.

Bennewitz handed over some 2,600 feet (8,400 m) of 8mm film and photographs he had taken of UFOs around Kirtland and other nearby military bases – particularly the nuclear weapons storage facility at Monzano and the Coyote Canyon weapons testing area – along with tapes of the "alien" signals.

Intriguingly, Doty's report – which has been made public – states that: "After analyzing the data collected by Dr Bennewitz, Mr Miller related the evidence clearly shows that some type of unidentified aerial objects were caught on film." However, Miller had not found any significance in the audio tapes, and the report concluded that there was no evidence of a threat to the bases' security.[35]

A further note of interest was in the comment: "Mr Miller has contacted FTD personnel at W-P [Wright-Patterson] AFB, Oh., who expressed an interest and are scheduled to inspect Dr Bennewitz' [sic] data."

Two weeks later Bennewitz was invited to a meeting at Kirtland AFB to present his evidence to an impressive line-up, including a brigadier general (William Brooksher) and four Air Force colonels besides Doty. Given that the object of the meeting was for Bennewitz to present evidence that he was "in contact with the aliens flying the craft", and that his own colleagues in the UFO community had begun to doubt his state of mind, why was there such interest on the part of the USAF?

It was soon after this that Bennewitz's paranoia began to increase. He paraded around with guns in case of alien attack and accused colleagues such as Stringfield of being in league with the CIA. He kept a cache of guns and knives around his house, where they could be easily grabbed if the aliens came. He became insomniac and paranoid. More worryingly – and possibly tellingly – he claimed that aliens were coming through the walls and injecting him with drugs. At the end of 1985 he had a breakdown and was institutionalized for a time.

Clearly, Bennewitz was in a bad way psychologically – but had he been driven to this state by the enormity of what he had uncovered, or had he been delusional at the outset? If the latter, why did the USAF take him so seriously?

In March 1983, William Moore received a call from one of his contacts in USAF intelligence instructing him to travel from his home in Arizona to a motel in upstate New York. Here he was handed a sealed envelope, told that he had exactly 19 minutes to do whatever he wanted with the contents, and then return it. Taking the envelope back to his room, he opened it to find a number of documents relating to the US Government's contact with aliens. He hurriedly photographed them – unfortunately, because of the conditions, his pictures were not of good quality and some were too blurred to read – before, exactly on time, the courier returned to collect them.

The documents seemed to be a part of top secret briefing from 1977, referring to the activities of a project designated "MJ-12",[36] and various other projects under its control that were responsible for different aspects of the ongoing alien contact. It included references to projects with the codenames Aquarius, Sigma and Snowbird.

In April 1983 – less than a month after Moore's experience – TV film producer Linda Moulton Howe received an unusual and exciting call. As we have seen, Howe had been investigating cases of cattle mutilations and the previous year had produced a TV documentary and book on the subject, both called *Alien Harvest*. She was called by Sergeant Doty and invited to Kirtland AFB, where she had a remarkable three-hour meeting.[37]

The reason for the meeting, Doty said, was that her documentary about cattle mutilations had "upset" some people in Washington, adding, "It came too close to something we don't want the public to know." But then he proceeded to open up to Howe.

Doty claimed that the USAF had film of a UFO landing at Holloman AFB in 1964 – clearly the same event discussed with Emenegger and Sandler eight years before. Three UFOs came, one landing while the others hovered overhead. The purpose was in order that the bodies of aliens killed in an earlier crash could be returned, but Doty said that the they had given "something" in return. He also promised Howe this footage for a documentary – perhaps, with Watergate forgotten, the time was now right for such a revelation. But once again the promise was not kept.

Doty also showed Howe a document, although she was not allowed to copy it or to make notes. It was a briefing document prepared for the President on his accession to power. (Although she does not remember which one, or the date, it is generally thought by UFOlogists that the President in question was Jimmy Carter).

The document used the term Extraterrestrial Biological Entity (EBE) to describe the small, grey-skinned aliens – the first time the term, which has now acquired wide currency, was used. The bodies of a number of EBEs had been recovered from several crashes, and housed in various facilities, including Wright-Patterson and the Los Alamos National Laboratory. In a second crash near Roswell in 1949, one of the six EBEs had survived. Designated EBE-1, it lived in the Los Alamos facility for three years, before dying of unknown causes. (Again, this was the same information given to Emenegger and Sandler by USAF officers in 1972.)

Since then two more aliens had lived on Earth, both coming voluntarily as "ambassadors" from their race. There was a second group of aliens, the "Talls", who were apparently in conflict with the Greys.

The document said that a secret group known as "MJ-12" had been established with responsibilty for handling the alien contacts. (Doty told Howe that "MJ" stood for "Majority".) MJ-12 were in charge of a number of projects that oversaw specific areas of alien contact. Project Aquarius gathered information about alien life-forms. Project Snowbird developed new technology based on that given by the aliens or recovered from UFO crashes. Project Garnet investigated the evolution of life on Earth. Project Sigma dealt with communication with the aliens. (Aquarius, Snowbird and Sigma had appeared in the "briefing document" shown to William Moore a few weeks earlier.)

Some of the revelations contained in the document held implications that go far beyond even the momentous knowledge that mankind is not alone in the Universe, and raised profound philosophical and religious questions. From the alien ambassadors, the government had learned that the EBEs had genetically manipulated life on Earth – and that they had even created Jesus as a kind of intermediary between us and them . . .

After Howe had finished the document, Doty told her that he would contact her again, using the codename "Falcon". (Later, Doty told another researcher that, while he had indeed met with Howe, he had shown her no such document, and had not promised her any film footage. He also denied that he was "Falcon", although he said he represented a more senior agent with that codename.)

The documents shown to Moore and Howe were only the beginning. Just over a year later the famous (or notorious) Majestic 12 papers emerged. They were first leaked to a trio of UFOlogists consisting of William Moore, Stanton Friedman and documentary producer Jaime Shandera. In December 1984 an anonymous package, postmarked Albuquerque, was delivered to Shandera, containing a roll of 35 mm film that, when developed, showed two documents: the Majestic 12 papers. However, the trio decided not to make them public until they had been able to analyze and, if possible, verify them. In the event, they were to sit on this momentous discovery for more than two years.

A few months after Shandera received his controversial parcel, British UFOlogist Jenny Randles, who was working on a book about the UFO cover-up, was approached by an anonymous source who offered her copies of documents revealing the US Government's secret knowledge of UFOs and alien contact. Randles became suspicious, and after she expressed her reservations he broke contact. When the Majestic 12 papers were released a year later, she realized that the documents she had been shown were part of them.[38]

However, soon afterwards, the documents were offered to another British researcher, Timothy Good, who had no hesitation in accepting them. Good received them from one of his contacts in US Intelligence. He was also writing *Above Top Secret* (1987) – a bestseller that introduced the idea of a government cover-up to a British audience – building the Majestic 12 documents into it as a central feature. An article about them even appeared in *The Observer* Sunday newspaper in May 1987 – which prompted the American team of Moore, Friedman and Shandera to go public with their own copies.

The first document was a brief memo bearing the signature of President Harry S. Truman to Secretary of Defense James Forrestal, dated 24 September 1947. Stamped "Top Secret – Eyes Only", it read: "As per our recent conversation on this matter, you are hereby authorized to proceed with all due speed and caution upon your undertaking. Hereafter this matter shall be referred to only as Operation Majestic Twelve.

"It continues to be my feeling that any future considerations relative to the ultimate disposition of this matter should rest solely with the Office of the President following appropriate discussions with yourself, Dr Bush and the Director of Central Intelligence."

The second document was dated 18 November 1952, and was a briefing document from CIA Director Roscoe Hillenkoetter to the recently elected President Dwight D. Eisenhower, stamped with classification TOP SECRET/MAJIC, and headed "Operation Majestic-12 Preliminary Briefing for President-Elect Eisenhower".

It referred to the establishment of the Majestic-12 group as a result of Truman's executive order, naming the members. They included Hillenkoetter himself, Dr Vannevar Bush (the top nuclear scientist), General Nathan Twining and General Hoyt Vandenburg. In a nice piece of irony, one of other members listed was astronomer Donald Menzel, who until his death in 1976 was the leading debunker of UFOs.

After a brief summary of the beginnings of the flying saucer phenomenon with Kenneth Arnold's sighting in June 1947, the document goes on to describe the Roswell crash, adding: "On 07 July, 1947, a secret operation was begun to assure recovery of the wreckage of this object for scientific study. During the course of this operation, aerial reconnaissance discovered that four small human-like beings had apparently ejected from the craft at some point before it exploded. These had fallen to earth about two miles [3.2 km] east of the wreckage site. All four were dead and badly decomposed due to action of predators and exposure to the elements during the approximately one week time period which had elapsed before their discovery. A special scientific team took charge of removing these bodies for study . . . The wreckage of the craft was also removed to several different locations . . . Civilian and military witnesses in the area were debriefed, and news reporters were given the effective cover story that the object had been a misguided weather balloon."

Apparent confirmation of the existence of Majestic-12 came when Jaime Shandera, following an anonymous tip, found a memo in the National Archives in Washington written from Eisenhower's Special Assistant Robert Cutler to General Twining in 1954, which referred to a briefing of the National Security Council by the "MJ-12" panel.

The controversy over the documents has raged ever since. Critics – both sceptics and believers in the UFO phenomenon – have pointed to various anomalies that suggest that the documents are fakes: the briefing document did not bear the page numbering standard to Top Secret documents, and the style of dating (07 July 1947) was not one in common usage in 1952. The designation "Roswell Army Air Base" (instead of "Air Field") is incorrect. However, Stanton Friedman has been able to show that other, genuine documents of the time bear similar anomalies – although no other document has so many as the Majestic-12 paper. The "Air Base" reference could also be explained by the fact that, by 1952, the designation had changed and therefore could be a simple error on the part of Hillenkoetter. Friedman also discovered that Donald Menzel had worked for the CIA and NSA, and held top security clearances – an aspect of his career that had not been known before.

The most damning criticism, however, came when Philip J. Klass was able to show that Truman's signature was identical to that on a letter written to Vannevar Bush in October 1947, suggesting that it had simply been photocopied onto the "Majestic-12" order.

When it came to the 1954 "Twining" memo, the National Archives pointed out several anomalies, and that there had been no meeting of the National Security Council on the date given. The

FBI began an investigation into what appeared to be the leak of a classified document, but abandoned it when it realized that the document was faked.

Since 1994 a number of other "Majestic-12" documents have been sent or given to UFOlogists by anonymous sources – the most recent crop being in 1999. Most of these have been directed at American researcher Tim Cooper, and include material relating to involvement of such illustrious figures as John F. Kennedy and Albert Einstein. However, it has been relatively easy to show that these documents are fakes, as is an alleged Majestic-12 "Special Operations Manual", giving instructions on methods of UFO crash retrieval, delivered to Don Berliner in 1994. But were these produced by a hoaxer, who had simply jumped on the bandwagon started by the orignal papers given to Shandera and Good, or were they misinformation designed to counter the original leak?

The controversy has raged ever since the documents were first release. Although some UFOlogists – most vociferously Stanton Friedman – accept at least the original documents as genuine, others believe them to be hoaxes. Timothy Good, who based much of his first book on them, now also admits his doubts, mainly because of the Klass discovery about the "Truman" signature, but still writes in *Alien Liaison* (1991) that: ". . . even if the entire MJ-12 document turns out to be fraudulent, I am convinced that the information contained therein, at least, is *essentially* factual."[39] [His emphasis.]

Belief in the MJ-12 group was given a significant boost in 1989 by Whitley Streiber in his novel *Majestic*. The novel – which is dedicated to Jesse Marcel – describes the Roswell crash and the setting up of the Majestic-12 operation, and links the story with the alien abduction phenomenon. Although told in the form of fiction, coming immediately after his two non-fiction works detailing his experiences as an abductee, *Communion* and *Transformation*, many readers took it as being based on inside information. They were helped in this by the introduction, in which the journalist hero is told by his "deep throat" source that if he wants to get his story published, ". . . your only hope is to publish your book as fiction" (so if the authorities take any action against him it will be tantamount to admitting it is true). Having been given this advice, the fictional narrator writes: "So this is fiction."[40]

The Majestic-12 documents, along with those shown to William Moore and Linda Moulton Howe, formed a consistent scenario involving the recovery of alien bodies from several crashes, including that at Roswell, and an ongoing dialogue between the US Government and EBE ambassadors. Towards the end of the 1980s, an important new element was added to this scenario: Area 51.[41]

The connection emerged in interviews by William Moore and Jaime Shandera with two of their contacts within the "Aviary", "Falcon" and "Condor". The two also appeared, in silhouette and with their voices electronically disguised, on a TV special, *UFO Coverup? Live*, in 1988.

Although "Falcon" was the name originally used by Sergeant Doty when he met with Linda Moulton Howe, and Doty was a contact of William Moore's, there is some dispute about whether he is the "Falcon" of these interviews. Doty told other researchers that his codename was not "Falcon" – although he said that this name was used by one of his superiors. Moore and Shandera have stated that their "Falcon" is not Doty. On the other hand, one of the producers of the TV special said that the Falcon interviewed *was* the AFOSI sergeant. Of course, if Doty was, as he claimed, a whistle-blower using the name to protect his identity, he would hardly be likely to admit it publicly. (And UFOlogists who tried to prove that he was Falcon seem to have had a curious disregard for their informant's well-being.)

The identity of "Condor" is easier to establish. He is almost certainly Robert Collins, a retired plasma physicist at the Sandia National Laboratories, which is attached to Kirtland AFB.

According to Falcon and Condor, the second alien, EBE-2, came to live on Earth voluntarily. It took a year to learn how to communicate with him, but they were then able to learn from him, and the second "ambassador" EBE-3, more about the aliens. They come from a binary star system in the constellation of Zeta Reticuli. They are around three-and-a-half feet (1.05 m) tall, with large heads and huge, insect-like eyes. They are extremely intelligent, with IQs in excess of 200, incredibly acute senses, and a lifespan of 350-400 years. They have a religion based on the existence of a Supreme Being. The EBEs were vegetarian, and could eat Earthly fruit and vegetables – but one of Falcon's revelations on *UFO Coverup? Live* about their eating habits threatened to make the subject ridiculous in viewers' eyes, when he said that their favourite dish was strawberry ice cream.

Falcon made one more revelation in the 1988 TV special: that eventually an agreement had been made by the aliens that allowed them to establish their own base in a designated area in the United States. He said, 'It's in the state of Nevada, in an area called Area 51, or 'Dreamland . . .'."

AREA 51: DREAMLAND

The now-legendary Area 51 is a vast tract of the Nevada desert given over to the military for the development of new aircraft and weapons systems. Officially it does not even exist – although its reality was briefly acknowledged by President Clinton, but then only to exempt it from Federal health and safety legislation. The area is highly restricted, and overflying is not allowed on pain of swift, even terminal, action. Strictly speaking, "Area 51" is the top-secret test area around Groom Dry Lake, about 80 miles (128 km) to the north of Las Vegas, but other sites, such as the huge Nellis Air Force Range and the Nevada Nuclear Test Site are also included. (Nellis AFB does have a facility called the Alien Technology Center, but this – apparently – refers to the evaluation of foreign aircraft.)

Support for Area 51's extraterrestrial connection was soon forthcoming. In 1989 Robert Lazar, who claimed to be an ex-employee of the top-security base, came forward, and in television programmes, lectures and interviews with UFOlogists, including Tim Good, told of his strange experiences there.[42]

According to Lazar, from 1982 he was working for the Weapons Division of the Los Alamos National Laboratory, on particle beam weapons for the Strategic Defense Initiative (SDI, or "Star Wars"). (When Lazar first started making his public statements, Los Alamos denied that he had ever worked there. But when a copy of their internal telephone directory was produced that showed Lazar's name, they changed the story to say that he had only worked on "non-sensitive" projects.) In 1988, with the assistance of the eminent physicist Dr Edward Teller, Lazar got a job at a part of the Groom Lake facility known as S-4.

Between December 1988 and March 1989 Lazar spent a total of six or seven days at the S-4 site, in underground hangars. He was taken there blindfolded, and internal security was tight, with severe restrictions on where any individual could go. Even conversations between the technicians were forbidden. Lazar said he was issued with a security badge bearing the code "MAJ", and that the project used money siphoned off from the "black budget" and funds allocated to the SDI programme, and that Congress – even the President – was unaware of its existence.

On his first day, Lazar was given a number of briefing papers to read. They detailed aspects of the project's involvement with UFOs, and autopsy reports, which included photographs, on dead aliens of the typical Grey variety. The briefing papers stated that the United States had gained possession of several intact UFOs, which were being test flown in Area 51. On his second visit Lazar was shown one of the craft – a circular machine 30 feet (9 m) in diameter and 15 feet (4.5 m) high. On another occasion he was allowed inside the craft, in which everything was rounded – "like it's made out of wax and heated for a time and then cooled off" – with seats that seemed to be made for children. He was told that it was powered by an "antimatter reactor", which utilized a new element called Element 115. On another visit he saw nine such craft in their subterranean hangar.

Lazar said: "It was definitely extraterrestrial. I don't know the history of them, but they're certainly alien craft, produced by an alien intelligence, with alien materials. We were trying to see if we could duplicate what was there, with earthbound material and technology."[43]

He was only to see one "flight" of the craft, in which, apparently under remote control, it lifted some 20-30 feet (6–9 m) off the ground and then settled back down.

Lazar's claims are, to say the least, controversial, and rest entirely on his personal testimony. It has been impossible to corroborate some of what he says about his background, in particular his claim to have obtained his masters degree in physics from the Massachusetts Institute of Technology. His supporters claim that this is part of a campaign to discredit him. On the other hand, researchers have located his wage slip for five days work at an unnamed government site in Nevada, which bears the budget code "E-6722MAJ"[44].

Since Lazar's claims, a number of other people claiming to be ex-employees at Area 51 have come forward, all anonymously. One female radar operator at nearby Nellis AFB tells how she saw a flight of between ten and fifteen UFOs, and was afterwards given mind-clouding drugs to make her forget the experience (obviously unsuccessfully).

BRITAIN

In the United Kingdom there is no equivalent of the Freedom of Information Act, and a culture of official secrecy exises that has been zealously protected by generations of civil servants. Given that situation, is there any way of telling what the British Government does or does not know about UFOs?

The official British line is that essentially the military are not interested in the UFO phenomenon. As far as we know, there was no British equivalent of Project Blue Book. However, several researchers – most significantly Timothy Good and Nicholas Redfern – have argued that this is simply evasion to hide the fact that the British Government, too, is covering up what it really knows. However, there is very little documentation – certainly no equivalent of the thousands of pages unearthed by American researchers – to challenge the official British line. But as Good, Redfern and others point out, this is hardly surprising given the British obsession with official secrecy. Most of their claims are based on anonymous inside informants within the armed forces and security services. So too are the claims of another leading advocate of the British Government conspiracy, Tony Dodd.

Probably the most debated British UFO incident is the Rendlesham Forest Incident of December 1980, although as that involved a USAF base any cover-up is once again down to the US Government. We have seen that there were claims in 1955 that a UFO had crashed in Britain ten years before, and that this was covered up. In his book *Cosmic Crashes* (1999) Nicholas Redfern has listed seven possible

UFO crashes in Britain, although most can be put down to meteors, debris falling from aircraft, or crashes of experimental aircraft. The crash of a twin-tailed aircraft on the runway of RAF Boscombe Down, Wiltshire in 1994, which was seen shrouded in tarpaulin and shipped to the USA a few days later, seems to fall into the latter category.

Redfern lists the 1945 crash, and another that he was told about by Leonard Stringfield that allegedly happened in Staffordshire in 1964. A UFO crashed in two parts, one in Britain and the other in Germany, and both parts were shipped out to Wright-Patterson AFB. There are no more details.

The most intriguing British UFO crash case is an event that occurred in the Berwyn Mountains of North Wales in 1974. As we have seen, on 23 January a formation of green lights were seen from many parts of north-west England streaking across the sky. Shortly after 8.30 p.m. a huge explosion and earth tremor shook the ground in the Berwyn Mountains of North Wales, recorded as reaching 4 on the Richter Scale in Edinburgh. Local people reported seeing a glowing object plunge to the ground. At first it was feared that there had been a plane disaster, and the emergency services were alerted.

A local nurse, Pat Evans, was among the first on the scene, alerted by a phone call from the local police that her services may be in urgent need. Driving as near as she could to the scene, Evans walked across the fields to be greeting by an astonishing sight. A UFO "the size of the Albert Hall" had crashed into the side of a mountain, and the ground was strewn with wreckage and bodies. Sensationally, it was claimed that she saw one close up, which was not human. Then a military team arrived and ordered her away from the scene.

In the last few years new claims have surfaced about the incident: two bodies were recovered from the site and taken to the chemical warfare establishment at Porton Down. These claims, promoted by Tony Dodd and Nicholas Redfern, are based on the testimony of an anonymous witness who claims to have been part of the retrieval team.

Although Parliamentary Questions have occasionally been asked about specific UFO incidents, the highest profile UFOs ever achieved politically in the United Kingdom was when they became the subject of a debate in the House of Lords in January 1979. The debate was initiated by the Earl of Clancarty, a UFOlogist who wrote under the name of Brinsley le Poer Trench, who proposed that the government set up an enquiry into UFOs. Nothing came of the debate, which seemed to get bogged down in the relationship of UFOs to Christian dogma, the Bishop of Norfolk expressing his concern of "the danger of the religious aspect of the UFO situation leading to the obscuring of basic Christian truths", and Lord Trefargne arguing that the notion of extraterrestrial life was incompatible with Christian faith.

Officially, all UFO reports made to the Ministry of Defence are filed with its Secretariat (Air Staff) section 2 – Sec(AS)2 – in Whitehall. Enquiries by UFOlogists to the Ministry of Defence usually elicit the response that the Ministry's policy, such as it is, that UFO reports are only to be examined from the perspective of whether or not they present a threat to national security. If it is judged that they're not – which usually seems to be the case – they are not investigated further.

However, Good and Redfern believe that the more significant site – where UFO reports are investigated more seriously – is at Rudloe Manor in Wiltshire, the headquarters of the Royal Air Force's Provost and Security Services (the RAF's military police force). According to Good, basing his claims on an inside informant in the Ministry of Defence, the reports are sent from there to the Provost Marshal's Office in London, and then on to the appropriate Ministry of Defence department.

Nicholas Redfern has located declassified documents from the 1950s concerning RAF UFO investigations. A 1953 instruction is that "sightings of aerial phenomena" by RAF personnel are to be sent to the Air Ministry's Deputy Directorate of Intelligence (Technical) for examination, and that personnel must not communicate their sightings to anyone else without authority. As with USAF and CIA documents from the same period, this shows that the RAF were more concerned with the flying saucer enigma than they admitted – but not that they knew what they were.[45]

Two members of the Ministry of Defence's "UFO desk" in Sec(AS)2 – or, as it was designated in its earlier incarnation,

Department S6 – have spoken about their work. The first was the late Ralph Noyes, who was in charge of S6 in the 1950s, and who went on to become a prominent parapsychologist, and Chairman of the Society for Psychical Research. Noyes' time with the section convinced him – from reports both from the public and RAF pilots – that there was a real, unknown phenomenon at work. However, he saw no evidence to suggest an extraterrestrial explanation, seeing UFOs as a rather more transient, more paranormal phenomenon. Shortly before his death he compared UFO sightings to rainbows, or the aurora borealis – something that, although insubstantial, was nevertheless real. The problem comes, he argued, when too much is read into such things, such as a rainbow being a sign from God, or UFOs being from outer space . . .[46]

In the early 1990s, the "UFO desk" in Sec(AS)2 was manned by Nick Pope, part of a routine three-year civil service posting. Pope had no previous interest in the subject, but his three years on the desk, and his personal investigations into the reports he received, convinced him that UFOs were real and that the only explanation for them was extraterrestrial. He disagreed with his Ministry's policy – or perhaps "attitude" would be a better word, as, according to Pope, no formal policy decision had ever been taken – that, as UFOs were deemed to pose no security threat they were unworthy of serious attention. For him, it was self-evident that intrusions by unknown craft into British airspace presented at least a potential security risk.

Pope's view was strengthened by the wave of "flying triangle" reports that came in from across the country on the night of 30-31 March 1991. He was especially concerned about the incident in which personnel at an RAF base in Shropshire had reported a huge triangular craft – described as the size of a jumbo jet – hovering over the base and beaming light at the ground. Pope reasoned if this did not present a security threat, what did? After trying in vain to persuade his superiors of this, he decided to go public, and in 1996 published *Open Skies, Closed Minds*, which argued for an extraterrestrial origin for UFOs and pressed the case for the British Government to take the subject far more seriously.

Coming from a Ministry of Defence official, these claims attracted a great deal of interest and publicity. Pope was even interviewed on the BBC's prestigious *Newsnight* programme, which is seldom given to coverage of "fringe" subjects. He also appeared on many other TV and radio programmes, and lectured widely.

Pope presented something of a paradox for those UFOlogists who believed in a British Government cover-up. Here was a Ministry of Defence official, who had investigated UFOs on behalf of the government, and who was not trying to debunk the subject. But on the other hand, he maintained that there was no cover-up, and that the official British policy was rather one of burying its head in the sand and pretending that the problem didn't exist. As Pope told me, "Anything that looks like conspiracy is probably bureaucracy".

Researchers such as Nicholas Redfern could only explain this contradiction by arguing that Sec(AS)2 was simply window dressing, and that Pope was not aware of the government's real interest in UFOs, the serious investigation of which were undertaken at Rudloe Manor. It has also been argued that, whereas Sec(AS)2's function is simply to receive and file reports from the public, sightings by military personnel are handled at Rudloe Manor. Pope disagrees, saying that instructions are that military reports are also sent to Sec(AS)2. The big question even if Pope was deliberately kept in the dark, is if the British Government are really trying to prevent the public learning the true nature of UFOs, why did they allow him to publish a book arguing the case for an extraterrestrial origin? Pope was, and still is, a Ministry of Defence employee and subject to the Official Secrets Act. Had they wanted, his superiors could easily have prevented him writing and publishing the book. In the event, although they vetted it to ensure that no official secrets were given away, they allowed its publication.

ELSEWHERE

In several countries, the official attitude to UFOs has been more open compared to that of the USA and UK. In France, the main UFO study group, GEPAN, is part of the government-funded National Centre for Space Studies. In Australia, the Royal Australian Air Force – which had investigated UFOs since 1960, working fairly openly with the public – stopped receiving UFO reports from the public in 1984, as they considered them to have no relevance to national security.

Although no head of state has yet declared his or her belief in UFOs, and no government has made an official statement on the matter, in recent years government or military officials in several countries – including Belgium, Mexico and Russia have publicly stated that they believe UFOs to be extraterrestrial. And, as we saw earlier, in China, newly receptive to the UFO phenomenon after the repressive Communist regime relaxed its grip, the President himself gave the opening address at a UFO Congress, which at the very least indicates that Chinese officialdom is not averse to being associated with the UFO phenomenon.

WHY THE COVER-UP?

If there is an official cover-up, what is being covered up? As we have seen, early theories alleged that governments were aware that UFOs were real and extraterrestrial, but feared the consequences of making this public in case it created mass panic. Other researchers, such as John Keel, argued that the cover-up was really down to ignorance – the US Government and military had no idea what UFOs were, but, since their job was to protect the American public, they dare not admit it.

The recent claims, centred on the Majestic-12 papers and the revelations about Area 51, maintain that the US authorities – or a secret group within them – are actually engaged in an ongoing dialogue with the aliens. This is being covered up either out of fear of the consequences, or because they have entered into an unholy alliance with the EBEs.

There is, however, a new twist to the "government conspiracy" theory that has been explored more seriously by UFO researchers in the last few years – most publicly by Gregory Kanon in his book *The Great UFO Hoax* (1997). This is the idea that there *is* a conspiracy on the part of the military and intelligence agencies, but not one aimed at suppressing belief in alien contact. Its object is actually to *foster* such a belief.

Kanon and other researchers point to the level of clandestine involvement in major UFO cases, and suggest that the belief in the extraterrestrial origin of UFOs has been deliberately constructed as a "smokescreen" for secret research projects, such as the development of the Stealth bomber.

This view is shared by William H. Spaulding, Director of Ground Saucer Watch and one of the pioneers of the use of the Freedom of Information Act, to find out what the CIA and other agencies really know. The documents he uncovered convinced him that those agencies *had* been keeping their involvement in the UFO phenomenon secret – but not because they knew that UFOs were extraterrestrial. Spaulding developed the "Federal Hypothesis", in which the CIA and USAF had *created* the UFO phenomenon, or at least kept it going after the initial wave of interest in the 1940s, as a cover for secret experiments.[47]

There is a certain amount of evidence to support this view. In 1997 CIA papers were released under the Freedom of Information Act which revealed that, in the 1950s and early 1960s, that agency allowed sightings of top secret spy planes such as the U-2 and SR-71 ("Oxcart") by members of the public to be recorded as "unexplained" UFOs. In other words, the authorities knew what the real explanation of these objects was, but labelled them "unidentified" because of their secret nature. Some estimates suggest that as many as *50 per cent* of "unexplained UFOs" during this period could be accounted for this way.

The news was something of a bombshell to UFOlogists, as at the very least they had to reconsider the statistics of "explained" as opposed to "unexplained" UFO reports. Some, such as Dr Bruce Maccabee, dismissed the CIA report as being of no significance. Others, however, are not so sure.

The CIA documents only refer to the period up to the early 1960s, but the revelation has prompted many to ask whether a similar strategy has been in operation ever since. After all, UFO reports could supply an effective smokescreen for all kinds of top secret projects – not to mention otherwise unacceptable blunders by the armed forces.

One case uncovered by William H. Spaulding neatly illustrates his point. In 1975, a USAF radar unit in California had detected an aerial object in the vicinity of Edwards Air Force Base. Suddenly – within one sweep of the radar – the object vanished. It had seemed to accelerate instantaneously from around 450 to over 2,000 miles per hour (720–3,200 km/h), the only way it could have moved out of range so quickly. The event was logged as a possible radar contact with a UFO. However, it was later revealed that this was an early test of a Stealth aircraft (which had been in development since the mid-1960s). The object had not gone away, but had merely become invisible to the radar. The test had been deliberately set up for un unsuspecting USAF radar team to see how they reacted – and who did not have a high enough clearance to know about the super-secret Stealth technology. As Spaulding asked: how many other such tests have there been, perhaps using civilian radar operators and air traffic controllers?[48]

In fact, the theory was not new. As we have seen, *True* magazine editor Ken Purdy, when he commissioned Donald Keyhoe to write his first article about flying saucers in 1949, had come to the conclusion that they were a "gigantic hoax" designed to cover up official secrets.

The earliest proponent of the Federal Hypothesis was Leon Davidson[49] in the late 1950s. Between 1949 and 1952 Davidson was a chemist at the Los Alamos National Laboratory, and subsequently worked for the Atomic Energy Commission and IBM, holding a high security clearance for more than seventeen years. He was one of the few to be allowed to read the CIA's Robertson Panel report when it was first produced.

Davidson undertook his own investigations into UFOs, and concluded that *every aspect* led back to the intelligence agencies, particularly the CIA; writing that the agency was "solely responsible for creating the Flying Saucer furor as a tool for cold war psychological research". Davidson, like Jacques Vallée many years later, believed – and produced a certain amount of evidence to support his contention – that George Adamski and other early contactees had worked with the full backing and support of the CIA.[50]

Although Davidson did go on to make some rather extreme claims – such as that the symbols seen on the Socorro UFO were a stylized version of the initials "CIA" – he also produced more cogently argued and compelling evidence. UFOlogists in general, however, simply refused to take it seriously, simply because it was not what they wanted to hear.

We have seen that, in the early 1950s, the CIA's main preoccupation with UFOs was their potential for psychological warfare –

both for offence and defence. Could it be that some of the major UFO "flaps" were created as experiments into the mass psychology of belief? In fact there is evidence that such a strategy was at least being considered.

Jacques Vallée is the only major UFOlogist to draw attention to the "Pentacle Memorandum". This is a secret memo from January 1953 (the same month, significantly, that the Robertson Panel met) written by the Air Technical Intelligence Center at Wright Field, discussing the simulation of a UFO flap in a remote area of the US. It recommended that "many different types of aerial activity should be secretly and purposefully scheduled within the area" in order to generate UFO reports from the public. This document has been in the files of UFO research groups since the 1960s, and yet has attracted very little interest – especially when compared to more contentious documents such as the Majestic-12 papers.

Then there is Donald Keyhoe and NICAP. As we have seen, Keyhoe was the first popularizer of the concept of the UFO cover-up, and the lobbying of the US Government for public disclosure of its UFO files was a major part of the policy of the organization he created, NICAP. However, it is also clear that his conviction that UFOs were extraterrestrial *and* that there was a cover-up largely came from his contacts within the USAF. And the head of Blue Book wrote an endorsement for the cover of one of Keyhoe's later books. All this is very odd behaviour from the very organization that Keyhoe was accusing of *denying* the reality of UFOs to the public – and specifically from Blue Book, which he believed was a major part of the coverup. Another informant of Keyhoe's was none other than Admiral Roscoe H. Hillenkoetter, the first Director of the CIA (1947-50) – an alleged member of the original Majestic-12 – who told Keyhoe that the Agency had been monitoring the UFO phemenon, and the USAF investigations, since the very first reports.

Keyhoe's organization, NICAP, also repays deeper investigation. One of the founding board members was Admiral Hillenkoetter. Jacques Vallée has revealed that no less than five board members of NICAP were former or serving officers of the CIA.[51] This close relationship between NICAP and the CIA is very strange, since NICAP was committed to exposing the supposed official cover-up – a cover-up in which the CIA is believed to play a major part.

Such suspicions extend into other aspects of the UFO phenomenon in the 1950s. Jacques Vallée has written that the seminal contactee George Adamski had several connections with intelligence agencies.[52] And another leading contactee of the 1950s, Howard Menger, later stated that he had been part of a CIA experiment to test public reaction to the idea of ET contact.

We have also seen that, in the early 1970s, USAF and other officers seem to have been actively encouraging individuals – usually film producers – to pursue their investigations into the government's alien contacts. Robert Emenegger and Allan Sandler were promised film footage of UFO landings by the USAF. Robert Carr and Leonard Stringfield – the two leading promoters of the "Hangar 18" scenario – drew their information from contacts in the Pentagon and Washington.

And what of AFOSI agent Sergeant Richard Doty – "Falcon" – who seems to have a hand in all the development of the most popular "government cover-up" scenario, based around the Roswell Incident, the Majestic-12 papers and Area 51? It was Doty who showed Linda Moulton Howe the "Presidential briefing paper" at Kirtland AFB in 1983. He was a major informant for William Moore and Jaime Shandera, and "Falcon" (whether or not this was Doty, he was certainly connected with him) who linked EBEs with Area 51.

(According to AFOSI, Doty was sacked from that department in 1986 for allegedly faking reports – although it appears that he was reinstated after a few months. He left USAF entirely two years later, although he is still on the Air Force Reserve list.)

In July 1989, William Moore – one of the first "revivers" of the Roswell story and among the initial recipients of the Majestic-12 papers – dropped a bombshell at a UFO conference in Las Vegas when he revealed that, for several years, he had been working as an informant and disinformation agent for AFOSI. He had, in fact, told this to some of his colleagues, including Karl Korff, as early as 1982, but had sworn them to secrecy.[53] Moore claimed that the reason he had agreed to work in this capacity was that he had been told that in return for giving AFOSI details of his colleagues' work, and giving misinformation to throw them off the scent, he would be allowed access to genuine information about the US Government's ongoing contact with aliens. However, the obvious alternative is that the information that Moore was given by his AFOSI handlers was the real misinformation.

As we have seen, Doty's first appearance on the UFO scene came when he investigated the claims of Paul Bennewitz. In September 1980, Doty asked Moore to act as an informant on Bennewitz – in exchange offering him the privileged information. Although Moore denies that he was responsible for feeding Bennewitz any of the misinformation that was to drive the unfortunate man to the brink of madness, if not beyond, he was aware that the USAF were doing so. But why? What had Bennewitz done to deserve such treatment?

Bennewitz first attracted AFOSI's attention when he began to claim that he was monitoring signals from military installations in New Mexico. Peter Brookesmith argues convincingly that Bennewitz *had* picked up such signals, but that they were nothing to do with aliens, and everything to do with classified activities at those bases. (Brookesmith points out that the ELF waves Bennewitz detected are not suitable for carrying detailed information, and have a very limited range in space.) Faced with this, the USAF decided to feed him misinformation, so that he believed in an increasingly extreme alien scenario.

In Brookesmith's words: "AFOSI then decided that it would feed Bennewitz a mass of misleading information about aliens, subterranean bases and anything else they could think of – so that, should he leak any technical details of their work, he could and would be discredited as a crank by all and sundry."[54]

Did AFOSI know that the result of this would be to make Bennewitz unstable – or did they help the process along? Remember that he claimed that aliens were entering his house at night, and giving him injections . . .

William Moore, as he now admits, was a participant in this operation to mislead and discredit Bennewitz.

Brookesmith has assembled a convincing case that the whole saga of Majestic-12 stems from a USAF operation to discredit Paul Bennewitz: the documents were Moore's "pay-off" for helping AFOSI, and were faked by that agency to appear that they had kept their part of the bargain – and which have successfully fooled the UFO community and many others besides ever since.

However, other aspects of the unfolding "alien contact" scenario seem to suggest an operation on a wider scale than the discrediting of one individual. For example, how does the briefing paper shown by Doty to Linda Moulton Howe – another TV producer – fit into the Bennewitz affair? This not only supported what AFOSI officers had told Emenegger and Sandler eleven years earlier about the 1964 Holloman landing, but also the documents shown to William Moore a few weeks earlier. And the Majestic-12 papers were also "leaked" to researchers in Britain: if they were contrived simply to pay off for William Moore, why?

Significantly, some of the project code names mentioned in the documents shown to Moore and Howe have subsequently been found to be those of genuine, classified, programmes – but ones that have nothing to do with UFOs or aliens. Sigma was a project to develop laser weapons systems, Aquarius to develop detection systems for sea-launched missiles and low-flying aircraft, and Snowbird (the least certain) a new type of weapons platform. Whoever devised those documents knew at least of these projects' existence – even though they were, at the time, highly classified. In other words, if they are fakes, they were faked from *within* the military intelligence community.

When Moore made his admission of working for AFOSI he said that he knew of several other leading UFOlogists who were also working for that agency, but refused to name them. It is a fact that several leading American researchers have links with intelligence agencies. Kevin Randle, another leading advocate of Roswell, is a former USAF intelligence officer; Karl Pflock is a former CIA employee, as is Derrell Simms, the researcher responsible for recovering implants from alleged abductees. Dr Bruce Maccabee admitted in 1993 that, for the previous fourteen years, he had regularly briefed the CIA on developments in the UFO field at their Langley, Virginia headquarters. Maccabee – who has claimed that he has knowledge that the CIA are sitting on thousands of secret documents relating to UFOs – asserts that he was simply briefing the CIA at their own request. But if so, as many of his colleagues have asked, why did he keep this so secret for so long?

Many UFOlogists base their claims of secret alien contact largely on their own – usually anonymous – contacts in the intelligence community. They include Leonard Stringfield, Robert Carr, William Moore, Timothy Good and Nicholas Redfern. In relation to claims of a British cover-up, Kevin McClure has identified what he calls the "Unknown Soldier syndrome", which he defines as "the spreading of disinformation by ex-military or pseudo-ex-military personnel".[55] And while sceptics may deride such anonymous informants as fictitious – a convenient excuse for researchers not to produce evidence to back up their claims – it is clear that at least some of them are genuinely who they claim to be. But have these researchers been the targets of a deliberate campaign of deception? It has been pointed out that UFOlogists never believe anything that comes from government, military or intelligence sources – unless those sources are telling them that UFOs are real, from outer space, and in contact with the US Government.

The idea that the intelligence community – or a faction within it – are actively *creating* the belief in alien contact and a government conspiracy does help to explain the paradoxical behaviour on behalf of certain individuals. We have seen that AFOSI officers seem to have been actively courting TV producers, although always failing to deliver the promised earth-shattering proof. And there are several examples of prominent members of the military intelligence community who should be denying everything actually lending credence to the contact claims.

We saw one example of this when discussing the Roswell Incident in the earlier chapter, in the testimony of Brigadier General Arthur Exon. This former commanding officer of Wright-Patterson AFB told Roswell researchers that, although he was not involved personally, he had been informed about the Roswell crash, the recovery of alien bodies and their transfer to Wright-Patterson, and of the formation of a Majestic-12-style group to oversee their study and to guard the secret of their existence. Why should someone who was at one time in charge of the very base where the aliens were supposedly preserved have volunteered this information to UFOlogists?

Another case is that of Admiral Bobby Ray Inman, a former Director of the the Office of Naval Intelligence (ONI) and Deputy Director of the NSA, CIA and DIA. He left government service in 1982 to become President of Scientific Applications International

Corporation (SAIC), which carries out research into anti-gravity propulsion systems. In 1989 Robert Oeschler, a former NASA scientist, recorded a phone conversation with Inman in which Oeschler asked him about back-engineering from "recovered craft", and whether the results would be made public. Inman, while not actually admitting that the US Government had such craft, failed to deny it either, saying that it was unlikely that the information would ever be made public. Perhaps Inman was speaking hypothetically and Oeschler – and Tim Good, the main promoter of the story – simply misunderstood.

In the documentary *Dreamland* (shown in Britain on Sky television),[56] presenter Bruce Burgess phoned Inman who said that while at NSA, DIA and CIA he had asked whether there was any credible evidence of UFOs and always received the answer "No". (However, it may be significant that he made no mention of the answer he received while at ONI . . .) On the other hand, his Executive Assistant, Tom King, did explicitly tell Burgess that Inman "asks that you do not quote him or use his name in any manner without his prior approval", so presumably Inman *did* approve the broadcasting of his comments.

Inman confirmed that the US Government *had* recovered crashed UFOs – but warned that, as they dealt with national security issues, the tapes could not be used without official approval. Yet Oeschler has several times allowed them to be broadcast in television documentaries – and no action has been taken either against him or the TV companies responsible.

On this reasoning, the feeding of information about EBEs, Hangar 18, Roswell, Majestic-12 and Area 51 was part of a concerted effort to construct a consistent scenario of alien contact – and of government conspiracy. Why?

Invoking Spaulding's "Federal Hypothesis", was this to provide a convenient cover for a variety of classified military projects? For example, is the belief that alien spacecraft are being piloted over the skies of Area 51 a handy way of diverting attention from very real *terrestrial* aircraft being developed there?

Could the "UFO" label also be used to cover up military accidents and mishaps that have endangered the public? Consider the Rendlesham Forest "landing" for example. It has only recently been revealed that this USAF base at Woodbridge in East Anglia was actually *the* major storage centre for America's nuclear arsenal in Europe at the time of the "landing". The more famous cruise missile base at Greenham Common in Berkshire – the focus for anti-nuclear protestors and the famous Women's camp – was, in fact, largely a decoy to distract attention from Rendlesham. Bearing this in mind, could the event of December 1980 not have been the landing of an alien spacecraft, or the misidentification of a lighthouse, but a story designed to cover up some kind of accident, perhaps involving the leaking of nuclear material? After all, apart from the conflicting evidence of the eye-witnesses, the only "hard" evidence for something out of the ordinary taking place in Rendlesham Forest was the above-average radiation count at the alleged landing site. What better way to discourage the interest of "serious" investigative journalists who might get to hear of the story than relegating it to the "lunatic fringe"?

The behind-the-scenes involvement of agencies such as the CIA and AFOSI is so obvious that many UFO researchers have acknowledged it, and proposed theories to explain it. One is that it is all part of a "softening-up" strategy to gradually accustom the public to the idea of alien contact, and which is building up to some momentous revelation. Another is that there are groups of whistle-blowers in the intelligence community who are opposed to the official policy of silence. A third is that these apparent leaks are in fact misinformation, designed to discredit the whole notion of alien contact in the public mind. (If so, such an exercise has been singularly unsuccessful, as if anything, belief in UFOs and the government cover-up is increasing in the developed world.)

Another solution that has been proposed is that a fake alien invasion is being created in order to unite the world against a common enemy. Those who propose this theory cite a statement made by President Reagan during a speech to the United Nations in the declining years of the Cold War: "I occasionally think how quickly our differences worldwide would vanish if we were facing an alien threat from outside this world. And yet, I ask you, is not an alien force already among us?"

But there are more sinister theories to account for the apparent manipulation of the UFO phenomenon by official agencies. Going beyond the use of UFOs as a convenient security cover for classified aircraft and weapons tests, it has been suggested that the alien scenario – in particular alien abductions – are being used as a cover for covert psychological warfare experiments, or what is now known as mind control. That the US military and intelligence agencies have conducted such experiments on innocent and unknowing subjects is beyond dispute. Most notoriously, the CIA's MKULTRA project of the 1950s used drugs such as LSD, together with hypnosis and electroshock therapy in experiments into the modification of an individual's behaviour.[57] (It was Admiral Hillenkoetter, the later member of NICAP, who authorized the first such experiments – then known as BLUEBIRD – in 1950.) The MKULTRA research, and similar projects by the US Army and Navy, caused a major scandal, and led to an official enquiry under Vice President Nelson Rockefeller, when they were exposed in the mid-1970s.

In recent years, many researchers have become interested in a sub-set of the abduction experience in which the abductee claims that *human* scientists have been involved. The term MILAB – "military abduction" – had been coined for this phenomenon. (The Bennewitz affair, it will be remembered, began with an investigation into just such an abduction). Dr Helmut Lammer, a scientist with the Austrian Space Research Institute, who has carried out a detailed study of MILAB claims, sums up these claims: "Researchers in the field of mind control suggest that those cases are evidence that the whole UFO abduction phenomenon is staged by the intelligence community as a cover for their illegal experiments."[58]

Lammer does not agree with this extreme position, arguing rather that the military have taken advantage of the abduction phenomenon as a convenient cover. His statistical study showed that, whereas alien abductions have been reported since the 1960s (although they have become much more frequent in the last two decades) and from across the world, MILAB reports are confined to the USA and have only been reported since about 1980.

A typical MILAB account recounted by Lammer involves a woman known only as "Michelle". When she came to the attention of alien abduction researchers, Michelle had initially reported the "classic" Grey alien experience. However, she began to experience flashbacks to what appeared to be a MILAB. Under hypnotic regression, she told of an experience that happened in 1970 while out camping with her boyfriend near Montauk Point, New York – then the site of a military base, which, since its closure, which has become the focus for some extreme UFO-related claims. Michelle and her boyfriend were kidnapped by a military team and taken to the base, where she was stripped and examined by a team of five or six people, including one woman, who were dressed in white gowns. She was given an intravenous injection which made her lose consciousness, and woke up later on a beach with her boyfriend. In later regressions, she remembered being placed in an isolation tank – a known method of brainwashing using sensory deprivation. It is significant that this early experience of hers had none of the "alien" trappings that became an increasing part of the MILAB phenomenon after 1980.

Many abduction researchers refuse to consider MILAB reports, since they are, they believe, self-evidently not "real" abductions. Others argue that these memories are implanted hypnotically by the aliens to cover their tracks. Some believe that the MILABs show that the aliens are working in co-operation with sinister government agencies. However, there is also the possibility that the "screen memory" idea is the other way round: that the *alien* experience is being used to cover the tracks of the military. Indeed, one MILAB victim, Katharina Wilson, even remembers holding a rubber Grey alien mask during her experience.[59]

Dr Lammer points out that neuroscientist Michael Persinger has demonstrated that mystical experiences, OOBEs and typical abduction experiences can be produced artificially by stimulating the temporal lobes with magnetic fields. (See the Explanations Chapter.)

In *The Stargate Conspiracy*, co-written with Clive Prince, I took a close look at one of the most popular movements based on apparent psychic contact with advanced extraterrestrial intelligences: the Council of Nine, or simply the Nine, who claim to be the gods of ancient Egypt who are about to return. In investigating the near fifty-year story of the Nine contacts (which began in 1952) we discovered that the driving force behind them, from their inception in 1952 until the late 1970s, the famed parapsychologist Andrija

Puharich, led a double life. At the very time of the first contacts Puharich was also working for the US Army on a joint project with the CIA that was part of the MKULTRA project. Moreover, the private research foundation at which the Nine first appeared was, in reality, largely financed by military and intelligence agencies. The involvement of the CIA in the story of the Nine continued for at least two decades, leading us to the conclusion that the whole thing was part of a long-term experiment – but with what aim?

The Stargate Conspiracy also looks at the CIA's role in manipulating New Age beliefs, particularly those concerning the imminent return of god-like extraterrestrial beings who were worshipped as the creator gods in ancient times. We concluded that this was part of a programme to create a new belief system, in which those who claimed to be in contact with the "space-gods" would, like the prophets of old, be in a uniquely powerful position.

If the same intelligence agencies are manipulating belief in alien contact, is this part of the same programme? And, if so, does this mean that the whole UFO phenomenon – at least in its most recent manifestation – was essentially manufactured with this aim in mind?

There is another possibility, one that has also been explored by Jacques Vallée, although most thoroughly in fictional form in his novel *Fastwalker* (1996). In this scenario, the effort to persuade the public that UFOs and their occupants are extraterrestrial is a smokescreen to prevent their real origins being known. If Keel's "Ultraterrestrial" hypothesis, and Vallee's own exploration of the parallels of UFOs with fairy lore and demonology are correct, then the "aliens" do not come from "up there" but have always been living alongside us, invisibly. But now they have the ability to influence us even more by persuading us that they come from the stars, and thereby accept their presence more easily. Have the Ultraterrestrials entered into what British researcher Geoff Gilbertson calls an "unholy alliance" not with extraterrestrials, but with the beings from myth and legend that have always sought to manifest in our material world?

Fastwalker concerns a super-secret Majestic-12-type agency called Alintel that is aware that the origins of UFOs lies in another, parallel world – but who encourage the idea that they are extraterrestrials. One of its scientists says: "In the last forty years we have conditioned the US public, indeed, the world masses, to expect visitors from outer space."[60]

Alintel's ultimate aim is power: to convince first the President, and then the world, that alien invasion is imminent – and that only they have the means to stop it.

And would such a scenario also work if the contact was benign rather then hostile? If an Alintel-like group could convince the world that only they had the secret of communication with powerful extraterrestrials . . . ?

As Vallée wrote in his 1979 study of the covert manipulation of the UFO phenomenon, *Messengers of Deception: "The group of people who will first manage to harness the fear of cosmic forces and the emotions surrounding UFO contact to a political purpose will be able to exert incredible spiritual blackmail.*" [His emphasis."][61]

Notes

1. Brookesmith, p. 29.
2. See Brookesmith, p. 22 for a reproduction of the memo.
3. Sachs, p. 92.
4. Sachs, p. 269.
5. Sachs, p. 269.
6. Klass, p. 53.
7. Brookesmith, p. 38.
8. Sachs, p. 7.
9. Klass, p. 47.
10. See Redfern, *The FBI Files,* (London) 2000.
11. Brookesmith, pp. 80–83.
12. Barker, p. 195.
13. Fontes, 'Report From Brazil'.
14. A French translation of Fontes' letter to APRO can be found on the RRO website (*www.multimedia.com/rro/fontes/html)*
15. See Nick Redfern's *The FBI Files,* (London) 2000.
16. See Redfern, *Cosmic Crashes.* (London) 1999.
17. *Ibid.*
18. Harbinson, *Projekt UFO,* p. 77.
19. Harbinson, 'The UFO Goes To War', p. 748.
20. Harbinson, *Projekt UFO,* pp. 71–78.
21. See Vesco and Childress, *Man-Made UFOs 1944–1994.*
22. Quoted in McClure, 'More Lies Than Secrets'.
23. In the video *UFO: Secrets of the Third Reich.*
24. McClure: 'Nazi UFOs – More Secrets Than Answers'.
25. McClure: 'More Lies Than Secrets'.

26. *Ibid.*
27. Sachs, p. 70.
28. J. Allen Hynek, *The UFO Report*, p. 282.
29. Good, *Alien Liaison*, p. 58.
30. *Ibid.*
31. *Ibid*, pp. 101–103.
32. Klass, p. 9.
33. Nickell, p. 7.
34. Brookesmith, p. 108.
35. Doty's report is reproduced in *Ibid*, p. 109.
36. Good, *Alien Liaison*, pp. 108–110.
37. *Ibid*, pp. 104–107.
38. Randles, 'UFOs Worldwide'.
39. Good, p. 113.
40. Strieber, *Majestic*, p. 5.
41. See Phil Patton, *Travels In Dreamland*, (London) 1997.
42. Good, *Alien Liaison*, p. 178.
43. *Ibid*, p. 155.
44. *Ibid*, p. 178.
45. *Ibid*, pp. 15–16.
46. Letter to *Magonia* on-line magazine, April 1998. (*www.magonia.demon.co.uk/ethbull/ethbull2.html)*
47. Brookesmith, pp. 91–92.
48. *Ibid*, pp. 93–94.
49. See Davidson, *Flying Saucers: An Analysis of Air Force Project Blue Book Special Report No. 14*.
50. Vallée, *Forbidden Science*.
51. Vallée *Messengers of Deception*, p. 202.
52. Vallée, *Dimensions*, p. 248.
53. Korff, pp. 166–167.
54. Brookesmith, p. 114.
55. McClure, 'More Secrets Than Lies'.
56. A two-part documentary produced and directed by Bruce Burgess for Transmedia and Dandelion Productions, 1996.
57. See Marks: *The Search for the Manchurian Candidate*.
58. Lammer, 'Preliminary findings of Project MILAB'.
59. *Ibid.*
60. Vallée *Fastwalker*, p. 196.
61. Vallée, *Messengers of Deception*, p. 157.

CHAPTER 8

EXPLANATIONS

"There never was an explanation that did not itself require an explanation"

Charles Fort

Few UFOlogists are content to amass data of sightings and alleged abductions without attempting to find a pattern, message or meaning in them. Many investigators even begin with preconceptions, but more usually one particular belief is seized upon relatively early and vehemently defended no matter what evidence may come to light that might challenge it. Like Cinderella's Ugly Sisters cutting off their toes in order to make their feet fit the magic glass slipper, they attempt to distort the facts to fit their preferred option, or ignore them altogether when they are too uncomfortable; too "damned" to be of service in their chosen cause.

Polarized at either end of the spectrum are the ETH enthusiasts – those who remain convinced that all alleged aliens are truly extraterrestrial in origin, and that all UFOs are nuts and bolts craft from the stars – and the professional debunkers to whom no amount of evidence will persuade them that there is even an *unknown*, let alone a paranormal, otherworldly or extraterrestrial explanation for the phenomenon.

In between those two extremes hover many of the world's UFOlogists, who are happy to admit that the phenomenon sometimes seems to be one thing and sometimes another, or find that new evidence substantially realigns the bias of certain cases towards other explanations. (Usually, but not exclusively, towards a mundane explanation such as a deliberate hoax, as in the case Kenneth Arnold's sighting in 1947, which now seems to have been simply a case of genuine misidentification.)

MISIDENTIFICATIONS

Every UFOlogist recognizes – frequently with a sinking heart – the moment when a promising UFO sighting dies on its feet and is relegated, terminally, to the Identified Flying Object (IFO) file, when the mysterious light in the sky is revealed to be a plane carrying football supporters to France, or the strange saucer-shaped object captured unwittingly on a holiday snap turns out to be that most deceptive of natural phenomena, the lenticular cloud – or a street-lamp, road-sign or light bouncing off a particularly large bird. In fact, almost all UFO sightings prove to have a rational explanation, although there is a small defiant percentage that remains to challenge, even affront, the reductionists, sceptics and debunkers.

As we have seen, a simple (but understandable) misidentification was probably behind Kenneth Arnold's classic 1947 sighting of "flying saucers". As James Easton eloquently – and convincingly – argued (see page 45), it appears that Arnold was bemused by the sight of a flock of large American White Pelicans catching the sunlight as it bounced off ice-covered mountains. And it is salutary that Jenny Randles, who fought for years to have the UFO sighting of Rendlesham Forest taken seriously, should be the one to acknowledge that, in the end, the mysterious lights could have been those of the local lighthouse (although this is a case that may yet hold some surprises). Equally disappointing is the fact that Betty Hill's ongoing, and virtually nightly, UFO sightings are usually – possibly always – merely misidentified aircraft flying overhead.

An example of hope triumphing over fact was indicative of a rash of UFO sightings, mysterious lights and so on, which emanated from the Irish Centre for UFO Studies, (ICUFOS), led by Eamonn Ansbro and Alan Sewell from their base at Craigavon, Co. Armagh. They, at least, were in no doubt about the nature of UFOs, telling sceptical researcher Dave Walsh[1] that "surveillance is being carried out on Earth by vehicles from extraterrestrial sources",[2] although they did admit to inviting them in, by transmitting "thoughts through crystals in an effort to make contact with aliens."

Bombarding Dave Walsh with "a collage of MIBs, suicides, black helicopters, abductions, harassments and so on" they almost began to persuade him that they had a case, but not quite. When they took him on a skywatch to Coomhola, 3 miles (5 km) north of Bantry, he noted that not only was it hardly an ideal place to look for UFOs (the landscape is too complicated with a confusing panorama of lights from islands and peninsulas), but also that the apparently mysterious red and green flashing lights they saw actually belonged to navigational buoys.

Walsh also cites "another sighting on 20 August [1997], by a couple driving on the Killarney-Mallow road (N72). . . .[3] I've been told by someone who lives in the area that a local nightclub uses a strong revolving outdoor light, which is generally pointed [in the] direction where the 'sighting' took place."

Of course the nightclub light may not have been the cause of the phenomenon, but as with every apparently watertight rationalization, a little bit of the magic dies.

Another fertile source of certain types of UFO sighting – the mysterious fireballs that chase across the sky – is the re-entering into the Earth's atmosphere of space junk. NASA has estimated that since man went into space in the late 1960s, some 100,000 artefacts – ranging from pens, screwdrivers and cameras to bits of rocket and solar panels – between a quarter of an inch (6 mm) and four inches (10 cm) in diameter, are still orbiting the planet at a rate of five miles a second. The *Daily Mail* of 19 August 2000 said, "Now NASA is working on the space-age equivalent of the broom to get rid of it, because the junk is becoming a major problem for the international Space Station being built hundreds of miles above the Earth. As the station grows, the odds are increasing that it will be hit by a piece of debris."

The article goes on: "While it has a shield against items under a quarter of an inch [6 mm], and large objects can be tracked and avoided, anything in between could be catastrophic.

"At the least, an object could cut through the station like a bullet, destroying everything in its path. To try to make sure this never happens, Nasa is testing a £100 million laser which will either sweep the junk out into deep space or pull it towards Earth where it will burn up in the atmosphere."

If the tests prove effective, the laser will be mounted on a space shuttle and clear up the space garbage bit by bit over a period of about two years. But in the meantime, pieces of debris will no doubt continue to be reported as intruders from distant planets.

Yet only a hardened sceptic would deny that there are still *unidentified* flying objects, not to mention experiences of high strangeness associated with them that remain to puzzle, annoy – and ultimately perhaps even to shame – today's keenest researchers.

EARTH LIGHTS

One of the more persuasive explanations for the sighting of simple lights – without the associations of a structured craft or entities – is Paul Devereux's "earth lights" hypothesis. The much-respected British researcher, formerly the Editor of *The Ley Hunter*, (a post he held for many years), argues that many spookily unexplained lights can be explained by the shifting of geological strata. Friction is caused by layers of rock rubbing against each other, which in turn release electrical discharges in the form of balls of light.

Scottish researcher Brian Allan (of Strange Phenomena Investigations) writes of the UFO hotspot, Bonnybridge:[4] "One thing worth noting is that there is a geological fault beneath the village, a belt of an ultra hard quartz material. This fault created major problems with mining operations in the area and caused the tunnels to be re-routed to avoid this layer of almost impenetrable material. It is feasible that this substance is prone to the 'piezzo crystal' effect, generating electrical charges when subject to the mechanical stress of tectonic movement. This phenomenon helps lend credence to a current theory that many UFO sightings have their origin in highly unusual, natural, geo-magnetic anomalies that create an electro-magnetic field, which in turn affects the temporal lobe in certain sensitive people."

Brian himself has observed something akin to this phenomenon in the neighbourhood of Bonnybridge. He says, "On a skywatch held on a clear, cold night in October 1998, I witnessed a brief flash of bright blue light that seemed to emanate from the ground. The light appeared as a single ball, then flashed outwards from the ball in two opposing arms. While this sighting only lasted for a fraction of a second, it was seen by eight or nine people . . ."

Sometimes what are known as "plasma balls" are also seen – and sometimes filmed – hovering over crop formations, although a watertight cause-and-effect has yet to be established. Although many people believe the balls of light, which appear to be either intelligent or under intelligent direction, actually create the designs in the crops, it may be that these "mini UFOs" have a natural affinity with any kind of crops, whether smoothed into giant patterns or simply left to grow and make money for the farmer in the usual way. After all, no one films fields without crop formations, so who knows?

The plasma balls also seem to congregate around electrical pylons, as Brian Allan notes, adding: "The large number of . . . pylons that march across the hills and moorland around Bonnybridge may also be a contributing element in this enigma. There are numerous reports of balls of light being seen clustered around the insulators mounted on top of the pylons . . . is this another phase of electromagnetic phenomena? Are these glowing spheres the result of a previously unexpected facet of high voltage physics? Or, could it be that the impressions of intelligent control are the subjective opinions of witnesses adversely affected by localized electromagnetic fields and their attempts to rationalize what they're seeing?"

As we have seen repeatedly, electricity plays a major part in UFO sightings and abductions, and is also associated with the onset of psychic abilities. Many gifted sensitives, including Uri Geller – whose strange talents kicked in after receiving an electric shock from his mother's sewing machine as a child – acknowledge that their paranormal careers began after being struck by lightning or having a close encounter with an electrical discharge.

Does this accidental "shock treatment" somehow change the brain/mind interface, opening up the gateway to the Otherworld? Does it make us see what others can't by creating a shift in our mental furniture, just as if it moved a sofa that was blocking a window? Or does it create a subtler form of Temporal Lobe Epilepsy, creating wonderfully magical worlds that have no reality on any level, but are merely the fairy gold of experience, simply empty illusions? Can the occasional shifting of layers of rock produce little bursts of electricity that actually change human perception, if only for a split second?

The ancient peoples who built the great and mysterious monuments such as Britain's Stonehenge, Avebury stone circles and the pyramids of the Giza plateau in Egypt seemed to use certain types of stone for a purpose, which can only be dimly guessed at today.[5] Inside the Great Pyramid, made of 2.5 million tons of limestone blocks, is the King's Chamber – which is lined with *granite*. Totally empty, except for a lidless and slightly broken "sarcophagus", also made of granite, the purpose of the Chamber is unknown. (Indeed, despite the airy pronouncements of Egyptologists, no one knows for sure what the pyramid itself was for.) Perhaps the builders and the priests who designed and used the Great Pyramid knew about the special properties of granite. Perhaps they knew that when great pressure is applied – under the subtly-shifting weight of 2.5 million tons of limestone, for example – it emits ghostly lights and some power that makes mere men see visions of gods . . .

The area around North Berwick in East Lothian, Scotland, is also a possible candidate for the piezzo electrical theory. Rich in granite, which gives off a steady stream of background radiation, it is a place of ancient fairy lore, of tales of being "pixie led" by dancing fairy lights – and now, of seeing UFOs, especially on the strange pyramid-shaped volcanic hill, known as Berwick Law. Local hotelier Dr Stephen Prior (see page 174) inclines to be sceptical about the tales, suggesting that, "My own suspicion is that they may be connected with coloured gas coming from the granite." Dr Prior, who is also a parapsychologist, acknowledges that *something* strange has been reported in that area for centuries, but the way it is perceived seems to have changed with the prevailing cultural expectations.

The world today is an electromagnetic spaghetti junction. Within approximately 100 years we have moved from polluters of the air with highly visible noxious fumes and smoke to polluters and destroyers with invisible tidal waves of electromagnetism that permeate the most remote areas. There is no hiding place from the endless radio waves, emissions from various telecommunications masts, microwaves and secret and semi-secret experiments such as the US Government's High frequency Active Auroral Research Program (HAARP), which is designed to bombard the ionosphere with high-energy, high-frequency, phased array radio waves with the ultimate purpose, so they say, of researching the natural phenomenon of the Aurora Borealis. Some, such as American activists Nick Begich and Jeanne Manning, authors of *Angels Don't Play This HAARP* (1995), accuse the Pentagon, who fund HAARP, of cooking the skies by beaming "more than 1.7 gigawatts (billion watts) of radiated power into the ionosphere, the electrically-charged layer above Earth's surface"[6], although the HAARP authorities counter by saying, "Mr Begich is off by three orders of magnitude"[7], which is apparently supposed to make us feel better.

The point is that *never* in the history of mankind, or indeed of the planet itself, has there been so much electromagnetism around. *No one* knows what effect it will have on humanity, although certain research into the use of mobile phones suggests that there may be escalating health risks. But what about changes at a subtler, psycho-spiritual level? Will there be an epidemic of psychic flashes, increasingly vivid dreams – even UFO sightings and encounters with aliens? Is the electromagnetic soup in which we are bathed already responsible for the thousands of abduction reports, almost exclusively from the developed world – and of that, almost exclusively from the United States?

Some people are more sensitive to electromagnetism than others: who knows what an overload will do? Or is it the case that this constant bombardment of electromagnetism will create a quantum leap in human evolution – just as the flash of lightning in the old movies jolts Frankenstein's monster into life – by giving us all a sixth sense? Is technology opening the stargate for us, whether we like it or not?

THE ELECTROMAGNETIC TRIGGER

British researcher Albert Budden, an engineer, caused a furore with his book *Aliens and Allergies* (1994), which argues that *all* paranormal phenomena – he includes UFO sightings and alien abductions in this category – are caused by the effect of concentrated electromagnetism on the brain of the experiencer. In an article for *Nexus* magazine in August 1999, he wrote: "If I were to encapsulate my case in a single general statement, I would say that in the understanding of the paranormal, electromagnetics are as fundamental as genetics are to biology."[8]

Budden's research seemed to reinforce his basic premise so repeatedly and emphatically that he said: "It almost became boring. Time and time again you find phenomena occurring on electrical hotspots. I've almost lost count of the times I retraced the steps of an experiencer and discovered criss-crossing power lines at the precise spot where their abduction took place. In the end I used to say, 'Just follow the power lines'. There's *always* a radio mast nearby or you discover that the witness has been struck by lightning – some major connection with electricity. I note that Whitley Strieber used to play around with electricity when young: he once fused all the lights in his house during one of his experiments as a sort of scientific swot, and he also encountered ball lightning at first hand."[9]

Another classic case that seems to support Albert Budden's claims is that of Dannion Brinkley, whose career as a New Age prophet began when he suffered a Near Death Experience (NDE) as a direct result of lightning exploding out of the telephone while he was in mid-conversation. Significantly, his otherworldly experience was markedly similar to Strieber's, its ontological shock apparently acting shamanically to open up his spiritual awareness. Unfortunately, Brinkley – like, to some extent, Strieber – is now something of a fallen prophet, his predictions for the late twentieth century having failed dismally.

However, although electricity seems to play *some* role in creating or triggering psycho-mystical experiences, Budden's theory fails to account for all the ghosts, poltergeists, fairies, demons, angels – and possibly aliens – that were reported prior to the late nineteenth-century discoveries of Edison and Tesla. Where did they come from in the days long before electricity was known? Yet Albert Budden is adamant, saying: "There are no beings, no spirits. In every case electricity plays a central role."[10] The debate goes on, but one suspects that while the majority of "beings" will continue to be seen where pylons and masts are at their highest concentration, a few will pop up in remote parts of Australia and South America where there are none . . .

HOAXES

This is a thorny topic. Hoaxing is a complex matter, as anyone in the crop circle community – on both sides of the great divide – will tell you. Motive is all: some hoaxers are in it for the potential profit, some for the fame (and the profit), some out of pure mischief, some to show up the credulity of the "experts", and yet others perhaps, more loftily, in order to create an art form.

Sometimes it is very difficult, particularly at this distance in time, to say for certain whether a case was a hoax or not. Was George Adamski a fake? And if so, what do we mean by that? Did he merely lie about everything – the UFOs, the meeting with the Venusians, the trips to the stars etc – deliberately and from the very beginning, perhaps in order to gain some status in the community, or even to create a new "religion"? Or were there subtler motives and forces at play? Perhaps he did see a strange light. Perhaps he had some kind of otherworldly encounter, but not quite as he later described it. It may be significant that his long-time secretary, Lucy McGinnis, left his employ because he had become a trance medium. Was he always inclined to mediumship? Was Adamski himself not the hoaxer, but the entities who used him?

And what about British schoolboy Alex Birch who hoaxed the infamous UFO photograph in 1962? As researcher David Clarke says: "Dismissing the whole saga [see page 72] as a hoax which got out of hand is easy, but ignores the important lessons the story can teach us about the complexity of human nature, the mysteries of the psyche and the creation of the UFO myth from very human origins."[11]

The Alex Birch case is perhaps reminiscent of the well-known case of the "Cottingley Fairies" in which two young cousins, Elsie and Frances, stunned – among many others – Sir Arthur Conan Doyle[12] with their photographs of winged fairies and a gnome apparently playing in the long grass near their home at Cottingley, near Leeds, in Yorkshire in the early years of the twentieth century. Over the years, as suspicions were voiced about the authenticity of the photographs – the professional debunker James Randi believed that Elsie and Frances had used invisible thread, for example – the increasingly elderly women said nothing. Then, in the early 1980s, one of the old ladies finally confessed,[13] declaring that all they had done was cut figures out of a book,[14] propping them up in the grass with hat pins. So far there is nothing to suggest anything other than Randi's straightforward hoax. But the whole point of the exercise, she claimed, was to show other people *what they genuinely saw* as a matter of course by the stream. In other words, the photographs were fake, but the fairies were real.

Yet, human nature being what it is, few people care to remember this most significant aspect, preferring to write the Cottingley Fairies off as a simple hoax.[15]

Elsie and Frances were in every other respect normal Yorkshire lasses with sound common sense, but they couldn't help it – they saw fairies. Alex Birch seems to have experienced a whole array of paranormal events as a child, which continued throughout his life, sometimes erupting into deeply unpleasant poltergeist activity, besides sightings of *genuine* UFOs. As a child he often witnessed lights darting about inside the house, "about the size of a ping-pong ball . . . always a bright flourescent colour with a green opaque colour:"[16] balls of light are reported – with very different slants – from all eras and countries, often tying in with local traditions of fairies and nature spirits, demons or, more recently of course, UFOs.

Alex also had one of the great initiatory and life-changing experiences: he underwent a Near Death Experience (NDE; see below) as a result of the diptheria he contracted when he was about six years old, spending several weeks on the danger list in an isolation hospital. He was lying, befogged by fever, when he saw a small but very bright light – which emitted "beautiful golden rays" – descend from the sky and enter his body. Then a being appeared, standing silently for many hours, simply smiling at him. That day he confided in his father, "Dad, I saw Jesus today." Just a week later, the child was on the road to recovery.

About two years after this extraordinary event, Alex observed a "dumb-bell" shaped UFO rising from the earth in a field near his home, which left no traces. As this was only about three years before he and his friends hoaxed the famous photograph, it reinforces the idea that the fake was very similar to the case of the Cottingley Fairies: in both instances the children saw something strange but knew that the only way they could show it to other people was through trickery. Perhaps it is telling that one of the Cottingley girls, when an old lady, maintained that *one* of their photographs *was* genuine after all . . .

THE EXTRATERRESTRIAL HYPOTHESIS (ETH)

Many UFOlogists, especially in the United States, are "ETHers", believing that UFOs are alien spacecraft and abductions are literal kidnappings by extraterrestrials. The ETH is based very largely on the fact that the UFOs exhibit a technology that cannot be matched by any known earthly craft, and the alleged declarations of the aliens themselves about their place(s) of origin. This has been reinforced in recent years by the allegations of some – such as Bob Lazar – that US scientists have retrieved crashed saucers and used them to "back engineer" their own versions (implying that some of the more recent UFOs may be of terrestrial origin) at secret locations such as Area 51 in the Nevada Desert.

However, the ETH is not without its problems. Over the years aliens have announced they have come from a huge variety of extraterrestrial locations, from the Moon to unknown planets in far distant galaxies. As John Keel says: "In recent years we have been informed by seemingly sincere contactees, several of whom have undergone psychiatric and lie detector tests and passed them with flying colors, that the saucers come from unknown planets named Clarion, Maser, Schare, Blaau, Tythan, Korendor, Orion, Fowser, Zomdic, Aenstria, and a dozen other absurd places."

He adds: "There have also been contactees who talk freely about the people of Venus, Mars, Jupiter, Uranus, Saturn and the Moon."[17]

George Adamski's Space Brothers allegedly came from Venus, but while they were reassuringly humanoid, we now know that they could not have survived in Venus's inhospitable conditions. Later, Adamski claimed to have travelled with them to Saturn.

It is significant that as modern knowledge of the stars progressed in leaps and bounds, the aliens changed the location of their home planets from near to very distant, out of the reach of our telescopes, or came up with star systems of which we have never heard and therefore cannot check. We are familiar with the name of the current favourite, Sirius, the home of – among other entities – the Council of Nine, but as the Nine are impostors, why should we take their claims to originate there seriously?

As for the idea that the UFOs are nuts and bolts machines from "out there", of course it may be true. Only a fool would say that there are no aliens anywhere in the Universe and that they could *never* reach the Earth (although, on our current understanding of interstellar travel it might take them a few generations to do so). In fact, recent discoveries have indicated that life may be inherent in space – but in what form? Microbes, green algae or strange microscopic bugs? Perhaps aliens are already here, in the form of the common cold or a host of other viruses that hitched a ride on space debris, planning their takeover of the world through deadly epidemics.

ETHers tend to categorize all unidentified flying objects as "spacecraft", besides presupposing that every anomalous light in the sky is either a sentient being or contains sentient beings. However, jumping to conclusions can have unfortunate repercussions, both for UFOlogy in general and the individuals to whom the ETH is king.

But are the flying saucers' occupants, as reported in the cases in this book (and elsewhere) really aliens? Do they truly come from other distant star systems, or do they hail from much closer to home? It is possible that while some are impostor spirits, like the Council of Nine, others may be creations of the CIA as part of some gigantic and unscrupulous experiment in social engineering, while others could be the product of an interaction with an invisible spirit world, be it the fairy realm or the old demons from the Pit in modern garb.

And what of the crashed saucers, for example at Roswell in 1947? Much as this is a beguiling idea, the sad fact is that *not one* of these stories has any checkable, objective evidence to support it. Saucers may have crashed on Earth, but these are not them.

Adherents of the ETH are adamantly opposed to the idea that there is any psychic or mystical component in the UFO saga. To them these are real, solid craft that do not need Spiritualist mediums or the services of card-carrying New Agers to be seen – which may well be true in some cases. UFOs are captured on radar, whereas (as far as we know) ghosts and fairies are not.

Yet in many – arguably even the majority of – cases, UFOs are witnessed by psychics, both trained and untrained, and it often emerges that contactees and abductees have a long history of experiences involving mysterious balls of light, poltergeists, extraordinary dreams, visions and Near Death Experiences. Alex Birch and Maggie Fisher were both familiar with an astonishing range of unexplained phenomena when they encountered their UFOs – whatever they may have been. The UFOs seemed to fit quite naturally into their scheme of things, and indeed they themselves thought of them as more paranormal than "nuts and bolts". Even George Adamski's fantastical stories may have owed more to his interest in philosophy and Eastern mysticism than a base somewhere on the planet Venus, and he may always have had a secret talent for trance mediumship. Like it or not, it seems that psychics find it easier to see UFOs than others, no matter how desperately the latter people may wish to do so. This would not be true if the objects were "real" in the sense that a jumbo jet is real, although even something as large and undeniably three-dimensional as that would be subject to personal interpretation if the witness had never seen or heard of one before, and had no frame of reference in which to put it.

TRICKS OF THE BRAIN

Much as it annoys the more mystically-minded and New Agers, scientists are always coming up with physiological or psychological explanations for visionary or unexplained experiences. Sleep paralysis is a favourite: the sensation of being awake but unable to cry out or move a muscle, while shadowy, and often malevolent, entities invade one's bedroom – and one's mind. A similar, and equally common, phenomenon of the twilight zone between sleep and the waking state, is the feeling of having a heavy weight on the chest, suffocating and terrifying. Because there is often an accompanying sense that this is a *being*, the old tales of the demon lovers – the incubi and succubi that haunted the troubled nights of the Middle Ages – are now dismissed as the by-products of a purely physiological condition, coloured by a sociological expectation and belief.

The connections between, and similarities with, modern stories of Grey aliens who invade the bedrooms – and lives – of the innocent are obvious. Instead of an evil spirit, a demon from the Pit, sitting astride the sleeper's chest with terrible lust, seeking not only sexual gratification but also our very souls, there have now come the more morally ambiguous Greys. Sometimes they take on the role of abusers like the demons of old, but increasingly pervade their victims' lives with a curious sense of belonging, inspiring intense feelings of being chosen, of a high destiny, and vivid sensations of rapture.

It is only too easy to dismiss the *form* these experiences take as merely the physical phenomenon of sleep paralysis in fancy dress, the specific garb chosen by the unconscious mind to reflect the contemporary folklore. Yesterday it was demons, today it is Greys, tomorrow it will be something else – but not, say the psychiatrists, psychologists and neurologists, something else *entirely*. Because it is, to them, essentially the same experience, it will always manifest in similar ways, no matter what kind of mask it wears.

It must be admitted that there is something to be said for the sleep paralysis school of thought, although perhaps not nearly as much as its scientific adherents would have us believe. There are many phenomena associated with sleeping that can be interpreted as "magical" although they are physiological in origin. For example, certain type of dream – notably those experienced just before waking – are dramatizations of sounds that filter through from the outside into the dreaming brain. An alarm clock going off in the next bedroom – in some cases even the next house, for the unconscious mind has very good hearing – can be turned into a fire alarm, or a car backfiring in the street can become a dramatic explosion in a First World War trench, in vivid and often horrifying dreams. Because those dreams happen just before waking, they are instantly memorable and may be seen as prophetic in some way, particularly if they were very vivid. Yet they were only the result of the brain making a huge drama out of a very ordinary event.

But does sleep paralysis really explain all the terrors of the night? Scientists may scoff, but those of us who have been unfortunate enough to experience the depredations of poltergeists (noisy, destructive and invisible ghosts) know only too well that, whatever their *cause*, their ultimate manifestation is certainly not "all in the mind". If only.

However, there are other mental conditions that can give rise to apparently otherworldly visions (apart from the time-honoured methods of deliberately inducing them, such as shamanic techniques including taking hallucinogenic drugs, whirling to a drum beat and fasting). One that is still relatively little-known is Charles Bonnet Syndrome (CBS), named after the Swiss naturalist who, in 1760, wrote an account of the "amusing and magical" visions experienced by Charles Lullin, his grandfather, as a result of poor vision following cataracts. What intrigued Bonnet most about this condition was the fact that Lullin evinced no signs of mental disturbance and was fully aware of the illusory nature of the visions.[18]

For 200 years CBS was thought of as something of a medical novelty, a rare condition,[19] until a study in 1989 of 500 patients with visual problems revealed that no fewer than 60 of them suffered from it. Similar surveys discovered that it is relatively common in all age groups, although the average GP is unlikely to diagnose it, and often refers the sufferer to a psychiatrist, believing the symptoms to be signs of developing mental illness. However, the visions induced by CBS are always harmless – one woman saw cows in a field when it was empty, for example[20] – and others may see anomalous but mundane objects such as articles of clothing or pieces of furniture. Where the condition impinges on this investigation, however, is when the sufferer sees more fantastical objects or creatures, such as angels – or perhaps little people, UFOnauts in their tight-fitting one-piece suits, or aliens. What Bonnet called "miniature spectres" are relatively common and can be amusing, like "the pair of tiny policemen seen hustling a midget criminal into a miniature police van".[21] As Fortean commentator David Hambling writes in *Fortean Times*:

> 'The recurring reports of little people in CBS visions are intriguing. They appear generally to old people living in socially isolated circumstances, most commonly at twilight; it is hard to avoid a comparison with the traditional behaviour of faerie folk.

> Some CBS patients report hallucinations which vanish when they look away, just like leprechauns.[22]

However, most CBS patients quickly realize the illusory nature of their visions – one patient said "these hallucinations have nothing to do with me" – although some have a more marked realistic quality. Hambling admits that "its cause remains mysterious", although there are suggestions that sensory deprivation may be involved. He goes on:"Some . . . are classed as bereavement hallucinations: in one study some 69 per cent of widows interviewed had seen their deceased spouse."[23] But are all visions of a deceased loved one merely effusions of the brain? Over the course of many years of listening to people who have had such experiences it has become clear to me that many of them contain what psychical researchers call "veridical" or evidential material – in other words the vision of the loved one gives the recipient some information that he (or more commonly, she) did not know beforehand, and was unlikely to have known even subconsciously. This does not include the common experience of being told where lost articles – such as insurance policies or wills – might be, for it is more than likely that the subconscious mind of the "visionary" may already know that, although they may have forgotten it at a conscious level. Some of the more thought-provoking cases that have come my way include a widow whose husband materialized to tell her that their daughter, living in Australia at the time, had been in a car crash the night before and had given birth prematurely as a result. He added that although the doctors were worried about the baby's prognosis, the child – who, he said, was a boy – would survive. All of this turned out to be true, but where had it come from? From the hard-wiring of the widow's brain – or the discarnate spirit of her husband? Or from another helpful entity *pretending* to be her husband? Sometimes the principle of Occam's Razor – the simplest and most obvious explanation is often the best – makes most sense.

Interestingly, one Neil Ogden of Wirral, Cheshire, wrote to *Fortean Times* in response to David Hambling's article:[24] "A short while after my father's death, he appeared to [my mother]. She said that rather than being scared, she felt immensely comforted by it, and it helped her through her grief. Hambling suggests that these 'bereavement illusions' are a type of CBS. I would argue that they are separate, and just a part of the natural grieving process."

Ogden adds: "Of course the other possibility is that it wasn't a hallucination, and that my father really did appear to her. For some reason, I find this possibility quite reassuring. Then again, perhaps the reassuring nature of such visions is the reason why the subconscious creates them in the first place."

Hambling suggests that where the vision is impaired, CBS may be caused by a "test signal" in the brain, which "originates from visual association cortex area 19 and is usually imperceptible.[25] In CBS patients where there is nothing to drown it out, the test signal produces 'release hallucinations' in the understimulated brain."

He also suggests that CBS may be a result of confabulation, where the brain supplies the missing details of a picture. You catch a glimpse of what looks like a body sitting in a chair and your brain adds a face – then you realize the "body" is just your coat. And there you were thinking you'd been visited by aliens (or your long-gone Grannie, depending on taste and inclination)!

No one doubts that the brain is an amazing organ, and that exploring its as yet largely uncharted territory is as exciting, if not more so, as finding a new continent or landing on Mars. Yet to study the *brain* in isolation, to ignore the evidence of its interaction with the *mind*, may be a serious mistake.

In recent years, several psychologists and neurologists have made much of the physical condition of "Temporal Lobe Epilepsy" (TLE), in which something akin to an electrical storm flashes through the temporal lobes at the front of the brain, causing massive epileptic fits – and extraordinary mystical or psychic experiences. In many cases the fits induce such a sense of universal connectedness and a rapturous awareness of the beauty of the world that the "sufferer" may seriously consider refusing long-term treatment, although some who experience this condition are also plummeted into the complete desolation of the abyss of hell.

One famous TLE sufferer was the artist Vincent von Gogh,[26] whose increasingly intense religious fervour – one symptom of the condition – was described as a "zeal almost scandalous". When in the grip of his TLE, he would visit three or four churches in one day. Religion consumed his whole life.

Unfortunately, because TLE presents scientists with apparent mystics – presumably in many cases the only mystics they will ever meet – this has prompted them to make a massive leap and declare that *all* religious or unusual sensations are the by-products of the synapses firing in a certain abnormal way. It has even been suggested that in discovering the effects of a massively-stimulated temporal lobe, they have found the "God spot" – that the experience of God is no more than a quick blast of electricity through a lump of pinkish-grey meat inside a small transient cave of bone, the individual human skull.

Professor Susan Greenfield, a neurologist of Oxford University has declared[27] that while it was once believed that thoughts somehow came in from outside (although this is a somewhat patronizing simplification of how our ancestors thought of the processes of cognition and inspiration), we now know that thoughts – and everything that makes us what we are – come from inside our brain. She claims that this is an exciting and uplifting concept, but while it may be to herself and her peers, there are some who find it deeply depressing that top modern scientists should ignore the huge amount of evidence for the power of the *mind* in favour of a mechanistic science that, while undoubtedly helping those with mental problems, is as limited in scope as the pulley-and-lever approach of Isaac Newton,[28] and arguably sets back our understanding of the human potential by many years.

One of Professor Greenfield's colleagues, Dr Michael Persinger of Laurentian University – incidentally, a former CIA employee – boasts how he can artificially induce almost any apparently otherworldly experience, from out-of-the-body experiences to alien abduction in the laboratory, simply through stimulating the frontal lobes of the brain with electromagnetic energy. He has successfully induced weird sensations in his volunteers, from a feeling of the presence of entities to a perception of tunnels and bright lights – just as in alien visitation, particularly of the "bedroom visitor" variety, and the typical Near Death Experience. However, Dr Persinger admits he has never yet succeeded in inducing a clear-cut religious experience, although he implies that it is only a matter of time before he does.

There is no doubt that he has had dramatic success in the

laboratory, but both he and Professor Greenfield seem not to notice the basic flaw in their logic. They claim that because almost any mystical, quasi-mystical or religious experience can be artificially induced by electrical stimulation, *that is always what they are*. That is like saying that because an electrode applied to certain areas of the brain will always – predictably – induce sensations of terror, that there are no *genuine* causes of terror in the real world . . . Putting Persinger and Greenfield in a cage with a hungry tiger and telling them it was only a by product of their brain chemistry would no doubt be a fairly dramatic way of persuading them to think again. Ours is not entirely a chemically induced virtual world – not yet.

Once again, however, believers and sceptics take up their time-honoured positions at the extreme ends of the spectrum, ignoring the possibilities – even the probabilities – that at least some of the answers may lie in the twilight world between them. True Believers declare that all extraordinary experience comes from "out there", from God or the angels without the contamination of flesh, while the materialist-rationalists believe that the evidence points to the brain as the seat of all experience, personality and what used to be called "soul". There are no fairies, no aliens, only abnormally functioning areas of the brain.

An interesting point was raised by a neurological nurse, Mark Glover from Bootle, Liverpool about the experience of TLE.[29] Mark pointed out that "there does not appear to be any evidence of typical CE4s [close encounters of the fourth kind – abductions] in people who are clinically classed as sufferers . . ."

He goes on: "As a nurse at a centre for neurology and neuro-surgery, I have worked with these patients and investigated to some extent the effects the condition can have on their perception of reality. I have as yet to find any indication that the perceptual aberrations the conditions can engender are anything remotely like the CE4 experiences."

Mark points out that he has had extensive experience of working with the mentally ill and of caring for "patients undergoing electro-convulsive therapy (ECT) . . . [which] involves a significant electric shock being applied to the brain via the temporal regions to relieve severe depressive disorders. Again, these patients do not appear to have the sort of perceptual disturbances that can be compared with CE4s."

Indeed, further research revealed that such patients do not generally experience *any* form of hallucination. Mark concludes: "One would expect that if TLE and ECT are instances of recordable electrical disturbances within the brain involving the temporal lobes, then at least some CE4 effects would be noticeable – yet this is definitely not the case. Unlike CE4 experients, patients with TLE who have hallucinations also seem to realize that their experience is just hallucinatory without any alteration to their view of reality."[30]

No doubt the sceptics would feel they can afford to ignore such inconvenient facts. After all, surely it is only a matter of time before new evidence comes to light that proves that full-blown and hugely detailed CE4s can be induced in the laboratory! But the concept of the personality and the seat of all experience residing in the brain is central to the materialist-rationalist-atheist. To them, there is no spirit, only electrical storms in the temporal lobes, therefore – and to them, this goes without saying – there cannot be an afterlife or any experience at a distance, for when the brain dies, that is the end of all experience.

Most scientists agree with that position, for they have no convenient intellectual, scientific or philosophical model that would allow them to approach the problem from a different angle while maintaining their academic respectability. However, one eminent British biologist, Dr Rupert Sheldrake[31] has sought to overcome this difficulty by providing an extraordinarily simple but elegant analogy to explain the difference between the *mind* and the *brain*, which goes like this:

Imagine intelligent people from a time before our modern era, such as the Victorians, when presented with their first experience of television. What would they make of the little moving pictures on the box in the corner of the living room? Having first satisfied themselves that the people they see on the screen are not actually inside the set, they might conclude that the pictures are actually *made* by some mechanical device inside. This seems a reasonable conclusion, particularly if they experimented by taking an axe to the set: a couple of swift blows and, if a picture still remains, it will be badly distorted, the controls will no longer work so that the channels are all mixed up, and in many cases, all that is left would be the hiss of static.

The television set is the brain. Brain damage will, of course, produce the human equivalent of getting the channels mixed up, the static resembling slurred or nonsensical speech, severe cognitive problems, coma and even death. Yet extending the television set analogy further demonstrates another dimension in understanding what may be the true relationship between the brain and the *mind.*

No one would suggest that in taking an axe to the television set actually damages the distant studios, programme controllers, executives, scriptwriters, producers, actors, crew and so on. Of course not. That would be ridiculous. They simply carry on as usual, totally unaffected by the damage wreaked on a piece of furniture many miles away. In this analogy, the ideas that take material form in pictures on the screen, and the people whose ideas they are, are the *mind.* The mind impinges on the brain, the television set, using it just as the output of the people in the studios creates pictures on the television. And just as human creativity will outlast the life of one television set, so the mind will continue beyond the death of the physical brain.

Of course an analogy, no matter how attractive, is simply a logical argument dressed up in semantics: for an argument to be totally persuasive, the basic premise has to be sound. It is at this point that the sceptics will impatiently wave away all analogies of television sets and distant studios, rightly demanding to be shown *evidence* that the mind is separate from the brain. In fact, this is already abundantly available, but not normally in university libraries. Look instead in the literature of the largely ignored and despised "Cinderella science" – parapsychology or, as it used to be known, psychical research. What is found there is so extraordinary that it challenges the very concepts of what we believe to be reality – and perhaps provides some insight into many of the phenomena associated with UFOs.

Once again, however, the evidence suggests that we are dealing not with some external spiritual or psychic force, but with some realm or dimension that requires *both the mind and the brain in order to manifest.* It is a partnership between the true phenomena from "out there" and an internal process of creativity, or perhaps the creative force is necessary in order to produce the gateway through which the phenomena can materialize. This is a subtle, two-way, dynamic process that may underpin much of what we call artistic inspiration at one end of the scale, and at the other provide the means of contact with the gods, of opening the "stargate".

However, there are still many questions to be asked, some of which carry very disturbing implications. Extrapolating from the mind/brain/television set analogy, one is left wondering who are *our* programme planners? Who writes our scripts?

THE UNKNOWN POWERS OF THE MIND

In the early months of 2000, there was a massive UFO flap over Italy: unexplained lights and strange objects being spotted all over the country, culminating on the evening of 23 February. Hanging on the western horizon, and astounding hundreds – perhaps thousands – of witnesses, were two bright lights "like halogen lamps" (or "like a car parked in the sky with two headlights on"). They were caught on camera and videotaped many times over, and reports of their appearance were submitted to air authorities, the media and the police. The Italian Centre for Ufological Studies (CISU) was overwhelmed with the number of calls from the public.

However, exciting though this was, as *Fortean Times* pointed out:[32] "The lights were indeed extraterrestrial in origin, though there was nothing mysterious about them. That evening saw a spectacular conjunction of Jupiter and Venus, visible just after sunset throughout Europe. Saturn was also visible nearby. In Italy, the combined brightness of the two planets was amplified to a startling degree by clear skies, and the effects of atmospheric refraction".

Yet despite this all-too-neat explanation, the rare visibility of the planets does not explain all of the UFO sightings reported in Italy during the early months of the year 2000. It is as if the misidentifications made the populace more open to genuine phenomena. Indeed, this is seen time and time again not only where UFOs are concerned, but in other areas of the unexplained: misidentifications and mistakes seem to usher in a wave of apparently genuine phenomena, or at least less easily explained events. But the misidentifications have to be wholeheartedly accepted for this to work: *obvious* fakes do not inspire mass paranormality.[33]

We have seen how even the godfather of all modern UFO cases – the Kenneth Arnold sighting in 1947 – may have been nothing more than the misidentification of a flock of large white pelicans, but there is often another, more elusive factor, which continues to make the UFO debate tantalizing and not infrequently maddening for the researcher. In the case of the mass Italian sightings of UFOs, could it be that, deluded though it was on this occasion, the mistaken belief had itself actually kick-started a genuine phenomenon – or at least created mass hallucinations for which there was no easy rational explanation? Although everyone knows how easily mass hysteria can spread, and the harm it can do – witness the deliberate use of such psychological tools by the Nazis when whipping up anti-Jewish hatred – what is less well known is the usefulness of a little artifice in the creation of genuine paranormal effects. Dubbed "artefact induction", this simple procedure was discovered in the 1970s by the late British parapsychologist Kenneth J. Batcheldor when he accidentally nudged a table with his knee during an experiment into table turning (where a table moves, sometimes violently, apparently of its own accord under the light touch of the "sitters").

A handful of experimenters sat in the dark waiting for the table to move, but when Batcheldor accidentally knocked it – and it moved – he had no time to apologize before the others expressed their amazement and joy that the experiment was working. The table had successfully moved all by itself! Sensing that this mistake was important, Batcheldor sat tight, waiting to see what would happen. Immediately after his colleagues "oohed" and "aahed" with wonder and pleasure at the apparently paranormal table-tilting, *a genuinely unexplained movement occurred.* In other words, his accidental nudge seemed to have induced authentic psychokinesis (PK, or mind over matter) in the group. Batcheldor soon came to realize that a little deliberate cheating oils the mysterious wheels of the paranormal magnificently, perhaps simply by triggering the psychokinetic powers of the group through the release of *belief*, as if showing what it *would* be like actually makes it happen.

Batcheldor gradually worked out a simple formula for inducing paranormality. He discovered that there are two major requirements for successful unexplained events to occur: the people involved must have, as intensely as possible, *belief* and *expectancy*. They must believe that events that challenge the laws of physics are possible, and must expect that those things are imminent. In other words, intellect will get you nowhere where the capricious world of the weird is concerned. It is an almost atavistic excitement in the presence of *magic* that wins through. (This is one of the main reasons why few amazing events happen to sceptics, although it is also true that there are none so blind as those who won't see – literally.)[34] With those twin psychological requirements firmly in place, and all doubts banished – Batcheldor encouraged his colleagues, instead of wondering whether some happening was paranormal to decide that it definitely was[35] – wonders were free to take place. Tables would rock, rapping sounds would come from all over the room, strange lights float about. It seemed as if Batcheldor had stumbled upon the secret of magic – and although in one respect it *is* "all in the mind", it is a child's mind, with its uncompromising belief and readiness for miracles, and the thrill of making magic happen that is the best for creating miracles.

Research[36] has shown that children are indeed considerably more psychic than adults, and the younger the child the more psychic he or she is. Perhaps the great English Romantic poet William Wordsworth summed it up perfectly when he described children coming into the world "trailing clouds of glory".[37] This notion of the child as magically knowing has been celebrated by many authors, but with particular success by J.M. Barrie[38] with the ultimate story of the magical powers of the child, *Peter Pan*, in which adulthood is seen as a curse, the loss of the magic, a sort of psychic crippling. (Grown-ups can no longer fly. Worse, they can't even remember that they used to fly.) Another author who, less seriously – but even more successfully, at least in material terms – encapsulates the magical potential of the child is, of course, the publishing phenomenon of the early twenty-first century, J.K. Rowling, with her Harry

Potter series. (Although the child/adult divide is not so clear cut in these books as in *Peter Pan.*)

Anyone with any experience of young children knows that they *claim* to see things that are not seen by adults. However, most grown-ups dismiss the stories of invisible best friends and fairies and so on as "just imagination" and actively discourage such anti-social fantasizing, certainly as the child grows older and rational thinking, in the form of formal education, is the order of the day. There are good and sound reasons for this: those who defy the conventions of society tend to have a tough time of it, as do their families and associates. In the modern western culture, it is rarely a good idea to be the neighbourhood mystic or visionary – even in societies where religion still reigns supreme, as in rural Ireland, those who claim to be on more intimate terms with the divine than the priests are looking for trouble. But in the rush to knock "all that childish nonsense" out of the youngsters' heads (sometimes only too literally), the baby is often thrown out with the bathwater. Made to conform and stop "telling lies" about the spaceman in the wardrobe or the strange little people in the hedge, the message is forcibly conveyed that magic does not happen (or at least, only in Harry Potter books) and the laws of physics are intractable. What goes up must come down. Nothing in the material world can be moved, shaped or created *directly* by the power of thought alone (although of course every great invention began with a thought). The individual on the whole determinedly puts away childish things, only to be disturbed and confused by glimpses of that other realm in dreams, Near-Death Experiences, paranormal happenings such as hauntings, and strange things seen in the sky. Perhaps the more psychic the person and the harder such characteristics have to be suppressed, the greater the potential to experience weird happenings in later life. It is as if the otherworld must get through at all costs. It will find any way to communicate its one central message: simply that there are more things in Heaven and Earth than are permitted in your world-view, so *admit it.*

True Believers, the mainstay of the nuts-and-bolts UFOlogical community, would argue that being psychic, or open to the paranormal at least, has absolutely nothing to do with seeing UFOs. To them, they are real, solid craft that anyone can see. Perhaps this is so, or at least on certain occasions. Yet even if all UFOs are as real as a tax demand, it is likely that only some people would see them, or that what was perceived would vary wildly from person to person, depending on their expectations and beliefs. Seeing even what is indubitably *there* is not always an easy matter, as can be seen from an elegant – and amusing – experiment carried out by Professor Arthur Ellison, a lecturer in electrical engineering at a North London college in the late 1970s. He had arranged that, as he was giving a talk to an extra-mural group on the subject of psychical research (as a former Chairman of the prestigious Society for Psychical Research), a bowl of flowers on the table in front of him would rise into the air, hover, then slowly return to its position on the table. He would not notice, but carry on speaking as if nothing had happened, keeping his eyes on the audience reaction. The only magic involved was that of cunning electromagnetic wizardry, but the audience did not know that.

Professor Ellison gave his talk, and, as planned, the bowl of flowers rose into the air, hovered, then returned to its place. At the time there was no discernible reaction from the audience, but at the very end of question time someone timidly put up a hand and murmured hesitantly that they had seen the flowers do something odd during the talk and had anyone else noticed anything? Once the ice had been broken, another member of the audience spoke up, saying that they saw the bowl rise into the air, at which the resolute sceptics among them howled with derision, claiming that nothing had happened at all, although one of their number did admit that he thought the bowl had "wobbled a bit". At the other end of the spectrum, someone alleged that they had seen "spirit hands" lifting the flowers up!

This perfectly illustrates the problem with perception. Human beings – even educated human beings, or perhaps *especially* highly intellectual people – do not tend to see the unvarnished truth that lies before them, but transmute it instantly and totally without conscious input into something that fits their world view[39] – and in that may well lie the secret of many cases of interaction with UFOs and their alleged occupants. (Perhaps there are hundreds of them darting about the skies all the time, but only certain people under specific conditions, are able to see them. Perhaps, too, the heavens have always been replete with UFOs of one sort or another, but until the age of space exploration humans have not been equipped with the *belief* that such things are possible, nor with the *expectancy* that they could happen at any time. The UFOs may have always been there, but because no one believed such things could exist, they remained invisible.)

However, the phenomenon is manifestly not merely one of misperception, the bloody-minded refusal to face the facts and acknowledge what's really there. Interwoven with the sightings of apparent craft are dream-like scenarios, absurdities and experiences of extraordinary surreality, as vivid as life itself, as if the boundary between the conscious and unconscious mind has been temporarily breached. Because of this element, a study of UFOlogy cannot successfully be undertaken in isolation: like it or not, the strangely capricious world of the paranormal has to be brought into the equation. True UFOlogy is not for the faint-hearted or the close-minded, although while there are perhaps few of the former among the dozens of UFO groups worldwide, there are certainly enough of the latter.

Yet what do we mean by the "paranormal"? It is the category of unexplained phenomena that is magical today, but which will probably become the mainstay of tomorrow's science. Two centuries ago peasants in France ran terrified to tell local scientists that stones had fallen out of the sky, but were arrogantly dismissed because everyone knew that there are no stones in the sky. Yet had the scientists bothered to get out of their armchairs and take a look, they would have had a head start in the understanding and scientific evaluation of *meteorites* . . . Similarly, Victorian explorers of Africa refused to take seriously travellers' tales of strange hairy humanoids of enormous strength – until they encountered gorillas for themselves.

IMAGINATION: HUMANITY'S SECRET WEAPON

To most people the word "imagination" conveys something unreal, unworthy of notice, even mendacious, but at its most powerful, imagination is the ultimate creative force. There is abundant evidence that the human mind can actually *create* visible phenomena, not only somewhat pointlessly bending metal, but also seeing at a distance, bestowing consciousness on inanimate objects . . . and *creating hallucinations that are seen by others*. This may help explain the peculiarly changeable nature of some UFOs, and the bizarre capabilities of the "Visitors".

In the early years of the twentieth century, the great explorer Madam Alexander David-Neel went to live for some months with the monks of Tibet, high in their mountain temples, to learn the secrets of their legendary magical powers. While there she witnessed extraordinary phenomena, including the creation of *tulpas*, or projected thought-forms that were visible to others, and determined to try to create one for herself. She decided on a monk, and, according to the prescribed rites, concentrated and meditated upon his image, vividly imagining his every detail so that she saw him fully formed in her mind's eye. Then, after six months of this intensive procedure, she became aware of a vague monkish outline, like smoke, drifting about. Gradually this filled in as she continued to concentrate on his image with all the power of her mind, and in a short time there, standing in front of her, was her monk, happy and jolly. She had created a man with imagination alone. But having created him, she forgot to maintain him with her mind-power. Separated from her, he took on a life of his own, becoming so malevolent that she had to destroy him, dissolving him gradually with her mind. The importance of this story is not so much that Madam David-Neel achieved this remarkable feat, but that *other people saw and interacted with him*.

Interestingly, one Mark Sheridan, of Auckland, New Zealand, sent this information to *The Unexplained* in 1981, after having read an article about *tulpas*, which may shed further light on the phenomenon: "Thomas Bearden, an aerospace engineer in Huntsville, Alabama, USA, believes that if people do not wake up when having terrible nightmares they would die. Dream monsters and other psychic hallucinations, he says, can acquire weight, volume and even a will of their own."

Mark Sheridan goes on to say that Bearden bases his belief on the fact "that the human brain is full of electrically charged particles that are analogous to 'mind stuff'. As he puts it: "When a person concentrates, this mind stuff condenses, getting thicker and thicker, until it becomes solid stuff, or real matter.

"Thus he believes that dream monsters may take on their own detached reality in a way that parallels the formation of *tulpas* through intense concentration, building the form bit by bit into his physical reality. Consequently, Bearden believes that the Tibetan yeti (or Abominable Snowman) is an example of a *tulpa* created by the inhabitants of a particular area in order to protect their sacred land."[40]

The creation of *tulpas* is a well-known magical discipline, and has underpinned many of the dark secrets of occultists over the centuries, especially the alchemists, those medieval and Renaissance scientists who dared to try to play God long before the advent of Dolly the sheep and the panoply of genetic engineering. The overwhelming ambition of many alchemists was not so much the fabled search for the Philosopher's Stone,[41] which would, they believed, make them immortal, nor the quest to turn base metal into gold, but the creation of life in the laboratory. These were the real Dr Frankensteins, although we know little about their success rate, if any, largely because they were wary of committing their experiments to paper – for theirs was a forbidden practice, and if caught they faced the most agonizing of deaths as heretics and blasphemers. Yet it may be significant that there were many rumours of success, however ephemeral, in animating *homunculi*, or little men said to be made out of anything from a mish-mash of human remains to clay or wood. Sometimes the *homunculi* were believed to look like aborted foetuses (possibly for very good reasons), kept in glass jars, and given life by the master alchemist in order that they become his servitors, his soul-less slaves. It does not take much imagination to see how such piteous creatures could look like today's cultural icon of the Grey alien, often described as having no souls, and seemingly the servitor of some unseen but greater power. Also it may be significant that photographs of the mummified foetuses buried with the boy-king Tutankhamun look disturbingly almost identical to the big-eyed, large-headed but spindly Greys.

It may be argued that the alchemists of old were merely deluded at best and charlatans at worst, and that their activities have no bearing on today's sophisticated world. But perhaps we should not be too quick to relegate alchemists and their works to the garbage of history. Not only were they the forerunners of today's cutting-edge scientists, but there are very good reasons for believing that they *may* indeed have created a form of life – which could explain some of the creatures now identified as aliens – for there are others, apart from Madam David-Neel, who have stumbled upon the secret of doing so.

Sometimes – albeit very rarely – this astonishing gift comes naturally, rather than as a learned technique, even to adults. In the 1970s an American woman known simply as "Ruth" visited the psychiatrist Dr Morton Schatzman, a fellow countryman, in London where they both lived. She was deeply distressed by the activities of her father, who had sexually abused her as a child: he followed her everywhere, even when she took a bath. Yet she knew he was not "really" there at all, because his physical self was thousands of miles away in the United States, but she could still see him. He was so real that, as she said, "I can smell him . . . I can even count his teeth. . . .[42] Life was becoming a living, waking nightmare for her, so she consulted Dr Schatzman in the hope he could cure her of this extraordinary aberration, thus beginning one of the most thought-provoking patient-doctor relationships on record.

He discovered that Ruth had an incredible talent. She could summon up absent people with the power of her mind, with such realistic detail that, for her, it was as if they were really there. Dr Schatzman advised Ruth to accept the visionary presence of her father, but be assertive and coolly tell him to go. Instead of being frightened by him – acting as his victim – she was to take control. This worked perfectly: on one occasion when he turned up while she was in the bath she asked him to pass her the towel! Soon her father had gone, but the psychiatrist was too fascinated by Ruth's wild talent to let her go without investigating it further.

It appeared that Ruth could conjure up, to her, totally real doubles of people already in the room – she did this on several occasions with Schatzman himself – holding conversations with them. In fact, the psychiatrist encouraged her to develop her gift without fear or shame, which he considered to be an unusually pronounced form of creativity. Soon she manifested "copies" of her best friend when she was lonely, and once or twice her husband when he was away. (She reported that the sex was the best ever!)

However, on at least one occasion the double she had summoned up *was seen by other people.* It appears that Ruth had a natural ability to perform what others consider to be a magical act – that of creating the illusion of life.

The truly exciting – or scary – aspect of this technique is that it is not necessary to travel to Tibet or be born with a curious gift, for research has demonstrated that, with a little trial and error, some knowledge, and a lot of Batcheldor's *belief* and *expectancy*, almost anyone can animate the inanimate.

In the 1970s in Toronto, Canada, psychical researcher Dr A.R.G. Owen and his wife Iris, together with a group of fellow researchers, set out on what was to be the most astonishing experiment in the history of parapsychology – arguably, one of the most important in the whole history of science. This was the Philip Experiment, a master class in how to create a ghost . . .

The aim of the group was to create a shared hallucination of a character they invented, whom they called Philip. Knowing that it was important to have all his details crystal clear in all their minds, they spent time agreeing on the details of his life – in seventeenth-century England – and an artist among them drew his portrait, so they could see him more clearly in their minds. Once they had him firmly ingrained in their imaginations, it was time to put him to the test.

Sitting around a little card table, the group asked the invisible Philip basic questions about himself, to which he responded by laboriously rapping out the answers – one rap for yes, two for no, or once for the letter "A", twice for "B" and so on. They had made contact with a non-existent ghost.

The phenomena associated with Philip seemed to work best when the group was being light-hearted, telling jokes or singing rousing songs, but imagine their amazement when the card table around which they sat began to jump up and down in time to the music! From that moment on, when "Philip" became synonymous with the table, rather than the fabricated English gentleman, events became extremely surreal. "Philip" (the table) began to act like a puppy, jumping up at people and even chasing Dr Owen to the door on one memorable occasion . . . Admittedly, this story never fails to crash the "boggle barrier",[43] but that, from all accounts, is actually what happened. There was more to come.

Over the months Philip the Table had become something of a *cause célébre* in the Toronto area so it came as no surprise when the group was invited to take him to the Toronto Television studios for a demonstration of his abilities before a panel and a live audience. Placed in the body of the studio he began by obediently rapping out answers to questions but soon, apparently, became bored, taking it into his little wooden head to "walk" around, as a table might walk in a cartoon. To the gasps of the audience, Philip even managed to cope with the three steps up to the platform, where he basked in his moment of glory.

That was the end of the Philip Experiment, which, despite its astonishing outcome, was not – at least technically – a success. The group had not succeeded in creating a shared hallucination of Philip the invented ghost, but what they had done was astounding enough. They proved that with the right attitude and a great deal of patience, a humble card table can become a star, walking, "talking" and even seeming to have the beginnings of a personality. Once, when the group were enjoying a session around the table, which was behaving boisterously, jumping up and down too energetically, one of the people said, "We only made you up, you know." At that all phenomena stopped abruptly. Philip was sulking. They had to work very hard at bringing the level of belief and expectancy back up to an effective level once again before he deigned to talk to them once more.

These jaw-dropping cases from the annals of parapsychology and – in the case of Ruth – abnormal psychology, reveal what must surely be the tip of the iceberg of human creative potential. One day perhaps these feats will be considered impossibly crude and embarrassing, but at the moment they should be celebrated as giant leaps for mankind in the evolution of the human mind.

These latent abilities may create certain aspects of UFO-related phenomena, reifying or making manifest in the "real" world cultural expectations and personal beliefs. Perhaps an unexplained light in the sky triggers this ability, creating temporary aliens and structured craft, some of which even leave traces behind like burn marks. (Patrick Harpur calls this phenomenon "Supernatural Branding."[44]) All the mind has to do is follow the guidelines provided by the classic cases, perhaps creating the physical marks by a process akin to that which produces stigmata, or the apparent marks of Jesus's crucifixion.

Stigmata have long been recognized as hysterical symptoms rather than – as many supposed – signs of personal holiness, yet they are a very significant psychokinetic talent. Those affected show a variety of symptoms, but characteristically bloody holes appear through the feet and hands (although it is impossible to crucify someone in this way – the palms would tear. Jesus was almost certainly crucified with nails through his wrists), just as depicted in statues and pious paintings. They may also present a slit wound in the side, as if pierced with a spear, marks of flagellation or the crown of thorns. In most cases this is extremely painful: it is as if the stigmatic is living through the Passion, and often there is a great deal of blood. Yet they are *not* actually being crucified. *Nothing* is being done to them, yet time and time again priests and independent observers have witnessed these marks appearing on smooth unblemished flesh. It seems that stigmata are the result of a form of self-hypnosis, brought on by intensive meditation on Jesus's suffering, and usually they appear at a specially meaningful time, such as on Good Friday, when Jesus was crucified. In other words, the stigmatic is evincing in the most dramatic and grisly form, albeit unwittingly, Batcheldor's two requirements for bringing about the "impossible" – belief and expectancy.

But has the matter-manipulating human mind, working either solo or collectively, really created the alien abduction scenario? Almost certainly a great many cases, particularly once the myth became established, can be ascribed to psychokinesis, triggered by the desire to be part of something extraordinary, to be famous, even to be a victim, rejoicing in the negation of responsibility for life's ills.

Perhaps, too, after the first wave of reported encounters – whether they were "genuine" or due to other causes – the phenomenon took on a life of its own, like a *tulpa*, or like the little card table called Philip. When thought-forms are built up by the concentrated mind power of a small group – or, in the case of Madam David-Neel, just one individual, they can become obstreperous, if not downright nasty. Maybe Dr Owen's Toronto team called a halt to the Philip Experiment just in time: who knows what would have happened if they had continued feeding "him" with the energy of their belief and expectancy? Philip could have become the only delinquent card table in history!

The aliens began as the Space Brothers of the Contactee Era, transmuted into the equally benign Greys of *Close Encounters of the Third Kind*, transmogrified into Whitley Strieber's morally ambiguous "Visitors" before taking on the role as coldly calculating abductors and rapists on a mass scale. Did a real race of beings really evolve so fast – or were they creations of the Collective Unconscious, becoming what we most hoped and feared, whichever emotion was felt with the most intensity? For after the extreme terror came another morally ambiguous phase, which has now almost completely been reformed into the aliens as *saviours*, making Chosen Ones of their erstwhile "victims". The Visitors have not changed, it is asserted, but our understanding of them has evolved. Certainly, the magical or occult concept of thought-forms imbued with the life of belief and expectancy does seem to fit the bill, but there still remain questions about the underlying nature of the scenario, the original experience that gave rise to the basic template upon which the thousands of other cases are based.

However, encounters with non-human entities did not begin with Whitley Strieber's archetypal experience, or with George Adamski, or with the aeronauts in their mysterious airships.

ABDUCTION AS SHAMANIC INITIATION

As we have seen, it has not escaped many of the more open-minded UFOlogists and abduction researchers that the core experience is often very similar to the classic initiatiory crisis in which the shaman (witch doctor or tribal healer) pushes through personal pain to a new level of spiritual reality. Abductees report escalating terror as they are repeatedly taken, often from childhood, by alien creatures who come uninvited into their homes or cars, destroying their peace of mind and, in some cases, pushing them to the very edge of sanity. Yet in the end, many abductees report not only a growing love for their abductors, but also a quantum leap in spiritual awareness as if a gateway has opened in their souls. Just as in the terrifying ordeal of the shaman, in which he is (in spirit) torn limb from limb then magically reassembled before climbing the tree to the realm of the gods, the abductees are put through hell – suffering physical pain and humiliation at their alien captors' hands, and the unimaginable horror of being removed from not only their family and friends but also their own *species*. This is a desolation that can only be guessed at.

Yet the shaman's agony is for a purpose: his ascent into the magical realm (which is invisible to ordinary, untrained tribal members) enables him to seek information from the strange beings that he finds there – bird- and animal-headed beings, similar to the gods of ancient Egypt.[45] However, as in fairy tales, the entities will only give him the answer to specific questions, such as how to cure a certain illness, or where to find water in a drought: no elaboration will be given or offered. And with each subsequent initiatory ordeal, the shaman will reach a higher level of the magical realm, where the information is "purer" – uncontaminated by his own mind. But even highly trained shamans know that the strange realms in which they move while entranced contain trickster spirits masquerading as benign helpers and guides, and sometimes they encounter downright evil entities who seek to take their souls. It is only through many years' experience and much personal suffering that the shaman learns true discernment.

In abductions, however, the experiencer is a raw recruit to a system of enforced initiation, thrown in at the deep end of an ordeal for which there is no preparation and no kindly elder to teach and protect. The road to spiritual awakening is rocky indeed, not least because there is no training provided that enables the individual to learn how to discern between the good, bad and merely fraudulent entities he or she encounters along the way. Perhaps this lack of training is why contactees' and abductees' "revelations" seem to begin so well then go off the rails: Whitley Strieber's adventures struck a chord with many people, but then he supported the idea of the Face on Mars and the UFO trailing the Hale-Bopp comet, although both have been disproved. Similarly, the prophecies given by alien beings to the likes of Dorothy Martin began persuasively enough, then ended with her complete humiliation when the world failed to end as predicted. Perhaps had these people realized the shamanic nature of their experiences – and sought help from those who know about such matters – they would not have brought such embarrassment on themselves, learning to sidestep neatly the traps of the shamanic realm in which they found themselves wandering.

It has been speculated – by British author Marisa St Clair, for example – that the Near Death Experience (NDE) is also an initiatory, shamanic ordeal, although in most cases it is low on negative content and high on rapture and enlightenment. It is simply the other side of the shamanic coin: the bliss of heaven goes hand in hand with the desolation and agony of hell, and they are both found in shamanic lore. Indeed, the similarities between the NDE and alien abductions are often striking: both involve seeing – and going towards – a bright light, encountering strange entities and being given mystical enlightenment that can include personal and global predictions. Of course there are differences: in the NDE the individual is not "taken" as such, although they have little or no choice in what happens to them, and they are actually clinically dead (if only for a matter of seconds). Yet the core experiences are very similar – both of which are now firmly ingrained in our cultural expectations. Only the "window dressing" is different: we encounter Grey aliens on the one hand, and religious figures on the other. Both occur either as a result of a personal crisis or actually *induce* a personal crisis, as if the person concerned needs a sudden jolt, an ontological shock, in order to make a quantum jump into another level of being.

It is significant that one of Dr John Mack's abductees, Edward Carlos, attributed the teaching of the "light beings" who took him to a form of shamanism, explicitly stating, "I am a shaman/artist/teacher", and adding, "They are teachers, [but] they are really interested in learning of us."[46]

Carlos believes that the "abduction" process is a transformation of the spirit, saying, that the shaman, "uses techniques to alter the psyche, and what the shaman is doing is playing with the emotional discourse between teacher and community, between shaman and student, between the person who travels and the person who remains or who lives a life here. I teach by emotion and experience."

Another of Dr Mack's cases was that of the thirty-nine-year-old Dave, a karate student whose spiritual quest largely centred on the Pemsit Mountain in Pennsylvania, sacred to Native Americans, with whom he felt a strangely strong kinship. As Dave's tangled story of repeated abductions – complete with humiliating examinations and the taking of sperm samples – emerged from several sessions of hypnotic regression, he began to see a pattern in all the agony. This was intensified when Dr Mack took him back to two past lives, one of which was lived as a Native American, and Dave realized that the familiar female alien he met during his abductions had always been with him. Everything began to fall into place. Dr Mack says: "Native American spirituality, shamanism, strange powers of nature, altered realities, Chi [the life force, which Dave defines as "the force which pervades the universe from which reality arises"], karate, the mastery of dreams, UFO abductions, past life experiences, and a multiplicity of synchronicities are all part of a mysterious puzzle for Dave whose pieces . . . he is learning to put together."[47]

Dr Mack points out that "the abduction phenomenon cannot be considered in isolation", and stresses the relationship between the unfolding abduction scenario and its roots – at least in this case – in Native American shamanism. Perhaps it is particularly significant that whereas today many locals think of Pemsit Mountain as a UFO base, previously it was the place where the tribal shaman would go to seek visions from the spirits. Is it the centre for a phenomenon of the spirit that changes only in the outward trappings, such as alien abduction?

Perhaps those who favour the nuts and bolts ETH may be missing a large part of the jigsaw: the transformative power of the UFO/abduction phenomenon seems to be the key to the whole experience. To leave what fundamentalist ETHers condemn as "mysticism" out of the equation may be missing the point.

Perhaps the alien abduction phenomenon is part of human evolution, not literally, as claimed, in the sense that half human and half alien hybrids are being created, but because the experience itself is moving us on to our next level. Humanity, especially in the West, is profoundly materialistic and even our religions are stripped of personal revelation or *gnosis* (the individual's knowledge of God at a profound level) that takes us beyond the confines of dogma into something much bigger and more intensely magical. Indeed, most religions frown on such experiences, even accusing those who are in touch with their shamanic selves as heretics or blasphemers. Narrow religions are often, in this way, worse than no religion at all, for they put constraints on the imagination, which may result in an explosion of angry, threatening phenomena . . .

Only by seeing the UFO experience as part of a much wider spiritual reality will any real progress be made in the understanding of the human condition as a whole. As Marisa St Clair says of the NDE: "Because the NDE is new to us, it may be a mistake to . . . attempt to make it fit any known category of human experience. It is now known that there are many different sorts of 'altered states of consciousness' (ASCs), such as dreaming, trance and hypnosis. The NDE, while bearing some similarity to certain aspects of those, appears to be another, quite distinct and separate ASC. Perhaps the time will come when death itself is seen as merely another altered state of consciousness!"[48]

Interestingly, although John Mack's friend, the abductee Carlos, also had an NDE, the abduction was far more transformative, saying that abductions give "access to the bliss, and the near death experience is . . . a momentary place in between. It is a soul place, to gather up." He sees the abduction as a completely healing experience on many levels, adding, "You are diseased and then you are healed. With each healing, the emotional growth is established and connected in the human realm and I can go and utilize that towards teaching others."[49]

It is a mistake to encourage abductees to think of themselves as either victims or Chosen Ones, when in fact they are travellers on a journey who are greatly in need of signposts, in quite another country.

THE MAGICAL KINGDOM

Until recently humanity believed that it shares the Earth with other beings, whose territory overlaps ours, but whose lives and destinies are separate. Woe betide any mere mortal who crosses the great divide between this reality and theirs, for although they can be kind, even generous, they are ultimately unimpressed with us and have a strong tradition of being mischievous or downright hostile. They are, of course, the fairies, otherwise known as the Good People, the Gentry, elves, sylphs, leprechauns, dwarves, gnomes, *fees*, korrigans and by many other names throughout the world. And although the fairy kingdom is now usually relegated to twee "flower fairy" designs on greeting cards or children's stories, we are well advised not to ignore certain traditions associated with the magical realm, for they have strong, even disturbing, similarities with modern myths – particularly that of alien abductions.

In the early years of the twentieth century, an American folklorist, Walter Wetz, gathered together a huge amount of data about strange magical beings in his thesis on Celtic traditions,[50] including this curiously matter-of-fact and detailed description of the "Gentry" from a native of Ireland:

> The folk are the grandest I have ever seen. They are far superior to us and that is why they call themselves the Gentry. They are not a working-class, but a military-aristocratic class, tall and noble-appearing. They are a distinct race between our race and that of spirits, as they have told me. Their qualifications are tremendous: 'We could cut off half the human race, but would not,' they said, 'for we are expecting salvation.' And I knew a man three or four years ago whom they struck down with paralysis. Their sight is so penetrating that I think they could see through the earth. They have a silvery voice, quick and sweet.
>
> The Gentry live inside the mountains in beautiful castles, and there are a good many branches of them in other countries . . . The Gentry take a great interest in the affairs of men . . . Sometimes they fight among themselves. They take young and intelligent people who are interesting. They take the whole body and soul, transmuting the body to a body like their own.
>
> I asked them once if they ever died and they said, No; 'we are always kept young . . . Once they take you and you taste food in their palace you cannot come back. They never taste anything salt, but eat fresh meat and drink pure water. They marry and have children. And one of them could marry a good and pure mortal.

> They are able to appear in different forms. One once appeared to me and seemed only four foot high . . . He said, 'I am bigger than I appear to you now. We can make the old young, the big small, the small big.

Several commentators, including Jacques Vallée, Hilary Evans and Patrick Harpur, have drawn attention to the striking similarities between the abducting, morally ambiguous and shapeshifting fairies and the aliens who menace, abduct but also often charm their human victims in the late twentieth and early twenty-first centuries. In his masterly *Passport to Magonia*, Dr Vallée cites the case of sixty-year-old chicken farmer Joe Simonton who lived close to Eagle River, Wisconsin as a classic story of encounters with the fairy folk – even though it is seen as proof of extraterrestrial contact.

At around 11 o'clock on the morning of 18 April 1961, Simonton heard a strange noise outside, which he compared to the sound of "knobbly tyres on a wet pavement". Going outside to have a look, he was confronted with a saucer-shaped object of a silvery hue that was "brighter than chrome", about 12 feet(3.6 m) high and 30 feet (9 m) across, which hovered above the ground. Suddenly a hatch opened revealing three men inside the craft, one of whom was wearing a black two-piece suit, but all of whom were about 5 feet (1.5 m) tall, smooth-skinned and "resembled Italians" with their dark hair and swarthy skin. The other two were dressed in knitted helmets and turtleneck tops.

One of these entities held aloft a jug and gestured to Simonton, apparently indicating that he needed water, so he took the jug into his house, returning with it full. Then he saw that one of the "men" was "frying food on a flameless grill of some sort", taking care to observe as much as he could of the interior of the craft. "The colour of wrought iron", the inside boasted instrument panels that gave off a slow whining noise, like an electrical generator. Simonton motioned to the men, indicating that he would like to try their food, and was rewarded with the gift of three biscuits, about 3 inches (7.5 cm) wide, with a scattering of small holes.

Then one of the UFOnauts attached a sort of belt to a hook on his person and the hatch closed so smoothly there appeared to be no join. At that, the UFO rose vertically into the air before zipping off southwards in a direct line, causing nearby pine trees to bend with the force of the blast.

Local sheriff Schroeder despatched two deputies to the scene, but they found no trace of the visitors. However, having known Joe Simonton for fourteen years, they had no reason to suspect that he was lying. Then the US Air Force became involved, concluding lamely that the witness had suffered a waking dream that his unconscious mind had inserted into reality. (However, at roughly the same time as Joe Simonton was having his weird experience, one Savino Borgo, an insurance salesman was driving along Highway 70, a mile from Eagle River when he saw a "saucer" rise into the air and fly off, parallel with the road.[51])

Jacques Vallée notes: "I understand several psychologists in Dayton, Ohio, are quite satisfied with this explanation, and so are most serious UFOlogists. Alas! UFOlogy, like psychology, has become such a narrow field of specialization that the experts have no time left for general culture. They are so busy rationalizing the dreams of other people that they themselves do not dream any more, nor do they read fairy tales. If they did, they would perhaps take a much closer look at Joe Simonton and his pancakes. They would know about the Gentry and the food from fairyland."[52]

Wetz's book features the tale of Pat Feeney, a well-off Irishman, who had been visited by a little woman who asked for some oatmeal: "Paddy had so little that he was ashamed to offer it, so he offered her some potatoes instead, but she wanted oatmeal, and then he gave her all that he had. She told him to place it back in the bin until she should return for it. This he did, and the next morning the bin was overflowing with oatmeal. The woman was one of the Gentry."

Paranormal abundance, however, is only one of the characteristics of fairy food. As Wetz's informant said, the Gentry "never taste anything salt, but eat fresh meat and drink pure water" – and water was what the "aliens" wanted from Simonton, while the analysis of little pancakes they gave him, carried out by the Food and Drug Laboratory of the US Department of Health, Education and Welfare on the request of the USAF, certainly provides something of interest for the folklorist, and for this discussion, although it has been dismissed utterly by UFOlogists, scientists and the military.

Simonton tried one of the pancakes and declared that it "tasted like cardboard", whereas the USAF's report stated: "The cake was comprised of hydrogenated fat, starch, buckwheat hulls, soya bean hulls, wheat bran. Bacteria and radiation readings were normal for this material. Chemical, infra-red and other destructive tests were run on this material. The Food and Drug Laboratories . . . concluded that the material was an ordinary pancake of terrestrial origin."

It is interesting that the food "tasted of cardboard": rural people used to put food out at night for the fairies, which although it was (usually) still there in the morning, had somehow lost its goodness, becoming tasteless, just like cardboard.

Also interesting is the fact that it was revealed that buckwheat hulls were cooked into the pancakes, for buckwheat was a staple of Brittany, where the *fees*, korrigans or *fions* had – and perhaps still have – a particularly strong presence. One Breton story tells of how one of the *fions*' cattle ruined a poor woman's buckwheat field: after she complained, they offered her compensation in the form of a spell that ensured that she would never run out of buckwheat cakes unless as long as she kept her mouth shut about their secret deal. All went well, and she and her family ate their fill of buckwheat pancakes, until the day that she gave some to a man to whom she blabbed about their paranormal origin. After that, the supply dried up.

Similar stories abound wherever there is a fairy tradition. One such tale from Wales contains several classic elements of both the small folk and alien abduction scenarios: "A man who lived at Ystradfynlais, in Brecknockshire, going out one day to look after his cattle and sheep on the mountain, disappeared. In about three weeks, after a search had been made in vain for him and his wife had given him up for dead, he came home. His wife asked him where he had been for the last three weeks. "Three weeks? Is it three weeks you call three hours?" said he. Pressed to say where he had been, he told her he had been playing his flute (which he usually took with him on the mountain) at the Llorfa, a spot near the Van Pool, when he was surrounded at a distance by little beings like men, who closed nearer and nearer to him until they became a very small circle. They sang and danced, and so affected him that he quite lost himself. They offered him some small cakes to eat, of which he partook; and he never enjoyed himself so well in his life."[53]

(This man was lucky to get away after three weeks: some are said to return home to find many years have passed, their wives are aged crones or dead, and they themselves are unrecognized, strangers in their own homes.)

Clearly these brushes with the fairies are strongly reminiscent of those with aliens, from the only-too-frequent absurdity of the whole experience to the phenomenon of missing time – although in the case of modern abductees it is usually measured in hours, not weeks or years.

British writer Patrick Harpur, in his extraordinary book *Daimonic Reality* (1994), urges the reader to consider the true nature of reality, which "is primarily metaphorical, imaginative, daimonic" – by which he does not mean "demonic", evil or occultly subversive, but the inspiration of the fluid, awesome and magical imagination. As we have seen, imagination is potentially the most potent power on Earth. Harpur considers that "daimonic reality" includes – or perhaps simply *is* – the fairies, saying that they put a spell on us, "altering our perceptions and making us see whatever they want . . . Fairy belief recognizes that enchantment lies sometimes more with us, sometimes more in the world."[54]

It is as if the fairies, recognizing the vagaries of human perception, deliberately play up to it, teasing us with their dazzling displays of UFOlogical cunning as apparently solid craft turn into amorphous blobs of plasma-like energy before our startled eyes – and so causing increasingly bitter divisions between mankind. It would be perfectly in keeping with the traditional *modus operandi* of the fairies to make nuts and bolts UFOlogists, rationalists and those who favour a more paranormal explanation waste their time and energy with ongoing hostilities. Fairies have always been the great *tricksters*, like agents of the great Cosmic Joker posited by Charles Fort. They thrive on human divisiveness, feeding off fear and confusion.

Everyone knows that fairy gold, attractive though it may seem, will turn into dust and ashes in due course in a horribly predictable fashion, perhaps simply in order to entertain and amuse the fairy folk who seem to like nothing better than to entice unwary humans with worthless goods or artefacts. Yet time and time again the objects left behind by non-human entities prove insubstantial or worse, liable by their very nature to cast doubt on the credibility of the witness. Joe Simonton's pancake made him the object of ridicule, but there are other objects, similarly "miraculous" that eventually prove highly unsatisfactory.

Photographs of UFOs are notoriously capricious, disappearing from cameras, developers' laboratories, locked drawers – anywhere – often prompting dark mutterings about conspiracies involving intelligence agencies. Yet, "Narratives of the theft of valuables from supernatural beings are found the world over,"[55] something to bear in mind when pursuing UFOs with expensive video cameras. Check your insurance first. Perhaps it is no accident that, as we have seen, there is not one unequivocal photograph of a UFO among the thousands, perhaps hundreds of thousands, now in existence. It could be that the "daimons" prefer it that way, living as they do off ambivalence. As Harpur says, "The 'remains' tease us and lead us on."[56]

One of the classic cases of disappearing paranormal artefacts is the story of the founder of the Church of Jesus Christ of Latter Day Saints (the Mormons), Joseph Smith, who as a farm boy in New York State in 1821 had a vision in which an angelic being called Moroni told him of the location of a box he was to dig up some six years hence. When he did so, in the presence of several witnesses, he discovered it contained bound brass and gold plates, covered in unknown hieroglyphics, which through a magical device called the "Urim and Thummim" he translated as the *Book of Mormon*. This purports to be the story of Jesus's post-resurrection mission to the New World. Although – despite the veneration of millions of Mormons worldwide – there are reasons to doubt the veracity of Smith's story,[57] the gold and brass plates did behave like archetypal paranormal objects because after they had been translated, *they disappeared*, thus placing Smith and his followers in a very difficult position. And although he founded one of the most successful of the nineteenth century's new independent churches, Smith lost his life to a mob at Nauvoo, Illinois in 1844, going the way of many whose lives had been touched by daimonic reality.[58]

(Perhaps the message is "beware of aliens bearing gifts", or at least try to avoid having visions. Hearing voices was not, in the end, the best career move for Joan of Arc, nor did finding the healing spring at Lourdes, as the mysterious lady directed her, do Bernadette Soubirous *personally* any good at all. She was hidden away in a nunnery, where she died of a tumour on the knee, still a young woman.)

As Harpur says sagely, "Daimons do not leave red herrings – they *are* red herrings".[59]

Daimons, or fairies, have a quintessentially contradictory nature. Indeed, part of their *vaison d'être* is undoubtedly to provoke disbelief, even in their manifestation. (A well-known British UFO sceptic once told me in complete seriousness about having been shocked to see a leprechaun by the side of a road on a rural Irish hillside.) Vanishing evidence is itself part and parcel of the fairy plan. Yet a little of their characteristic ambivalence often seems to rub off on those who come in contact with them, as if the enchantment never completely goes away. It is significant that the old word for a fairy spell was "glamour", perfectly encapsulating the *feel* of the experience, as can be seen in the strange story of the visionary Maggie Fisher and, of course, the archetypal glamourized abductee, Whitley Strieber. Once you have entered daimonic reality, nothing will ever be the same again, though you can try with all your heart and soul to make it so. You can run from the fairies, but you can never hide.

However, there is a slight variation between the fairies of old and the aliens of today, representing perhaps no more than a modernization, an updated PR exercise on the part of the shape-shifting daimons. As Patrick Harpur succinctly points out: "Unlike the fairies who belong in the past and appear only fleetingly in the present, UFOs and their 'occupants' appear in the present and belong in the future. They are not, like the fairies, always going but never gone; they are always coming, coming – but never here."[60]

Like the cults who constantly wait for the prophesied imminent advent of some supernatural figure, be it the Second Coming or the space gods, but who are always frustrated – as in the case of the Council of Nine and the "landing on planet Earth" of superior beings from Sirius – UFOlogists wait for the perfect case that will prove the reality of the phenomenon once and for all. It may be that they will have to wait a very long time, if the origins of the "real" UFOs (as opposed to *human* hoaxes or misidentifications) really do lie with the invisible mischief-makers with whom, it is said, we live side by side.

THE EMBODIMENT OF EVIL

As the stories of hundreds of traumatized abductees makes clear, encounters with aliens are not fun (at least not often).[61] Frail and credulous humans are snatched by repulsive Greys and subjected to disgusting, humiliating and often painful examinations – even surgical operations. They are raped, abused, often on several occasions spread over a lifetime, perhaps for the purpose of creating a hybrid alien/human race (as even the notably less traumatic case of Antonio Villas-Boas suggests). Even those who are not maltreated in this way, but merely witness UFOs, are often damaged by the experience with bouts of what appears to be radiation sickness, other wounds – and, in at least one case, death.

Such events go well beyond mere pranks. If creatures from the otherworld are truly tormenting us, sometimes unto death, then we are looking at pure evil on a massive scale. This is not a new idea, although sometimes it may not have been taken totally seriously.

"Crawling, Creeping, Unbelievable terror! See . . . the Night of the Green Horror! See . . . The Disembodied Hand that Crawls! See . . . the Earth Ravaged by Creatures from Hell!" cried the posters for the 1950s "B" movie *Invasion of the Hell Creatures* (X Certificate), summing up the hysterical hype common to films of that type and era, which must take the blame for the anti-alien phobia that inflicted many people in those days, despite the best attempts of the Space Brother cults to persuade them otherwise. That, even despite *Close Encounters of the Third Kind*, with its benevolent Greys, this attitude has not shifted noticeably can be seen from the gung-ho Republican movie, *Independence Day* (1996) in which the American President saves the world from giant crayfish through the cunning use of computerized special effects.

Of course no one doubts that the very thought of a *real* invasion from "out there" is utterly terrifying. To be attacked by fellow human beings is traumatic enough, but imagine being the target of creatures with less in common with us than bed bugs.

Thanks largely to the creative hype of Hollywood – and the insidious propaganda of the alien abductee grapevine, no matter how "real" their experiences – the idea that UFOs and their occupants are hostile is deeply embedded in the collective consciousness. Despite those who claim to be in touch with peaceable "space brothers" (such as the followers of the Aetherius Society), we view the idea of flying saucers as a threat to civilization as we know it, to all the things that humanity treasures: compassion, decency, autonomy and freedom. As our own history is a sorry catalogue of rape, pillage, invasion and genocide we imagine that beings from elsewhere must think the same way, viewing Earth as a nice little outpost of their Empire, once they have subjugated the natives.

It may be that such a cynical idea is justified. Even if not actually extraterrestrial, the occupants of UFOs do seem to behave with a callous disregard for human health and well-being, both mental and physical – even causing deaths, as in the case of the Brazilian ranch worker Inacio de Souza, who died of leukaemia, apparently as a result of his encounter with a UFO in August 1967. Others have been blinded[62], some claim to have been raped, while almost all have been traumatized in one way or another. Perhaps we are right to treat the concept of aliens on Earth with great caution. Given the unique circumstances of the arrival of the aliens – although the ancient astronaut school may think it anything but unique – we may well be justified in considering them guilty until proved innocent.

However, there are those who believe they can never be innocent. Many see the UFOnauts as nothing less than emissaries of evil, their craft as Satanic saucers. American authors Brad Steiger and Joan Whritenour wrote:[63] "Certain saucer cultists, who have been expecting the space brethren to bring along some pie in the sky, continue to deliver saucer-inspired sermons on the theme that the saucers come to bring starry salvation to a troubled world. The self-appointed ministers who preach this extraordinary brand of evangelism ignore this fact that not all 'saucers' can be considered friendly. Many give evidence of hostile actions. There is a wealth of well-documented evidence that UFOs have been responsible for murders, assaults, burning with direct-ray focus, radiation sickness, kidnappings, pursuits of automobiles, attacks on homes, disruption of power sources, paralysis, mysterious cremations, and destructions of aircraft. Dozens of reputable eye-witnesses claim to have seen alien personnel loading their space vehicles with specimens from earth, including animals, soil and rocks, water, and struggling human beings."

"Mysterious cremations" sound bad enough, among the catalogue of crimes against humanity, but Steiger and Whritenour's wilder claims, such as the plural "destructions [sic] of aircraft" seem rather too much to take. Science fiction writer Frederick Pohl will have none of it, asserting angrily in *True's New Report*:

> It's as false as false can be; there not only is not a 'wealth' of such evidence, there isn't *any*. The absolute best you can say in support of that claim is that there are many people who *think* such things happen, and a mass of circumstantial bits and pieces of events. There is no evidence at all for the assumption that the saucers are almost certainly hostile.

Steiger was to change his view on UFOs dramatically. Within a few years of writing that alarmist catalogue of alien crimes he penned *Gods of Aquarius* (1976), which was subtitled *UFOs and the Transformation of Man*, expounding the theory that "the UFO will serve as the spiritual midwife that will bring about mankind's starbirth into the universe" – a view seemingly aimed at the New Age market. As the eminent British UFOlogist Hilary Evans commented wryly:[64] "Cynics might suggest that Steiger, having milked the UFO-scare theme for all it can give, is now finding the positive approach more profitable . . ." He added ". . . but perhaps he has genuinely changed his mind."

While even the alien abductees are moving towards a gentler interpretation of their experiences, the idea that the saucers are satanic in origin and intention has never gone away, although it has several variations, some more "*X-Files*" than others in tone. As might be expected, however, Christians – especially Fundamentalists – are most likely to hold the view that the UFOs are demonic and should be treated as such.

In the 1950s, "flying saucers" were seen as solid, nuts and bolts craft, but Gordon Cove, writing in a privately published booklet, *Who Pilots the Flying Saucers?* (1954) posited a subtler, if even more disturbing, hypothesis. Although Cove agreed that the craft were real, he went much further, saying: "What we are suggesting is the possibility that Satan has seized one of the planets as his base of operations to attack the Earth. This thought . . . may seem fantastic: but upon cool meditation, does it seem so absurd? The first thing a military general seeks, when war is declared, is a convenient headquarters. Satan is the cleverest military genius ever known. Is it feasible that Satan, along with his principalities and powers, his wicked angels and demons, would continue to float airily around in the atmosphere for thousands of years, when there are literally millions of planets which would be well adapted for a headquarters?"

He goes on: "Satan is partially powerless unless he can get some willing instruments to work through. Therefore, if Satan wanted to manufacture some flying saucers in order to facilitate the flight of his evil hosts throughout the vast universe, it would also be to his great advantage to get a race of beings under his control who would manufacture them for him. Could he not inspire the Venusians, if such exist, with supernatural cunning and wisdom to make a fleet of flying saucers, and also show them how to pilot these supernatural machines?"

Cove also suggests that even the kindly "space brother" type of alien that contacted George Adamski, for example, are not what they seem, but "demon-possessed Venusians or Martians", whose sole intention is to deceive humanity and turn us away from good. At this point in his argument, however, matters become more complex: Cove concedes that *some* aliens may not be agents of Satan after all, but benevolent forces dispatched to fight the good fight. Quite how he proposes to tell the *faux* angels from the true ones is undetermined. Perhaps, like most people with a strong conviction of the reality of Satan, he believes those on the right side would know.

Cracks begin to appear in Cove's credibility – if, indeed, they have not already made an appearance – when he quotes a prophet called Hehr, who predicted in the early years of the twentieth century that an "older race on Venus" would "take measures to re-establish a new and better order in the shortest possible time", giving the dateline of 1965, which would, he asserted, be preceded by a global conflict and total anarchy. As with many UFO cults, the great mistake was to be specific. Dates can be checked: they come and go remorselessly – although of course, Hehr, prophesying in the 1900s, was safely distant from his all-important year of 1965.

A few years after Cove published his pamphlet, other cultists expressed their fervent belief that the UFOs heralded some great spiritual upheaval, an – if not *the* – Apocalypse. A good many of these True Believers put their faith in the saucers as signs and wonders of the coming of a New Age of love and wisdom, even redefining Jesus's background as otherworldly – literally. While the Aetherius Society, and others, now gave his origins as Venus, others contented themselves with seeing the saucers as messengers. Bob Geyer, of the Los Angeles Church of Jesus the Saucerian, said in 1970:[65] "Our conversations on the religious aspect of UFOs brought forth the conclusion that they herald the Second Coming of Christ."

Unlike Cove, who believed that the UFOs are piloted by devils in disguise, Geyer saw them as genuine extraterrestrials, albeit with a apocalyptic twist, explaining: "Satan, the old prince of darkness, and his legions of demons, are also beings from other worlds. They came down from another planet. Once, Satan was a member of God's astronauts. He became too greedy and too ambitious. He may have exploited the inhabitants of earth, or other planets. He may have tricked people into slavery."

Unsurprisingly, the view that UFOs are demonic in origin is most common among Fundamentalist Christians, particularly in the United States, although many less zealous American Christians also tend to take the Bible much more literally than their European counterparts. Zola Levitt and John Weldon are two commentators who see UFOs as signs that the coming of the Biblical Antichrist – the supernatural embodiment of evil – is at hand, writing in their 1975 book *UFOs – What On Earth Is Happening?*:

> Are the flying machines really up there? Maybe so; it's not that important. If the demons wish them to be there, they are there,

> and if they wish people to imagine they're there, then they are imagined to be there.

They go on: "We believe demons can induce a whole series of experiences that, in fact, never really happened, similar to the experiences Uri Geller and Dr Puharich found were induced by their extraterrestrial contacts. They can also, however, through various means produce "real" UFOs which are visible to anyone. With the powers we know demons have, they could theoretically transform a large chunk of rock into a UFO, assume human form inside of it, and land openly, thus 'proving' the existence of advanced intergalactic civilizations."

Perhaps, blinded by their fundamentalist fervour, those authors failed to notice the clues hidden in their own subtext. As we have seen, the experiences of Uri Geller and his mentor, Dr Andrija Puharich with non-human entities suggest other, but no less disturbing, interpretations . . .

Even apart from the hellfire-and-damnation school of thought, there is still something sinister about UFOs, something elusive and unsatisfactory. They appear to be solid, but then shapeshift into amorphous blobs of light, darting around the sky in impossibly capricious and super-manoeuvrable ways. They haunt missile bases, but rarely seem to interact with the personnel who work there (leaving aside the ongoing debate about back engineering at Area 51 and other similar sites); they have an inexplicable fondness for water – to the extent that there are several cases of USOs (Unidentified Submarine Objects), but few cases of fishermen or sailors having seen them; and they are apparently highly superior to us technologically, yet there are many cases of broken down UFOs. Sometimes secretive, sometimes witnessed by thousands of people (as in Mexico City or China), their aim is by no means clear. An overview of the UFO question often suggests that their purpose, if they have one, may well be simply to confuse and deceive – and it is by no means only the Fundamentalist Christians who believe that to be the case.

British authors Anthony Roberts and Geoff Gilbertson in their 1970s cult classic *The Dark Gods* observe: "The over-riding reality of UFOs will be seen to manifest on an archetypal, psychic level, as an attempted 'control syndrome'. This control syndrome tries to influence any civilization that aspires to come to terms with the cosmological patterns and purpose, and must therefore be viewed with great caution by sentient beings. The 'control' appears to work through a careful programming of any belief-system while conforming to a definite pattern and moulding the physico-spiritual awareness of the recipient culture or species."

They elaborate further: "The numerous phenomena manifest just outside any knowledge-framework of the receivers, and yet they seem to have 'natural laws' and formal interconnections of their own. These impinge on the reality consciousness of the various imbibers, rather like whiskey (or any drug) impinges on the neuro-circuits of human beings and drastically alters rational perspective. The UFO aspect of this control syndrome incorporates in man as psychic, magical, mythological and paranormal components as it does the more overtly sociological and technological paradigms. They add: "It seems to be a very comprehensive control system indeed if its true source is in the realms of the metaphysical, then the full cosmic connection becomes more readily apparent."

Roberts and Gilbertson build to their climax, warning that: "If the whole thing is a cosmic conspiracy of sorts, then all humanity have been pawns in the game of these 'dark gods' for millennia. It only follows that these same forces have directed human history."

We are not looking at light-beings or merely mischievous fairy folk here, but a deadly enemy. To those authors, the conclusion is inescapable: "This [hypothesis] holds that the course of recent (that is, the last few millennia) history has been secretly and ruthlessly controlled by small groups of self-appointed 'elitists' who are servants of evil and who seek to establish the rule of the Dark Gods in a hell-on-Earth over which they will preside. If the UTs [Ultraterrestrials] are exerting a 'control syndrome' through manipulation of paranormal phenomena and the illusory UFOs, then their elitist disciples are the earthly agents who handle the social side of the great game."

THE COSMIC JOKER

Charles Fort,[66] the great collector of "damned data" – bizarre and provocative evidence that science cannot or will not consider, such as mysterious falls of fish from clear blue skies – is often dismissed, even by his admirers, as a marginal character in the world of the weird, whose ideas are fun but not to be taken too seriously. Yet if any one individual has the right attitude to the vexed UFO/alien problem, it has to be Fort.

Fort wrote, as a sly dig at the pompous absurdity of scientific arrogance: "In the New York newspapers, September, 1880, are allusions to an unknown object that had been seen travelling in the sky, in several places, especially in St Louis and Louisville . . . Unless an inventor of this earth was more self-effacing than biographies of inventors indicate, no inhabitant of this earth succeeded in making a dirigible aerial contrivance, in the year 1880, then keeping quiet about it. The story is that, between 6 and 7 o'clock, evening of July 28th, people in Louisville saw in the sky 'an object like a man, surrounded by machinery, which he seemed to be working with his hands and feet.' The object moved in various directions, ascending and descending, seemingly under control. When darkness came, it disappeared. Then came dispatches, telling of something that had been seen in the sky, at Madison-ville, Ky [Kentucky]. 'It was something with a ball at each end.' 'It sometimes appeared in a circular form, and then changed to an oval.' " Fort goes on:

> These are stories of at least harmless things that were, or were not, seen over lands of this earth. It may be that if beings from somewhere else would seize inhabitants of this earth, wantonly, or out of curiosity, or as a matter of scientific research, the preference would be for an operation at sea, remote from observations by other humans of this earth. If such beings exist, they may in some respects be very wise, but – supposing secrecy to be desirable – they must have neglected psychology in their studies, or unconcernedly they'd drop right into Central Park, New York, and pick up all the specimens they wanted, and leave it to the wisemen of our tribes to explain that there had been a whirlwind, and that the Weather Bureau, with its usual efficiency, had published warnings of it.[67]

In his usual tongue-in-cheek manner, Fort satirized and joked his way through a huge assortment of unexplained phenomena, but underlying it all was his idea that we are the playthings of a Cosmic Joker, who delights in creating "anomalistics", evidence of High Strangeness, just to catch us out. Like all his many concepts, some completely contradictory, and all of them expressed in his energetic but highly quirky style,[68] this may not have been entirely serious, but it is worth considering (if only to excite conspiracy theorists). Perhaps momentously he wrote: "I think we're property . . . That once upon a time, this Earth was No-Man's Land, that other worlds explored and colonized here, and fought among themselves for possession, but now it's owned by something, all others warned off."

(Did this wonderful Fortean thought[69] transmogrify into the stultifyingly serious Council of Nine's agenda?)[70]

Fort may have been joking – although one is never sure with him – but it is still a disturbing thought that despite all our great human achievements and the nicely ordered world in which we live, with its allegedly immutable laws of physics, everything can be turned upside down by the Joker, *who is our owner*. He toys with us, perhaps creating amusing little scenarios with Grey aliens and terrified humans spread-eagled on examining tables, just to see us squirm, in the same way that a nasty little child will pull the legs off an insect, then immediately become bored and do something else.

There may be something serious underlying Fort's flippant words. Perhaps it is true that once upon a time beings came to Earth and "colonized" us. Perhaps, too, it is time for them to return to rule their Empire: as we have seen, there are those who believe this is precisely what is about to happen. Or maybe "they" do not actually own us, but find us marginally amusing and occasionally annoying – *as near neighbours*.

They may not actually be the fairy folk – although the latter probably also exist – but another, separate, civilization that inhabits a dimension that is best described as being literally *above* us. John A. Keel calls them the "Ultraterrestrials" – although they are not necessarily the downright evil beings, of the same name, of Roberts and Gilbertson's "Dark Gods" hypothesis – human-like beings living parallel lives that occasionally impinge on ours, when they succeed in baffling and terrifying. Certainly the idea that there are people "up there" – however simplistic and silly it may seem –

would explain the mysterious "falls" of a huge range of objects over the centuries, including an ornately carved stone pillar; unknown powders of various colours; oils; resinous matter; fresh, *cooked*, and dried fish; hideous pieces of raw human flesh; "rains" of blood; wooden crosses and blocks of ice – many years before the existence of wayward airline lavatories.[71] The most common sort of fishfalls are of sprinklings of examples of one species only, and the most common explanation (today as in Fort's time) is that they must have been swept up by a "whirlwind". However, the objections also remain the same: what kind of whirlwind picks up just one sort of fish, but leaves behind every other sort of debris? As Fort wrote: "Many other falls we shall have record of, and in most of them segregation is the great mystery. A whirlwind seems anything but a segregative force. Segregation of things that have fallen from the sky has been avoided as most deep-dyed of the damned . . ."

Then Fort adds gleefully: "But several days later, more of these objects fell in the same place."[72]

Keel suggests that the beings themselves may have given us the very occasional clue: remember that the tall people allegedly from "Zomdic" told willing contactee James Cook of Runcorn to *jump* on board their UFO because the ground was damp. They added that they could not operate in damp weather because their craft was enveloped in some sort of electrified field. But most tellingly of all – *if* they are to be believed – they told James Cook that " . . . *the saucers were used only in the vicinity of the earth and could not operate in outer space*".[73] But did this mean that they themselves inhabit some kind of invisible realm that is literally above our heads – an Ultraterrestrial domain? Or were they referring to a type of craft that they use only around the Earth, while perhaps keeping more robust and far-reaching ships for outer space journeys?

Bizarre and fantastical as it may seem, however, there may be something to be said in favour of the Ultraterrestrial hypothesis. After all, as it seems highly unlikely that such terrestrial-sounding objects as wooden crosses and carved pillars were jettisoned by beings from other planets, perhaps they came from much nearer home, from the dimension "up there" that may be inhabited by beings similar to, but in some ways different from, humankind.

Sometimes the objects that have fallen to Earth have a more obvious message. The great airship mystery of the late nineteenth century produced many odd tales of half-peeled potatoes falling overboard from one of the "ships" and even pieces of card inscribed with elegantly-written information about the aeronauts. But nothing seemed to add up: to this day no one knows who they were or what they wanted, where they came from or where they went. Perhaps that's how it's supposed to be.

TIME TRAVELLERS AND OTHER DIMENSIONS

Many people hypothesize that UFOs come from another reality that coexists with ours, but on another dimension, which explains how they appear and disappear so inexplicably. Perhaps they have learned how to pop into our reality for a quick tour of our world, gaining solidity through the belief and expectancy of those to whom they show themselves, in an endless dance of reciprocity: they show themselves, the witnesses believe in them and the belief gives them more "reality", more visibility, just as layer upon layer of belief creates a *tulpa*.

Maybe the UFOnauts need electricity or nuclear waste with which to power and maintain themselves in our dimension, hence the numerous occasions they have been observed around military installations and power stations.

Some suggest that the aliens are ourselves – the human race, evolved almost beyond recognition – from the future, coming back to warn of imminent ecological or natural disasters, trying to prevent the annihilation of the planet and all its life. If so, however, their knowledge of the immediate future is patchy and unreliable, for although they are given to dire prophecies about the end of the world, time and time again the appointed hour comes and goes without so much as a new conflict breaking out, let alone a global calamity.

Yet the idea of intruders from other dimensions, which may be described as *elsewhen*, is perhaps not so far off the mark. Quantum physics – to which all modern scientists pay lip-service, although usually remaining firmly in the old Newtonian-Cartesian mechanistic mould – provides the template for this possibility, positing the existence of many invisible dimensions, where something may be both here and there, depending on whether it is observed or not. The British cutting-edge physicist, Professor David Bohm, described a highly controversial view of the world in his seminal book *Wholeness and the Implicate Order* (London, 1980), arguing that while our separation of the world into apparently distinct objects has enabled us to control our environment very well, on a deeper level this distinction between "me", "you" and "that" is actually false. As the astro-physicist Professor Archie Roy of Glasgow University writes in his *A Sense of Something Strange* (1990), "Many of these objects, such as electrons and other atomic particles, are illusions. [Bohm] puts forward reasons for believing that the level of reality manifesting itself, the level that we study, is produced by the creative, flowing processes of a sub-world. These 'forms' and 'objects' thrown up do seem to have a certain stability, longer and shorter durabilities, their behaviour capable of being described by a set of laws deduced from observation. But because they are only a sub-set of phenomena projected or manifested from a deeper, more fundamental world of dynamic processes, certain anomalies or paradoxes reveal that, however deeply we believe we have come to grips with ultimate reality, the artifacts we are studying are projections, as it were, into a lower number of dimensions from a higher dimensional world of existence."[74]

In other words, seeing is not necessarily believing. The evidence of one's own eyes is almost always incomplete and flawed, being too dependent on a host of factors from physical health to prejudice and expectations, not to mention the fact that human beings can only perceive a limited area of the spectrum – as anyone realizes who has had the disturbing experience of watching a cat or dog follow something invisible around the room with their eyes. Not only is there *more* there than we can see, but what we do see is probably very different in its basic nature: neatly distinct chairs and tables are really, according to theoretical physicists, not really solid objects at all, but whirling masses of invisible atoms that happen to form the configurations that we recognize as tables and chairs. Before they were made, their atoms would have formed what we call planks of wood and before that, trees. Yet in a sense none of these things exist, for everything is in a constant state of flux, a universal dance of life, only some of which is perceived by mortal man.

With such limits on our perception, the only wonder is that more people are not aware of intruders from elsewhere, who are perhaps impatient to be recognized as fellow inhabitants of our world. As the ever-wise Jacques Vallée writes: "These unexplained observations need not represent a visitation from space visitors, but something even more interesting: a window toward undiscovered dimensions of our environment".[75]

He goes on to present the great challenge to our current belief system: "I believe that the UFO phenomenon represents evidence for other dimensions beyond spacetime; the UFOs may not come from ordinary space, but from a multiverse which is all around us, and of which we have stubbornly refused to consider the disturbing reality in spite of the evidence available to us for centuries."[76]

CONCLUSIONS

While many of the more recent UFO sightings – and even some of the abductions – may be the work of very terrestrial agencies, such as the Pentagon and the CIA, a hard core of experiences remain unexplained.

There are certainly tricksters involved, and not always shady characters who could have stepped straight out of *The X-Files*, but apparently busy spiritual entities, who may or may not come from the far distant stars they occasionally claim to be their home. They may be the beings that used to be called fairies, or more malevolent Ultraterrestrials, the invisible fellow-inhabitants of the Earth who delight in taunting, seducing and sometimes seriously damaging human beings, or they may be the henchmen of the Cosmic Joker, who operates from a nearby home, from where he throws strange objects, ranging from carved stone pillars to cooked fish at unsuspecting humans. It may be that the trickster beings are an ever changing, chameleon race that owes a little to all those ideas, with some added vampiristic tendencies. Perhaps inherently they have little real form or character themselves, but pick up life and colour from the belief, perception and expectations of the more solid and predictable human race. In this way the UFOs keep just one step ahead of modern technology, just enough to inspire umpteen conspiracy theories but keep the truth as elusive as ever.

Or perhaps there are many races of beings from other dimensions competing for our attention, love or simply our *energy* . . .

If such creatures exist, we may be sure that the world's governments, especially that of the United States, will have attempted to contact them, to make a deal, to offer our energy for their knowledge and protection. Perhaps that is the secret behind the MILAB hypothesis, which claims that the military are abducting innocent people, not instead of the aliens, but *on their behalf.*

Or maybe the simple truth is, after all, that the UFOs are extraterrestrial craft invading our airspace, and their occupants are abducting millions of individuals in order to create a hybrid race.

As I said in the Introduction, this phenomenon is so fluid that of all the competing theories, *no one is always right*, and admitting that may be the most sensible thing any UFOlogist – whatever their basic beliefs – can do. Researchers have wasted years through over-selectivity of the data, prejudice and mutual hostility. Surely it is time to call a truce and get talking with humility, a willingness to admit mistakes, and an openness to new ideas. For somewhere in the mass of data may already lurk, if not the answer, then some of the right questions – and that at least is a good enough place to start.

Notes

1. *Fortean Times* describes Dave Walsh as "an internet consultant, liar and author of the weekly online Fortean newsletter 'Blather". See *daev@-fringeware.com*
2. See 'Irish Skies Are Twinkling' in *Fortean Times*, No. 105.
3. Reported by Joseph Trainor in *UFO Roundup*. See *http://www.digiser-ve.com/ufoinfo/roundup/vo2/mdo2_35.shtml*
4. In an unpublished article that he kindly gave permission for me to quote extensively.
5. See *The Stargate Conspiracy* (London, 1999) by Lynn Picknett and Clive Prince.
6. See 'HAARPing It Both Ways' in *UFO Magazine* (US) vol 15 No. 7, July 2000.
7. *Ibid.*
8. See *http://www.attnews.com.au/nexus/Polter.html*
9. From a telephone conversation in London with the author on 30 August 2000.

10. *Ibid.*
11. See Jenny Randles, Andy Roberts and David Clarke, *The UFOs That Never Were*, (London, 2000), p. 136.
12. Sir Arthur Conan Doyle, 1859-1930, the Scottish doctor who became famous as the creator of the "Great Detective", Sherlock Holmes. After the trauma of the First World War, he became an ardent Spiritualist and was excited at the prospect of using the Cottingley photographs to publicize his book, *The Coming of the Fairies,* in the early 1920s.
13. This author, as Deputy Editor of the weekly publication *The Unexplained,* was the first person to hear the audio tape on which Frances confessed to Leeds-based psychical researcher Joe Cooper, a family friend, that they had faked it and how it was done. She said that they had been particularly gleeful at fooling such a great man as Conan Doyle, recalling that Elsie's mother had said, "Fancy, our Elsie impressing him – and her at the bottom of the class, too!"
14. *Princess Mary's Gift Book.*
15. According to Joe Cooper, Frances saw a fairy again shortly before she died in the early 1980s, and was horrified by the experience.
16. See p. 137 of David Clarke's chapter on the Alex Birch mystery in *The UFOs That Never Were*, co-authored with Jenny Randles and Andy Roberts. (London, 2000.)
17. John A. Keel, *UFOs: Operation Trojan Horse*, p. 213.
18. K. Gold, "Isolated Visual Hallucinations and the Charles Bonnet Syndrome", *Comprehensive Psychiatry* vol 30, No. 1, 90–98 (1989), quoted in David Hambling's article 'Phantom cows and little people' in *Fortean Times*, No. 125.
19. Only forty-six cases were discovered in two centuries, see *Ibid.*
20. *Ibid.*
21. *Ibid.*
22. *Ibid.*
23. D. Bresnahan, 'Charles Bonnet Syndrome', *Grand Rounds at Froedtert Hospital,* vol 4 No., 6 (Nov/Dec 1997), quoted in *Ibid.*
24. No. 127.
25. R.E. Foerster 'Hallucinations within the brain', *Journal of Neuropsychiatry* 35:677:706 (1976). Quoted in *Ibid.*
26. Vincent Willem van Gogh (1853–90), the Dutch Impressionist who painted, among other popular works of art, the famous *Sunflowers.* A sufferer from chronic mental illness, he cut off part of his ear in remorse for having threatened fellow artist Paul Gauguin with a razor.
27. On the series she presented for BBC2, 'Brain Story', first shown in July 2000.
28. Sir Isaac Newton, 1642–1727. While widely considered to be a classic example of the materialist-rationalist scientist, in fact he was deeply involved in the occult, and was a practising alchemist.
29. See the Letter page of *Fortean Times* No. 112.
30. *Ibid.*
31. Originally of Clare College, Cambridge. His first book, *A New Science of*

Life (1981) – which described the theory of formative causation, or the natural phenomenon whereby disparate members, and future generations, of the same species learn new techniques through a sort of collective unconscious – was vilified by the respected scientific journal *Nature* as "a book for burning".

32. No. 125.
33. The classic example of such a phenomenon is the wave of genuinely psychokinetic metal-bending that swept the West after the debut of the young Israeli psychic superstar in October 1973 (in Britain on the *David Dimbleby Show*). Whether the sudden upsurge of the ability to render cutlery useless on a massive scale in school playgrounds and the like was a result of "catching" the gift from a genuine psychic, in this case Geller, or was simply due to the fact that he presented them with a new and exciting idea, releasing dormant powers, may never be known. Yet the obvious *fakes*, the metal-bending conjurors, such as Geller's great critic (and later, grudging admirer), David Berglas, had no such effect on the public at large. No one ever bent metal at home after watching one of his shows. Perhaps the secret is to present oneself as genuine with as much passion and intensity as possible – then mysterious powers will follow.
34. Sometimes physical objects are blanked out by the brain – people simply do not see them – because they have no frame of reference for them. For example, when Captain Cook's tall ship first arrived on the shores of Australia many of the Aborigines failed to notice it.
35. Not "Is it?" but "It is!"
36. In particular the work of parapsychologist Dr Ernesto Spinelli.
37. William Wordsworth (1770–1850). The line is from *Intimations of Mortality*.
38. Sir James Matthew Barrie (1860–1937). *Peter Pan* was first produced on the London stage in 1904.
39. As Clive Prince and I discovered when researching our 1994 book *Turin Shroud: In Whose Image?* (Updated for the year 2000 paperback edition with the subtitle *How Leonardo da Vinci fooled history*.) The Shroud of Turin was, and in some circles, still is, believed to be the winding sheet of Jesus miraculously imprinted with his image when he was taken down from the cross. The object of enormous reverence and the subject of great passion, a vociferous lobby of True Believers, fronted by British author Ian Wilson (whose 1978 bestseller *The Turin Shroud* made the cloth world famous for the first time; before then only Catholics had taken an interest in it) finally persuaded the authorities to allow it to be carbon dated, believing that a scientific endorsement would force the world to take it more seriously. Yet their ploy backfired badly: in October 1988 the carbon dating results were announced. It was a medieval or early Renaissance fake . . . Immediately many of the very people who had put their faith in the technique of carbon dating began to find reasons to cast doubt on it, even to suggest some form of conspiracy. Our own research revealed that the Shroud is indeed a fake, a masterpiece by none other than Leonardo da Vinci, who created it using a primitive form of

photography, and using his own face as that of Jesus. So although it is worthless as a holy relic, the Shroud is a priceless photograph of the Maestro Leonardo. But of course the True Believers won't have any of it, even though they themselves have claimed that the image "behaves like a photograph" and that, if it were a fake – and obviously they don't believe it is – the only candidate for faking it would be Leonardo. To this day, they continue to overlook obvious clues on the image itself that demonstrate conclusively that it is a fake – the image is 6ft 8in (2 m) at the front and 6ft 10in (2.05 m) at the back, for example. (A projected image, like that of the Shroud, can be any size.) And instead they have begun to see curious images in the weave of cloth – including flowers, a hammer, nails – and, most ludicrous of all, a broom. Yet all there is to see, apart from the image itself, is the herringbone-style weave of the linen cloth.

40. See *Postscript*, in No.131 of *The Unexplained.*
41. For a discussion about the real nature of alchemy, see Picknett and Prince, *The Templar Revelation.*
42. See *The Story of Ruth*, by Morton Schatzman, 1977.
43. The phrase "crashing the boggle barrier" was invented by the late British psychical researcher Renée Haynes to describe the moment when one's mind skids on encounters with what appears to be the impossible.
44. Patrick Harpur, *Daimonic Reality*, Viking, 1994, p. 201.
45. See the Epilogue of Picknett and Prince's *The Stargate Conspiracy* for a detailed discussion of the relationship between the ancient Egyptian religion and shamanism.
46. *Abduction*, by Dr John E. Mack (New York, 1994), pp. 363–364.
47. *Ibid*, p. 289.
48. *Beyond the Light*, (London, 1997) by Marisa St Clair, p. 155.
49. Mack, p. 368.
50. Walter Yveling Evans Wetz, *The Fairy Kingdom in Celtic Countries, its Psychological Origin and Nature* (Oberthur, Rennes, 1909), also quoted on p. 25 of Jaques Vallée's *Passport to Magonia.*
51. Harpur, p. 144.
52. Vallee, p. 25.
53. Edwin S. Hartland, *The Science of Fairy Tales-An Inquiry into Fairy Mythology* (London, 1891).
54. Harpur, 1994, p. 137.
55. Hartland, p. 136.
56. Harpur, p. 141.
57. See, among others, *The Mormon Murders: A true story of greed, forgery, deceit and death*, by Steven Naifeh and Gregory White Smith (Weidenfeld and Nicolson, New York, 1988).
58. Even if he had, as some claim, fabricated his story of angelic instruction, it seems that Smith was no stranger to "occult practices" such as dowsing – see Naifeh and White Smith, for example – and the hill in which he unearthed the plates, Cumorah, had long been sacred to the Native Americans. Had Smith employed a sort of "psychic questing" to locate

the plates? And by doing so, had he disturbed the local hill spirits or fairies who were determined to get their own back by making him look stupid – or worse?

59. Harpur, p. 141.
60. Harpur, pp. 176–177.
61. Exceptions must include George Adamski, who (allegedly) travelled round the universe thanks to his Venusian friends, and Antonio Villas-Boas of Brazil, who had sex with a blonde female humanoid (although her habit of barking like a dog when excited must have been a bit off-putting). These cases, however, took place before the wave of alien rapes that existed in parallel with allegations of satanic ritual abuse.
62. See the case of Brazilian Almiro Martins de Freitas (*q.v.)*, a security guard, blinded after shooting at a UFO near the town of Italiania, Rio de Janeiro.
63. *Flying Saucers Are Hostile* (1967).
64. *The Unexplained,* p. 1322.
65. In conversation with writer Eric Norman.
66. Charles Hoy Fort of Albany, New York, 1874-1932, who inspired and gave his name to Britain's most astonishing magazine, the admirable *Fortean Times, The journal of strange phenomena* – see *www.forteantimes.com*
67. See Fort, p. 641.
68. In his extraordinary books *The Book of the Damned* (1919); *New Lands* (1923); *Lo!* (1931) and *Wild Talents* (1932).
69. Fort wrote – see p. 641 of *The Complete Books of Charles Fort* – "I labor, like workers in a beehive, to support a lot of vagabond notions. But how am I to know? How am I to know but that sometime a queen-idea may soar to the sky, and from a nuptial flight of data, come back fertile from one of these drones?"
70. See *The Stargate Conspiracy,* by Lynn Picknett and Clive Prince.
71. See *The Complete Books of Charles Fort.* (London, 1974).
72. *Ibid,* p. 49.
73. See John A. Keel's *UFOs: Operation Trojan Horse* (New York, 1971), p. 200.
74. Archie E. Roy, *A Sense of Something Strange* (Glasgow, 1990), p. 41.
75. Jacques Vallée, *Dimensions: A Casebook of Alien Contact,* (New York, 1988), p. 203.
76. *Ibid,* p. 253.

FURTHER READING

Books and articles

Arnold, Kenneth and Ray Palmer, *The Coming of the Saucers,* Amherst, 1952.
Barker, Gray, *They Knew Too Much About Flying Saucers,* University Books, New York, 1956.
Bender, Albert, *Flying Saucers and the Three Men,* Saucerian Books, 1962.
Berlitz, Charles and William L. Moore, *The Roswell Incident,* Berkeley Books, New York, 1988.
Blum, Howard, *Out There,* Pocket Star Books, New York, 1990.
Brookesmith, Peter, *UFOs: The Government Files,* Brown Books, London, 1996.
Budden, Albert, *Allergies and Aliens: The Visitation Experience – An Environmental Health Issue,* Discovery Times Press, London, 1994.
UFOs: Psychic Close Encounters, Blandford, London, 1995.
Butler, Brenda, Dot Street and Jenny Randles, *Sky Crash, A Cosmic Conspiracy,* Grafton Books, London, 1986.
Cassirer, Manfred, *Parapsychology and the UFO,* self-published, London, 1988. *Dimensions of Enchantment: The Mystery of UFO Abductions, Close Encounters and Aliens.* Breese Books, 1994.
Conroy, Ed, *Report on Communion, the facts behind the most controversial true story of our time,* Avon Books, New York, 1989.
Evans, Hilary, *Visions*Apparitions*Alien Visitors: A Comparative Study of the Entity Enigma,* Thorsons, London, 1984.
Alternate States of Consciousness: Unself, Otherself, and Superself, The Aquarian Press, Wellingborough, 1989.
Fiore Edith, *Encounters: A Psychologist Reveals Case Studies of Abductions by Extraterrestrials,* Ballantine Books, New York, 1990.
Fontes, Dr Olavo T., 'Report from Brazil', Project 1947 website (*www.project1947.com/fontes1.htm*)
Fort, Charles: *The Complete Books of Charles Fort (The Book of the Damned; Lo!; Wild Talents; New Lands),* Dover Publications Inc., New York, 1974.
Friedman, Stanton T., *Top Secret/MAJIC,* Marlowe & Co., New York, 1996.
Friedman, Stanton T. and Don Berliner. *Crash at Corona: The U.S. Military Retrieval Cover-up of a UFO,* Paragon House, New York, 1992.
Fuller, John G. *The Interrupted Journey,* Dial Press, New York, 1966.
Godwin, Jocelyn, *Arktos: The Polar Myth in Science, Symbolism and Nazi Survival,* New Leaf, 1996.

Good, Timothy, *Above Top Secret: The Worldwide UFO Cover-Up*, Sidgwick & Jackson, London, 1987
Alien Base: Earth's Contact with Extraterrestrials, Century, London, 1998.
Beyond Top Secret: The Worldwide UFO Security Threat, Pan, London, 1997.
Harbinson, W.A., *Genesis*, Corgi, London, 1980
Projekt UFO: The Case for Man-Made Flying Saucers, Boxtree, London, 1995.
'The UFO Goes to War', *The Unexplained*, vol 4.
Harpur, Patrick, *Daemonic Reality*, Viking, 1994.
Hobana, Ion and Julien Weverbergh, *UFO's Behind the Iron Curtain*, Bantam Books, London, 1974.
Hopkins, Budd, *Missing Time: A Documented Study of UFO Abductions*, Richard Marek Publishers, New York, 1981 *Intruders: The Incredible Visitations at Copley Woods*, Random House, Inc., New York, 1987.
Howe, Linda Moulton: *An Alien Harvest: Further Evidence Linking Animal Mutilations and Human Abductions to Alien Life Forms*, Linda Moulton Howe Productions, Huntingdon Valley, PA, 1989.
Hynek, J. Allen: *The UFO Experience: A Scientific Inquiry*. Henry Regnery Company, Chicago, 1972.
Jacobs, David M: *Secret Life: Firsthand Accounts of UFO Abductions, Simon & Schuster, New York, 1992.*
Keel, John A., *UFOs: Operation Trojan Horse*, Sphere Books Ltd., London, 1973.
Keyhoe, Donald E, *The Flying Saucers Are Real*, Fawcett Publications, New York, 1950.
Flying Saucers From Outer Space, Henry Holt, New York, 1953
The Flying Saucer Conspiracy, Henry Holt, New York, 1956.
Aliens From Space, Double Day and Company, New York, 1973.
Klass, Philip J.: *The Real Roswell Crashed Saucer Cover-up*, Prometheus Books, Amherst, 1997.
UFO Abductions: A Dangerous Game. Prometheus Press, Buffalo, 1989.
Korff, Kal K., *The Roswell UFO Crash: What They Don't Want You To Know*. Prometheus Books, Amherst, 1997.
Lammer, Helmut, 'Preliminary Findings of Project MILAB: Evidence for Military Kidnapping of Alleged UFO Abductees' CSETI website (*www.cseti.org/position/addition/milab2.html*) 1996.
Leslie, Desmond and George Adamski, *Flying Saucers Have Landed*, Werner Laurie, London, 1953.
Lindemann, Michael (ed.): *UFOs and the Alien Presence: Six Viewpoints*, The 2020 Group, Santa Barbara, CA, 1991.
Lorenzen, Coral and Jim: *Abducted: Close Encounters of a Fourth Kind*, Berkley Books, New York, 1977.
Mack, John E., *Abduction: Human Encounters with Aliens*, Simon & Schuster, London, 1994.
Marks, John, *The Search for the 'Manchurian Candidate': The CIA and Mind Control*. W.W. Norton & Co. London, 1979.
McAndrew, Captain James, *The Roswell Report: Case Closed*. US Government Printing Office, Washington, 1997.
McCarthy, Paul, 'The Missing Nurses of Roswell', *Omni*, vol. 17, No 8, 1995.

McClure, Kevin, 'More Lies Than Secrets: Continuing an Investigation into the Nazi UFO Legends', Magonia website (*www.magonia.demon.co.uk/abwatch/2secrets.html*) October 1999.
'Nazi UFOs – More Secrets Than Answers', Magonia website (www.magonia.-demon.co.uk/abwatch/aw10.html), June 1998.
Menger, Howard, *From Outer Space to You*, Saucerian Press, Clarksburg, WV, 1959.
Nickell, Joe, 'The Hangar 18' Tales – A Folkloristic Approach'. *Common Ground* No. 9.
Patton, Phil, *Travels In Dreamland: The Secret History of Area 51*, Orion, London, 1997.
Peebles, Curtis, *Watch the Skies! A Chronicle of the Flying Saucer Myth*, Berkley Books, New York, 1995.
Picknett, Lynn and Clive Prince, *The Stargate Conspiracy: Revealing the truth behind extraterrestrial contact, military intelligence and the mysteries of ancient Egypt*, Little, Brown and Company, 1999.
Pope, Nick, *Open Skies, Closed Minds*, Simon & Schuster, London, 1998.
The Uninvited, Simon & Schuster, London, 1999.
Pringle, Lucy, *Crop Circles*, Harper-Collins, London, 1999.
Randle, Kevin D., *The October Scenario: UFO Abductions, Theories about them and a Prediction of when they will return*, Berkley Books, New York, 1989.
Roswell UFO Crash Update: Exposing the Military Cover-up of the Century, Global Communications. New Brunswick, 1995.
Randles, Jenny, *The Complete Book of Aliens + Abductions*, Piatkus, London, 1999.
Investigating the Truth Behind MIB: The Men In Black Phenomenon, Piatkus, London, 1997.
'UFOs Worldwide', *The Unknown*, October, 1997.
Randles, Jenny, Andy Roberts and David Clarke, *The UFOs That Never Were*, London House, London, 2000.
Redfern, Nicholas, *A Covert Agenda: UFO Secrets Exposed*, Simon & Schuster, London, 1998.
Cosmic Crashes: The Incredible Story of the UFOs that Fell to Earth, Simon & Schuster, London, 1999.
The FBI Files: The FBI's UFO Top Secrets Exposed, Simon & Schuster, London, 2000.
Rimmer, John, *The Evidence for Alien Abductions*, The Aquarian Press, Wellingborough (Northamptonshire), 1984.
Roy, Archie E., *A Sense of Something Strange*, Dog and Bone, Glasgow, 1990.
Ruppelt, Edward J., *Report on the Unidentified Flying Objects*, Ace Books, New York, 1956.
Sachs, Margaret, *The UFO Encyclopedia*, Corgi, London, 1981.
Sparks, Brad, 'Colonel Philip Corso and William Birnes' Bestselling Book Exposed as a Hoax!' International Roswell Initiative website (*www.roswell.org*)
Spencer, John, *The UFO Encyclopaedia*, Headline Book Publishing PLC, London, 1991.

Spencer, John and Hilary Evans (editors), *Phenomenon*, Futura Publications, London, 1988.

St Clair, Marisa, *Beyond the Light*, Barnes & Noble, London, 1997.

Steiger, Brad, *Project Blue Book: The Top Secret UFO Findings Revealed*, Ballantine Books, New York, 1976.

Strieber, Whitley, *Communion: Encounters with the Unknown, A True Story*, William Morrow & Co., 1987.

Transformation: The Breakthrough, William Morrow & Co., 1988.

Majestic, Futura, London, 1990.

Breakthrough: The Next Step, Simon & Schuster, London, 1997.

The Secret School, Simon & Schuster, London, 1997.

Vallée, Jacques, *Passport to Magonia: From Folklore to Flying Saucers*, Neville Spearman Ltd, 1970.

Messengers of Deception: UFO Contacts and Cults, And/Or Press, Berkley, 1979.

Dimensions: A Casebook of Alien Contact, Ballantine Books, New York, 1988.

Confrontations: A Scientist's Search for Alien Contact, Ballantine Books, New York, 1990.

Forbidden Science, North Atlantic Books, Berkeley, 1992.

Vallée, Jacques with Tracy Tormé, *Fastwalker (A Novel)*, Frog Ltd., Berkeley, 1946

Vesco, Renato and David Hatcher Childress. *Man-Made UFOs 1944-1994: 50 Years of Suppression*, AUP, Stelle, 1994.

Weaver, Colonel Richard L. and Captain James McAndrew, *The Roswell Report: Fact vs. Fiction in the New Mexico Desert*. US Government Printing Office, Washington, 1995.

ADDRESSES

To contact the author, write to: Lynn Picknett, Box 184, 78 Marylebone High Street, London W1M 4AP, UK.

The British UFO Research Association (BUFORA) publishes *UFO Times and UFO Newsfile* bimonthly. For more details send a large SASE to: BUFORA (UMU), 1 Woodhall Drive, Batley, West Yorkshire, WF17 7SW, UK. Or log on to BUFORA ON-line at www site URL *http://citadel.co.uk/citadel/eclipse/futura/bufora/bufora.ht*

J.Allen Hynek Center for UFO Studies (CUFOS), publishes the bimonthly *UFO Reporter and the Journal of UFO Studies*, besides publishing files that once belonged to NICAP. Contact: J. Allen Hynek Center for UFO Studies, 2457 W. Peterson Ave., Chicago, IL Tel: 60659 (773) 271-3611, Website: *http://www.cufos.org/index.html*

IUFOPRA (Irish UFO & Paranormal Research Association), PO Box 3070, Whitehill, Dublin 9, Eire.

IUN (Independent UFO Network) *www.iun.org*

MUTUAL UFO NETWORK (MUFON), is the investigative UFO organization in the world. It publishes the monthly *MUFON UFO Journal*. Those interested in joining and subscribing to the magazine may do so by submitting $30 (USA) or $35 (foreign) in US currency to: Mutual UFO Network, 103 Oldtowne Rd., Seguin, TX 78155-4099. Tel: (830) 379-9216. Fax (830) 372-9439. Website: *http://www.mufon.com* Email address: *mufonhq@aol.com*

NARO (Northern Anomalies Research Organization) publishes *NARO Minded,* 6 Silsden Avenue, Lowton, Warrington, WA3 1EN.

OVNI Presence, BP 324, 13611, Aix-en-Provence, Cedex 1, France.

SUFOI (Scandanavian UFO Investigation) Postbox 11027, S-600 11, Norrkoping, Sweden.

UFORA (UFO Research Australia), PO Box 1894, Adelaide, South Australia 5001.

UFORIC (UFO Research Investigation Centre Canada) Department 25, 1665 Robson Street, Vancouver, British Columbia, V6C 3C2, Canada.

PUBLICATIONS

Abduction Watch, 3 Claremont Grove, Leeds, LS3 1AX.

Bulletin of Anomalous Experience, (BAE), 614 South Hanover St., Baltimore, MD 21230-3832, USA.

Flying Saucer Review (FSR), PO Box 162, High Wycombe, Bucks HP13 5DZ.

Fortean Times, The Journal of Strange Phenomena, edited by Bob Rickard and Paul Sieveking, is a monthly newsstand publication in the UK, price £2.70. For subscription enquiries, phone 01454 642458, general enquiries 020 75653130. US readers can discover their local stockist by ringing Eastern News toll-free on (800) 221 3148. .

Lumiéres Dans La Nuit, 5 Rue Lamartine, 91220 Betigny sur Orge, France.

Magonia, 5 James Terrace, London SW14 8HB.

Ohio Notebook, Box 162, 5837 Karrie Square Drive, Dublin, OH 43016, USA.

Skeptics UFO Newsletter (SUN), 404 N Street SW, Washington, DC 20024, USA.

UFO Afrinews PO Box MP 49, Mount Pleasant, Harare, Zimbabwe.

UFO Magazine (UK), Wharfebank House, Wharfebank Business Centre, Ilkley Road, Otley near Leeds, LS21 3JP.

UFO Magazine (US) (monthly), 5455 Centinela Avenue, Los Angeles, CA 90066, Tel: (310) 827-0505, Fax: (310) 827-6865.
To subscribe (at $24.95 for annual domestic subscription) call (888) 836-6242. Website: *www.ufomag.com*

WEBSITES

Albert Budden: *http://195.195.73.39/users/fasttrack/budden*
www.pharo.com

RESEARCH INTO ALIEN ABDUCTION

Academy of Clinical Close Encounter Therapists, 2826 O Street, Suite 3, Sacramento, CA 95816, USA.
BUFORA (see above).
Intruders Foundation, PO Box 30233, New York, NY 10011.
Program for Extraordinary Experience Research, 1493 Cambridge Street, Cambridge, MA 02139, USA.

INDEX

'AFB' indicates 'Air Force Base'.

Other titles available from Robinson Publishing

The Mammoth Book of Encyclopedia of Unsolved Mysteries **£7.99 [ ]**
Colin Wilson & Damon Wilson
Here in one terrific collection is a compelling and remarkable history spanning over two thousand years of the greatest mysteries known to mankind. Renowned authors and paranormal investigators Colin and Damon Wilson explore such diverse puzzles as the disappearance of Glen Miller to the Loch Ness Monster.

The Mammoth Book of Locked Room Mysteries and Impossible Crimes Ed. Mike Ashley **£6.99 [ ]**
For the first time ever, is the biggest collection of such stories. They include how a man can be stabbed in open countryside surrounded by plenty of witnesses who saw nothing, and the women who was killed from a theatre's stage when the only people on it are male strippers – with no visible weapons! These and 30 other stories will stretch your powers of deduction to the limits!

The Mammoth Book of Murder and Science Ed. Roger Wilkes **£7.99 [ ]**
Where traditional methods of detection have failed time and time again, scientific routes of investigation and analysis have led to the conviction of many dangerous crimnials. *The Mammoth Book of Murder and Science* examines the landmark cases and shows how forensic science has developed over the last 150 years.

The Mammoth Book of Best New Science Fiction Ed. Gardner Dozois **£9.99 [ ]**
Widely regarded as the one essential book for every science fiction fan, *The Mammoth Book of New Science Fiction* contains more than two dozen Sc-Fi stories from the previous year. This year's volume includes stories by David Marusek and Eleanor Arnason as well as many other bright stars of science fiction.

Robinson books are available from all good bookshops or direct from the publisher. Just tick the titles you want and fill in the form below.

TBS Direct
Colchester Road, Frating Green, Colchester, Essex CO7 7DW
Tel: +44 (0) 1206 255777
Fax: +44 (0) 1206 255914
Email: sales@tbs-ltd.co.uk

UK/BFPO customers please allow £1.00 for p&p for the first book, plus 50p for the second, plus 30p for each additional book up to a maximum charge of £3.00.
Overseas customers (inc. Ireland), please allow £2.00 for the first book, plus £1.00 for the second, plus 50p for each additional book.

Please send me the titles ticked above.

NAME (Block letters) . . . .

ADDRESS . . . .

. . . .

POSTCODE. . . . .

I enclose a cheque/PO (payable to TBS Direct) for . . . .

I wish to pay by Switch/Credit card

Number . . . .

Card Expiry Date . . . .

Switch Issue Number . . . .